Ergebnisse der Physiologie · Reviews of Physiology

Ergebnisse der Physiologie

Biologischen Chemie und experimentellen Pharmakologie

Reviews of Physiology

Biochemistry and Experimental Pharmacology

64

Herausgeber / Editors

R. H. Adrian, Cambridge · E. Helmreich, Würzburg
H. Holzer, Freiburg · R. Jung, Freiburg · K. Kramer, München
O. Krayer, Boston · F. Lynen, München · P. A. Miescher, Genf
H. Rasmussen, Philadelphia · A. E. Renold, Genf
U. Trendelenburg, Würzburg · W. Vogt, Göttingen
H. H. Weber, Heidelberg

With 63 Figures

Springer-Verlag Berlin Heidelberg GmbH 1972

Authors:

Jouvet, M., Prof. Dr., Laboratoire de Pathologie
  Générale et Expérimentale, Faculté de Médécine,
  8, Ave. Rockefeller, F-69 Lyon

Moruzzi, G., Prof. Dr., Istituto di Fisiologia della Università di Pisa,
  Via S. Zeno N. 29—31, I-56100 Pisa

ISBN 978-3-662-31272-8   ISBN 978-3-540-36627-0 (eBook)
DOI 10.1007/978-3-540-36627-0

© by Springer-Verlag Berlin Heidelberg 1972
Originally published by Springer-Verlag Berlin · Heidelberg 1972
Softcover reprint of the hardcover 1st edition 1972

Library of Congress Catalog Card Number 62-37142.

Universitätsdruckerei H. Stürtz AG, Würzburg

Neurophysiology and Neurochemistry
of Sleep and Wakefulness

Contents

Contents

The Sleep-Waking Cycle

Giuseppe Moruzzi*

With 39 Figures

Table of Contents

---

* Istituto di Fisiologia, Università di Pisa, e Laboratorio di Neurofisiologia del CNR, Pisa, Italia.

I. Introduction

To review the entire field of sleep physiology might be nowadays the task of a collaborative treatise, rather than the aim of an article for *Ergebnisse der Physiologie*. There is no urgent need, moreover, to carry on such a huge task, since several books, review articles, and proceedings of symposia, partially or wholly devoted to sleep, have been published during the last years.

Monographs have been written by Hess (1949b, 1968a, b), Bremer (1953), Oswald (1962, 1966), Magoun (1963), Kleitman (1963), Hassenberg (1965), Bonvallet (1966), Foulkes (1966), Hartmann (1967), and Koella (1967), while books with articles on sleep have been edited by Harms (1964), Bürger-Prinz and Fischer (1967), Webb (1968), Jovanović (1969), and Baust (1970a).

Symposia partially or wholly devoted to sleep have been edited by Hess (1950), Adrian et al. (1954), Wolstenholme and O'Connor (1958, 1961), Jasper et al. (1958), Jasper and Smirnov (1960), Passouant (1962), Moruzzi et al. (1963), Hernández-Peón (1963), Akert et al. (1965), Jouvet (1965a), Fischgold (1965), Eccles (1966), Granit (1966), Quarton et al. (1967), Schlag and Scheibel (1967), Kety et al. (1967), Nauta et al. (1967), Gastaut et al. (1968).

Reviews covering different aspects of hypnic physiology have been written by Kleitman (1929, 1957, 1967), Bremer (1937, 1938, 1951, 1954, 1960, 1968a, b), Hess (1954), Moruzzi (1954, 1960, 1962, 1963a, b, 1964a, 1965, 1966, 1969), Poeck (1955), Rossi and Zanchetti (1957), Hediger (1959, 1969) Dell et al. (1961), Akert (1961, 1965), Cordeau (1962), Cordeau et al. (1965), Hill (1962), Passouant and Cadilhac (1962), Dell (1963), Jung (1963, 1965, 1967), Hess, Jr. (1964), Rossi (1964), Ruckebusch (1964), Zernicki (1964, 1968), Luce (1965), Meulders (1965), Garcia Austt (1965), Palestini (1965), Jouvet (1965b, c, 1967a, b, c), Pompeiano (1965, 1966, 1967a, b), Tissot (1965), Hernández-Peón (1965), Bürgi (1966), van den Hoofdakker (1966), Deutsch and Deutsch (1966), Roffwarg et al. (1966), Routtenberg (1966, 1968), Monnier and Fallert (1967a, b), Passouant et al. (1967), Zanchetti (1967a, b), Hoffmeister (1968), Passouant (1968), Sterman and Clemente (1968), Jeannerod (1969), Weyn and Letasch (1969), Steriade (1970), Berlucchi (1970).

The reader will find an almost complete list of references in Piéron's book (1913) and in the second edition of Kleitman's treatise (1963), respectively, for the periods up to 1912 and up to 1962; in RECHTSCHAFFEN and EAKIN's bibliography (1968) for the years from 1962 to 1968; and, finally, in the *Sleep Bulletin* edited by EAKIN since 1968, to be completed from 1970 onwards with the *Sleep reviews*, edited by DEMENT, KALES, and RECHTSCHAFFEN. PIÉRON's book (1913, see pp. 365–422) and two recent essays (MORUZZI, 1964b; KRUTA, 1967) provide information on the history of the theories of sleep.

Thus the aim of the present introduction is not only to give a positive definition of our theme, the usual task of any foreword, but also to state clearly the areas of sleep physiology which are not going to be covered. In order to give to the reader the possibility to complete his information in the fields of hypnic physiology which are outside the scope of our review, the introduction will be followed by a short summary of several trends of sleep physiology, with references to a few key reviews or articles.

We shall be concerned here with the *neural mechanisms underlying the sleep-waking cycle*. The major sources of information and the main object of a critical appraisal will be such phenomena as the coma produced by midbrain transection and the related processes of recovery; the hypersomnias, the hyposomnias, and the lethargies produced by brain stem or cerebral lesions; and the sleep induced by electrical or sensory stimuli. Thus attention will be concentrated upon the experimental attempts to *modify* the sleep-waking cycle, rather than upon the endeavor to describe and to classify the phenomena occurring during natural sleep.

The *phenomenology of sleep* will remain, therefore, outside the scope of our article. The description and the classification of the phenomena occurring during physiological sleep has made great progress in the last twenty years. We are now able to combine the classical behavioral approaches with EEG and EMG recording, and even with refined analysis of single unit discharges led from cortical or subcortical neurons. These technical improvements have led to great advances in the study of sleep and waking behavior of the unanesthetized, free moving animal. The extent of this progress is shown by the expansion of the area covered by behavioral studies. Originally, such studies were exclusively concerned with the observation of movements and of postures, i.e. with the outcome of the integrated activity of skeletal muscles. We may nowadays talk of "behavior" of final common paths: motoneurons, preganglionic autonomic nerve cells, neurosecretory neurons. Our information is coming in fact not only from the study of the integrated activity of the effectors—striated muscles, visceral organs, endocrine glands—but also from the electrophysiological analysis of the spontaneous or evoked discharge of single units or large populations of neurons. We may easily get direct information concerning not only the final paths, but also the interneurons of the brain, and we may correlate these observations with the behavior of the free moving, unanesthetized animal. Notwithstanding this tremendous technical progress, this approach is still based on observation, description, and classification of *spontaneously* occurring phenomena.

The investigator tries in fact to observe what happens in physiological conditions, and carefully avoids producing major interferences with the occurrence of the natural event, at least with respect to the sleep-walking cycle. The mental attitude is therefore very much the same as that of the ethologist who observes, describes, and classifies movements and postures. The area of investigation has simply expanded, because of technical advances.

The difference between the observational and the experimental approaches was clearly defined by Claude Bernard in his *Introduction à l'étude de la médecine expérimentale.*

«On donne le nom d'*observateur* à celui qui applique les procédés d'investigations simples ou complexes à l'étude de phénomènes qu'il ne fait pas varier et qu'il recueille, par conséquent, tels que la nature les lui offre. On donne le nom d'*expérimentateur* à celui qui emploie les procédés d'investigation simples ou complexes pour faire varier ou modifier, dans un but quelconque, les phénomènes naturels et les faire apparaître dans des circonstances ou dans des conditions dans lesquelles la nature ne les lui présentait pas. Dans ce sens, l'*observation* est l'investigation d'un phénomène naturel, et l'*expérience* est l'investigation d'un phénomène modifié par l'investigateur.» (1865, p. 29.)

Both approaches are important, but the study of the neural mechanisms underlying the sleep-waking cycle cannot be made without «l'idée d'une variation ou d'un trouble *intentionnellement* apportés par l'investigateur dans les conditions des phénomènes naturels» (1865, p. 17).

Of course the temporary abolition or the striking alteration of the sleep-waking cycle by lesions or by electrical stimulations imply experimental situations which do not occur in physiological conditions. We must be aware of the limitations of this approach, particularly when we attempt to interpret the results. Undoubtedly our conclusions must also take into account the results obtained, along the observational line, from intact, free moving animals. However the historical importance of what Claude Bernard called the experimental approach is shown by the fact that our concepts of waking or sleep "centers", of activating and deactivating "systems" are based upon the results of stimulation or lesion experiments. These results would never have been attained solely by observation of phenomena occurring in natural conditions.

II. An Outline of the Phenomenology of Sleep

When it became possible to combine the classical behavioral studies made on the free moving, unanesthetized animals with practically any type of electrophysiological recording, the progress was bound to be tremendous. Two separate branches of science, ethology[1] and physiology, converged on a single

---

1 Ethology: the objective study of behavior (Tinbergen, 1955; see p. 1).

type of experimentation. Revolutions in science are frequently due to a major technical improvement. This was not an exception to the rule.

However the wealth of data which has been gathered in this way is mainly related to the phenomenology of sleep. The studies on the sleep-waking rhythm have utilized so far only a few of the "signs" of sleep which are now available. All of these signs should be known, since the data gathered by means of the phenomenological approach are likely to be extensively utilized, in the near future, for experiments on the sleep-waking cycle.

It should be stated at the outset that several contributions to the phenomenology of sleep were in fact experimental, not simply observational in nature. To test a reflex response or an evoked potential during sleep and wakefulness implies, undoubtedly, to carry on an experiment. However, whenever the sleep-waking cycle is not modified by the experimental situation, as in the majority of cases, we are still in the realm of the sleep phenomenology, at least with regard to the central theme of our review.

This chapter contains the outline of an huge inventory of hypnic phenomena, and provides also the references to a few review articles. The material is classified, following an approach which is both historical and methodological in character, into six separate sections, depending upon whether the results were obtained i) with the classical behavioral approach, ii) by combining EEG and EMG recording with behavioral observations, iii) by completing the second approach with microelectrode recording from single units; iv) by recording the motor responses to different kinds of sensory or central stimulation; v) by recording central evoked potentials, with particular attention to the problem of sensory transmission; or finally vi) by utilizing several types of recording of the visceral functions.

Before BERGER's discovery sleep studies were based on behavioral observations of postures and movements, and on the study of a few changes occurring mainly in the autonomic sphere; pupil diameter, blood pressure and heart rate, body temperature etc. The reader will find a complete review of the classical literature in KLEITMAN's book (1963; see pp. 8–67, 81–91), which also contains detailed information on depth and duration of sleep (l.c.; see pp. 108 to 121) and on its periodicity (l.c.; see pp. 131 to 184). For old literature PIÉRON's book (1913) should also be consulted.

The distinction between monocyclic and polycyclic animals (KLEITMAN, 1963; see p. 148), the discovery of the amazing differences in pattern and duration of sleep in different mammals (see HASSENBERG, 1965); the appearance of circadian rhythms (KLEITMAN, 1963; see p. 131); finally, KLEITMAN's evolutionary theory of sleep and wakefulness (l.c.; see p. 363) are the outcome of several observations made along the lines of the classical ethological approach. The technical procedure was extremely simple, and yet the power of observation and the intellectual insight of the ethologists and of the old physiologists led to important results, whose validity has remained unchallenged.

Reference must also be made to the results of the comparative[2] and developmental[3] studies on sleep. With only a few exceptions our review will be concerned with the sleep-waking cycle in the adult mammal. Most of our information on mammalian sleep comes from experiments on cats, but important data have been obtained on rats, rabbits, monkeys, and on man.

Berger's discoveries on the human EEG, with the demonstration of the striking changes occurring during sleep, opened a new era in the field of hypnic studies. The *correlation between EEG and behavior was the theme of several investigations*, first made on man and then extended to laboratory animals. Already in the 1930's Adrian's hypothesis on the synchronization and desynchronization of cortical neurons was universally accepted, while the existence of a close relationship between EEG patterns and the different stages of sleep or the state of arousal was regarded as a well established fact (see Hess, Jr.,1964, and Jung, 1967). The discovery (Klaue, 1937; see Jouvet, 1967a) that the *slow, synchronized* phase of sleep was interrupted by episodes characterized, paradoxically, by low voltage fast EEG activity, muscular relaxation, and tonic twitches—later called *fast, desynchronized* or *paradoxical* sleep—was of major importance for the specific theme of this review. In fact investigations with Tönnies' automatic EEG Interval Spectrum analysis have demonstrated that the desynchronized sleep shows different EEG features from the state of arousal (Tönnies, 1969), an important observation suggesting that the term "paradoxical" is not justified. The main point, however, is that for the first time the physiologist realized that besides the well known alternation between sleep and wakefulness, another alternation went on *within* each period of sleep. This new, irregular cycle was characterized by the alternation of slow and fast sleep. The fast, paradoxical sleep was bound to attract the attention of physiologists (see Moruzzi, 1963b; Jouvet, 1967a; Koella, 1967), psychologists (see Hartmann, 1967; Foulkes and Hobson, 1969), and clinical neurologists (see Gastaut et al., 1968) for many reasons, some of which are outside the scope of our review. The study of the correlates of fast sleep represents a quickly developing chapter in the area of sleep phenomenology. What is important for our theme is to realize that in normal conditions sleep is not a homogeneous state, but rather consists of a cycle between synchronized and desynchronized episodes.

---

2 See Hediger (1959, 1968, 1969) and Hassenberg (1965) for ethology; Ruckebusch (1963, 1964), Jouvet and Jouvet (1964), and Klemm (1966) for physiology; Foulkes and Hobson (1969) for recent preliminary notes.

3 See Valatx (1963) and Jouvet-Mounier (1968).

The EEG patterns have been correlated with changes of cortical D.C. potentials (see CASPERS, 1961; KAWAMURA and SAWYER, 1964; KAWAMURA and POMPEIANO, 1970) and of brain circulation (see KETY, 1967), and with records of the hippocampal activity (see PASSOUANT and CADILHAC, 1962).

Microelectrode studies on the discharge of single units during synchronized sleep and EEG arousal were initiated in the early fifties. As subjects curarized cats (VERZEANO and NEGISHI, 1961), "pyramidal" (WHITLOCK et al., 1953; VON BAUMGARTEN et al., 1954), and *encéphale isolé* (VON BAUMGARTEN et al., 1954; CREUTZFELDT and JUNG, 1961) preparations were used (see MORUZZI, 1954). Great progress was made when new techniques permitted experimenters to combine *single unit recording with EEG and behavioral studies*, in the free moving, unanesthetized animal. The results of unit recording from cortical neurons have been reported and reviewed by EVARTS (1962, 1964, 1965, 1967), and other investigations were reported later by CALVET and CALVET (1968), McCARLEY and HOBSON (1970) and by NODA and ADEY (1969, 1970). Single unit discharges were led also from the midbrain reticular formation (HUTTEN-LOCHER, 1961; BALZANO and JEANNEROD, 1970), and from thalamic (MUKHA-METOV and RIZZOLATTI, 1970; MUKHAMETOV, RIZZOLATTI and TRADARDI, 1970) and hypothalamic (FINDLAY and HAYWARD, 1969) neurons. Averaged multi-unit recordings from the cerebral cortex (SCHLAG and BALVIN, 1963), the pyramidal tract (ARDUINI et al., 1963; MARCHIAFAVA and POMPEIANO, 1964; ROUGEUL et al., 1966), the midbrain reticular formation (SCHLAG and BALVIN, 1963) and the diencephalon (GOODMAN and MANN, 1967) were made during the states of sleep and wakefulness.

The fourth line of investigation was made possible by the analysis of *spinal reflexes* and of *pyramidal movements* in the unanesthetized, free moving cat. Non-painful, electrical stimulation of sensory nerves combined with electromyographic recording of reflex responses, or of the movements produced by pyramidal stimulation, led to the discovery (GIAQUINTO et al., 1964a, b) that strong inhibitory volleys arise in the brain stem during the paradoxical episodes of sleep. These volleys block the motoneuronal responses through mechanisms of pre- and postsynaptic inhibition (see POMPEIANO, 1965, 1966, 1967a, b). The classical literature on reflex and postural changes during sleep was reviewed by TRÖMNER (1912) and TOURNAY (1934) several years ago.

The fifth line of work was concerned with the responses of single units or of neuronal populations to "test stimuli", including natural stimuli or electric shocks, applied either to peripheral fibers or to central structures. The results obtained with the technique of the evoked discharges have been reviewed by CORDEAU et al. (1965), ROSSI et al. (1965), and by STERIADE (1970).

Several data showing an inhibition of transmission of sensory volleys during paradoxical sleep have been reported or discussed by Pompeiano (1967a) in a recent review. A comprehensive review of the effects of sleep and wakefulness on the sensory system has been made by Koella (1967).

The last line of endeavor, the influence of sleep on visceral functions, had been extensively investigated by the old physiologists. The classical literature was reviewed by Piéron (1913) and Kleitman (1963). A revival of this approach occurred when it became possible to recognize the episodes of paradoxical sleep by combining EEG recording with cervical electromyography and electronystagmography (see Jouvet, 1967a). Perhaps the most impressive development during recent years is the discovery of a marked fall of blood pressure occurring during the paradoxical episodes (Candia et al., 1962; Kanzow et al., 1962), a phenomenon that appears strikingly enhanced after baroceptive denervation (see Zanchetti et al., 1966, and Kumazawa et al., 1969). Two reviews by Baust (1970b, 1971) should be consulted for the phenomenology of the autonomic manifestations of sleep.

III. Lesion Experiments

Theories may become obsolete even while the validity of the experiments on which they are based remains unchallenged .This statement is frequently made, and we are going to see that it is certainly true for some classical works of sleep physiology. However old concepts linger on in the original terms which are still routinely used, and they may subreptitiously be revived during a comprehensive review of the literature. It would not be easy to avoid such terms as activation and deactivation: these words might in fact be dispensed with only by adopting several circumlocutions, which would make reading cumbersome and possibly obscure. For these considerations we are going to start the chapter on lesion experiments with a short introduction on *terminology*. Our aim is to state clearly the position of modern physiology with respect to a few words which are too strongly entrenched to be avoided.

There is little doubt that terms such as activation and deactivation, fall or slackening of cerebral tone, facilitation and withdrawal of an energizing influence, imply concepts whose validity was unchallenged in the thirties, but which have become untenable as a consequence of two major progresses: the discovery of the paradoxical phase of sleep and the introduction of the techniques of microelectrode recording in the free moving, unanesthetized animal. We know now that the differences between wakefulness and sleep, and between physiological sleep and coma, cannot be expressed exclusively in quantitative terms, simply by postulating a progressive fall in the overall tonic activity of both the "energizing" and the "facilitated" structures, the ascending reticular system of the brain stem and the cerebrum, respectively.

The fallacy of regarding the ascending reticular system mainly as an energizer, as a structure which would be concerned merely with maintaining a given level of activity in the cerebrum, had already been pointed out (MORUZZI, 1958) even before methods for microelectrode recording from the free moving cat were available. Single unit investigations have been reviewed by EVARTS (1967) and by BERLUCCHI (1970), and the theme is outside the scope of our article. Suffice it to recall here that the main discharge frequency of neocortical (EVARTS, 1962, 1964, 1967; CALVET and CALVET, 1968), lateral geniculate (BIZZI, 1966; SAKAHURĀ, 1968; MUKHAMETOV and RIZZOLATTI, 1970) and reticular (HUTTENLOCHER, 1961) neurons of the cat—and of hypothalamic units of the rat (MINK and BEST, 1967) and of the rabbit (FINDLAY and HAYWARD, 1969)—is indeed usually lower during slow wave sleep than during wakefulness, but is actually the same or even higher during the paradoxical phase. Indeed the firing level of the large-sized pyramidal neurons of the monkey motor cortex may be even higher during slow sleep than during waking without movements (EVARTS, 1965). Only when measurements are made in the lulls between bursts of rapid eye movements, the rate of firing of single visuo-cortical neurons (VALLEALA, 1967; McCARLEY and HOBSON, 1970), or the integrated output of cortical neurons to callosal fibers (BERLUCCHI, 1965) may be low during the paradoxical phase, indeed even lower that during the slow wave period (BERLUCCHI, 1965). Moreover, the hippocampal neurons tend to become silent during the paradoxical phase of the rat (MINK and BEST, 1967) and of the cat (BELUGOU et al., 1968), and a decrease of firing may be observed in the same conditions in some units of the lateral geniculate body of the cat (BIZZI, 1966). All these data suggest that during sleep *some* units may strikingly decrease their activity and become even silent. However this is by no means true for *all* units. In fact the field is still largely unexplored and we have no direct information on the sleep behavior of small inhibitory interneurons (see EVARTS, 1964).

If one thinks that, besides the average frequency, the temporal patterns of discharge of single neurons and their interrelations are deeply affected by sleep (see EVARTS, 1967, and BERLUCCHI, 1970), one comes to the conclusion that it would be impossible to express sleep in purely quantitative terms, for example, as a fall of the overall tonic activity of the cerebrum.

The lesion experiments, crude as they admittedly are, provide information on the tonic influence exerted by the brain stem upon the cerebrum, and on the physiological significance of this ascending influence. This *integrated* picture of the interrelations between brain stem and cerebrum cannot be achieved even with the most sophisticated electrophysiological techniques available at the present time. The lesion experiments are so far irreplaceable. We shall deal at length with them. We believe that there will be no harm in using words of the old terminology, and some comparisons taken from much simpler situations occurring in classical physiology, provided our aims are clearly stated. We are interested only in demonstrating that a tonic influence is exerted by the brain stem on the cerebrum, and in attempting to construct an integrated picture of the effects on brain activity and on animal's behavior when there are changes in the ascending reticular discharge.

The term *activation* will still be used to represent the physiological processes leading from quiet sleep to relaxed or to alert wakefulness, while a process going in the opposite direction will be designed as *deactivation*. Our attention will be concentrated on animal's behavior rather than on the EEG patterns, which will be classified simply as *synchronized* or *desynchronized*. No attempt will be made to cover other details provided by EEG studies (see Ursin, 1968), since we are interested only in an alternation of activity: the sleep-waking cycle. The terms cerebral tone, facilitation, and energizing influence will be avoided, unless required by the historical treatment. The term *modulation* will be used to designate the multifarious influences exerted by the brain stem on the cerebrum. Finally we want to point out that the influence exerted by the ascending reticular system on the processes of sleep may not be due exclusively to withdrawal of a tonic activation. Other factors, such as the patterns of the ascending reticular discharge, may be important.

A. Complete Brain Stem Transection
1. The Acute *Cerveau Isolé*

When the midbrain is transected just behind the third nerves, the sensory afferents to the isolated cerebrum are interrupted, with the exception of the olfactory and optic nerves, while the outflow is mediated exclusively by the motoneurons and the preganglionic parasympathetic nerve cells of the oculo-motor nucleus.

This is the classical *cerveau isolé* preparation, first described by Bremer in 1935 (see Bremer, 1937, 1938). Soon after the transection, once the effects of ether anesthesia have disappeared ,the ocular behavior and the EEG patterns resemble those of a cat under slow wave sleep (Fig. 1). Throughout the survival time of an acute preparation, the alternation of sleep and wakefulness is abolished altogether. This behavior is entirely different from that of Bremer's *encéphale isolé* (see Bremer, 1937, 1938), since the sleep-waking cycle is present after spinal section at C_1, as shown by ocular behavior and the EEG records (Fig. 2). Thus either the abolition of the sensory inflow through the other cranial nerves, as originally suggested by Bremer (1937, 1938), or the abolition of an autochthonous influence arising from the brain stem, must be responsible for the striking differences between the *encéphale isolé* and the *cerveau isolé* preparations.

The *cerveau isolé* is characterized by an extreme myosis. Keller (1932) had previously reported that fissurate myosis was present both after pre-collicular decerebration, when any cerebral influence upon the Edinger-West-phal nuclei had been eliminated, and following a midbrain transection made behind the third nerves, as in Bremer's *cerveau isolé*. With chronic experiments Keller disposed of the irritative hypothesis, and made the interpretation quite easy for the experiments of precollicular decerebration. The myosis was obviously due to release of the preganglionic parasympathetic neurons in-nervating the pupils from a tonic inhibition arising from or driven by the cerebrum, possibly the hypothalamus. However, the extreme myosis observed after intercollicular transection was difficult to explain in the early thirties.

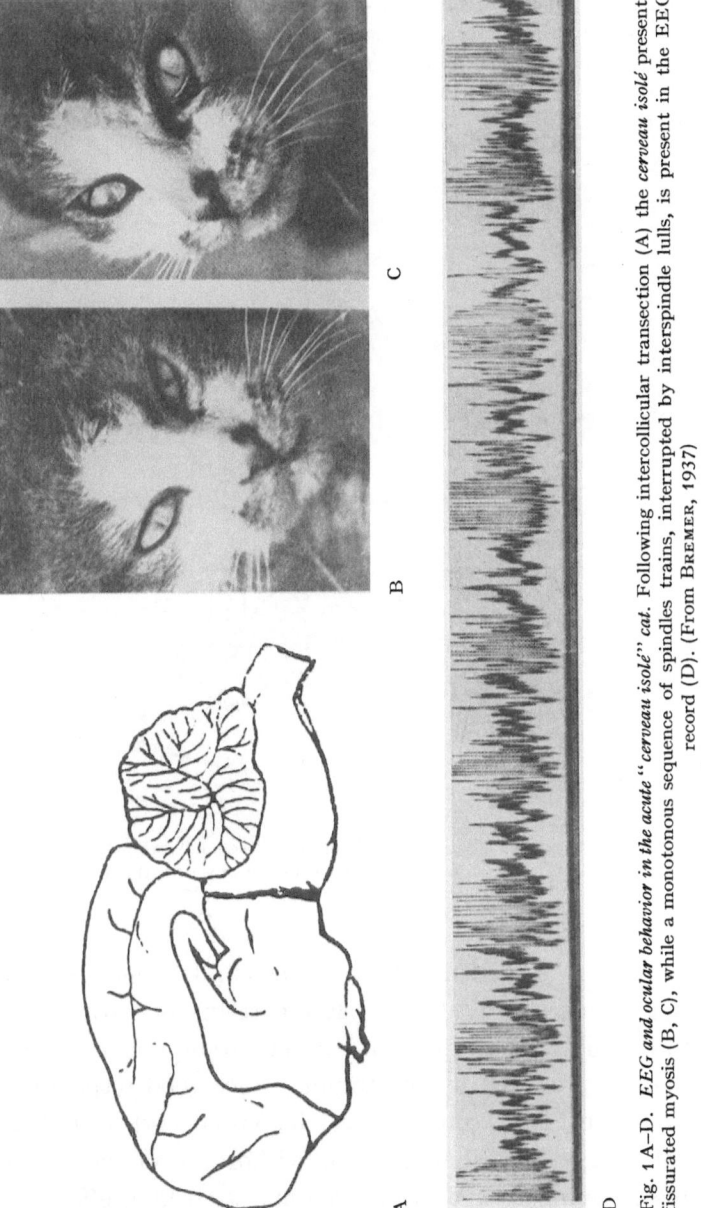

Fig. 1 A–D. *EEG and ocular behavior in the acute "cerveau isolé" cat.* Following intercollicular transection (A) the *cerveau isolé* presents fissurated myosis (B, C), while a monotonous sequence of spindles trains, interrupted by interspindle lulls, is present in the EEG record (D). (From BREMER, 1937)

Following BREMER's discovery a convincing interpretation of this phenomenon became possible. BREMER (1937) suggested that the tonic inhibitory influence exerted by the brain upon the pupillo-constrictor neurons was abolished as a consequence of the striking fall of cerebral activity produced by the inter-collicular transection. He pointed out that the descending inhibitory paths impinging upon the parasympathetic pupillo-constrictor neurons were intact in his preparation. Hence the release effect, the myosis, could be explained only with a fall of the tonic discharge of the inhibitory brain neurons. To quote:

«Cette suppression fonctionnelle se produit malgré l'integrité des voies corticifuges, oculomotrices et pupillo-dilatatrices. *Elle est l'expression d'une diminution de ce que l'on pourrait appeler le «tonus cortical et diencéphalique»*[4] (Bremer, 1937, p. 78–79).

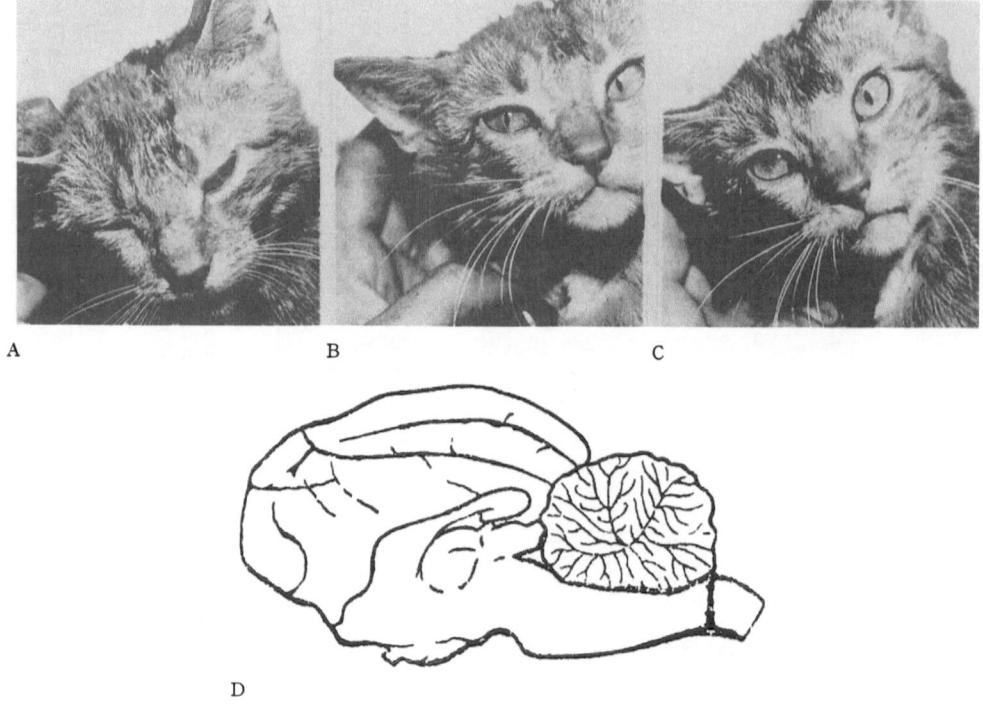

Fig. 2A–D. *Sleep and wakefulness in the acute "encéphale isolé" cat.* Following spinal section at C_1 (D) the *encéphale isolé* presents an alternation between ocular behavior of sleep (A) and of wakefulness (B, C). (From Bremer, 1937)

The hypothesis of a sudden fall of the cerebral "tone" was supported by the striking similarities with the EEG patterns of barbital anesthesia. Bremer pointed out that the cerebral deactivation was not due to local circulatory changes produced by the trauma, since good responses to photic stimulation could still be recorded from the visual cortex. Hence the fall of the cerebral tone was attributed to the withdrawal of the steady flow of sensory impulses impinging upon the cerebrum, a striking effect after inter-collicular transection.

«Sa seule explication possible est qu'il est le résultat de la suppression, par l'inter-ruption des voies corticipètes, de l'afflux incessant des excitations, d'origine cutanée et proprioceptive notamment, qui sont essentielles pour le maintien de l'état vigile du télencéphale, état vigile que les seuls appareils olfactif et optique ne suffisent apparem-ment pas à entretenir» (Bremer, 1937, p. 79).

Bremer suggested, as Kleitman (1929) had done before him, that a slight reversible sensory deafferentation was the cause of physiological sleep, and that the results obtained with the *cerveau isolé* could be taken as evidence

---

4 Italics are ours.

confirming this hypothesis. This was a bold extrapolation, since BREMER had in fact produced a state of coma, not one of sleep. However all SHERRINGTON's studies on postural tonus are based on an extrapolation of this kind.

MAGNUS had stated, in the book summarizing his life work (1924), that decerebrate rigidity is in fact the caricature of normal reflex standing. The extensor hypertonus is clearly a pathological phenomenon, one that should be regarded as the result of a distortion, rather than the sheer magnification, of the physiological mechanisms underlying postural tonus. Nonetheless the discoveries made by SHERRINGTON and by MAGNUS himself on the neural mechanisms underlying the regulation of postural tonus actually started from the simplifying hypothesis, one of major importance in the history of neurophysiology, that the neural mechanisms underlying normal posture can be studied in the decerebrate preparation. BREMER (1937) followed the same line of thought, when he assumed that the coma of the acute *cerveau isolé* could be utilized for the study of physiological sleep.

The main significance of the *cerveau isolé* and *encéphale isolé* preparations, at least from the physiologist's standpoint, is of course not the experimental production of a state of coma, but rather i) *the abolition of the sleep-waking cycle by the acute midbrain transection*, and ii) the *demonstration that the critical level of acute transection still permitting the alternation between sleep and wakefulness lies somewhere between the midbrain and the lower medulla*. In summary, the main conclusion that may be drawn from BREMER's experiments is that a tonic influence arising below the midbrain transection, but above the spinal segments, is necessary in order to maintain the sleep-waking cycle.

Although the coma produced by midbrain transection is obviously different from physiological sleep, the similarities between the two states deserve comment. The monotonous sequence of spindles occurring in the *cerveau isolé* can no longer be regarded as the electrophysiological evidence of the existence of a state of sleep, an assumption universally accepted in the thirties. We have learnt later that during behavioral sleep episodes characterized by low voltage fast activity, the paradoxical or fast sleep, alternate with those characterized by EEG synchronization, the slow sleep (see JOUVET, 1967a for ref.). However BREMER's interpretation of the fissurated myosis occurring during natural sleep and in the *cerveau isolé* has been fully confirmed by recent experiments.

A distinction should be made between i) the retinal component of the pupilloconstrictor tonus, which is due to the photic reflex, and ii) the extraretinal components, one of which is probably autochthonous in nature (KELLER, 1946; ZERNICKI et al., 1970). The decrease of the pupilloconstrictor tonus, which markedly contributes to the mydriatic response to darkness occurring during wakefulness, is due i) to abolition of the tonic component of the photic reflex and ii) to inhibitory influences exerted by the retinal dark discharge upon the Edinger-Westphal nucleus (KING et al., 1963). Of course the inhibitory influence of the dark discharge on the pupilloconstrictor tonus is abolished following visual deafferentiation, and the mydriasis that characterizes blindness may then be explained only by the absence of the tonic reflex excitation of the Edinger-Westphal neurons, i.e. with the abolition of the tonic component of the photic reflex. In this experimental situation, however, a *tonic* discharge reappears in the pupilloconstrictor neurons during sleep. The phenomenon is present in man and may be easily reproduced in the cat.

Old clinical observations (see ROSSI, 1957, for ref.) had shown that the mydriasis of the blind man was replaced by myosis during sleep; an observation fully confirmed, following visual deafferentiation, during the physiological sleep of the unanesthetized, free moving cat (BERLUCCHI et al., 1964b; see Figs. 3, 4) and in the *cerveau isolé* preparation (ROSSI, 1957). The difference between wakefulness and sleep is actually very striking when the contamination by the photic reflex is absent following visual deafferentiation.

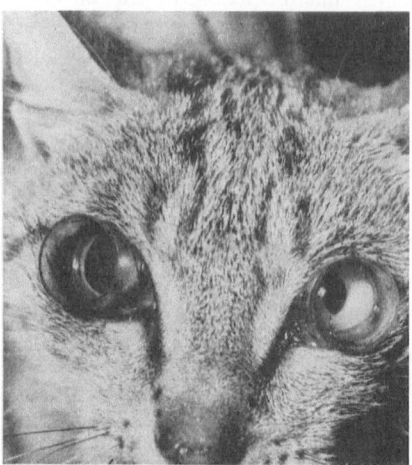

Fig. 3. *Mydriasis in the blind cat during wakefulness.* The left eye had been sympathectomized 12 hours before, so that pupillary dilation was due only to inhibition of the parasympathetic tone. Mydriasis was obviously greater on the right side. (From BERLUCCHI et al., 1964b)

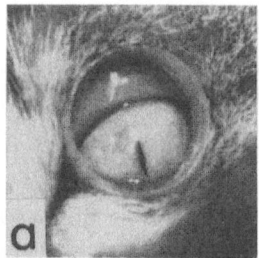 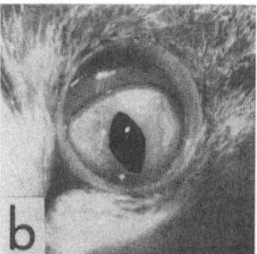

Fig. 4a and b. *Myosis in the blind cat during sleep with EEG synchronization.* a Myosis with downward rotation of the eye during physiological sleep. b Pupillary dilation and small upward movement of the eye following noise too slight: to produce EEG and behavioral arousal. (From BERLUCCHI et al., 1964b)

In brief, the parasympathetic pupillo-constrictor tonus of the Edinger-Westphal nucleus is regulated by i) its reflex excitation from the retina under light; ii) its inhibition by centers lying in the brain; and iii) by extraretinal components, in part autochthonous in nature. Pupils are myotic during relaxed wakefulness, in conditions of light adaptation, because the first mechanism overwhelms the second; the opposite situation prevails in blindness, but myosis reappears in sleep because brain inhibition decreases so that, in the absence of the photic reflex, the autochthonous components of the pupilloconstrictor tonus are released and become easily detectable.

The theory of the fall of the cerebral tonus during sleep might seem to be contradicted by the experiments of microelectrode recording from the cerebral cortex. In confirmation of HEBB's prediction (1949), EVARTS (see 1967) has shown—in the free moving, unanesthetized cat—that sleep is related with a generalized reorganization rather than with a fall of the neuronal discharges, at least with regard to those types of cortical neurons which have been recorded so far. However the experiments on the pupils can be best explained with a genuine fall of the tonic activity of well defined inhibitory centers. At least with respect to the regulation of the pupillo-constrictor discharge, the theory of the fall of a cerebral tone is still the more likely explanation of the myosis observed during physiological sleep and in the *cerveau isolé*. Thus the *cerveau isolé* reproduces events occurring during physiological sleep, as BREMER (1937, 1938) had predicted.

Several studies made on the acute *cerveau isolé* after BREMER's discovery suggest that a distinction should be made between the effects of precollicular (high *cerveau isolé*) and of postcollicular transections (low *cerveau isolé*). These two lines of investigation have been reviewed some years ago (MORUZZI, 1964) and will be discussed separately.

In the *low cerveau isolé* most of the midbrain is still connected to the cerebrum. Some degree of *tonic* activation is present in the cerebrum isolated from the pons by an acute postcollicular transection, as shown by the striking increase of EEG synchronization in the cat's neocortex, following reversible inactivation of the retinal dark discharge (BIZZI and SPENCER, 1962; see Fig. 5). The hypothesis may be advanced that some degree of a tonic activating influence still arises from the midbrain and/or the hypothalamus of the acute low *cerveau isolé*; an ascending tonic facilitation, however, that would not be adequate to maintain a waking state. The additional hypothesis may now be made that the critical level or the critical patterns required for the maintenance of wakefulness are gradually reestablished as chronicity progresses. Thus the results obtained in the chronic *cerveau isolé*, to be reported in the next chapter, would be explained by the gradual enhancement of a *latent or subliminar tonic* activation, one that was already present in the acute stage, soon after the postcollicular section. The latent activation of the acute low *cerveau isolé* might become supraliminar under the impact of sensory volleys, thus explaining the *phasic* EEG arousal produced in the cat's neocortex by olfactory stimulation (ARDUINI and MORUZZI, 1953). This hypothesis would permit an unitary explanation for the observations made in both the acute and chronic *cerveau isolé* preparations.

In summary, the acute low cerveau isolé preparation appears to be activated subliminarly with respect to the neocortex. This activation may be enhanced, phasically, by olfactory stimulation and, tonically, by recovery processes occurring only in the chronic preparation.

In the high *cerveau isolé* the midbrain is entirely separated from the cerebrum by a precollicular section. The analysis of the ocular behavior is obviously no longer possible and only electrophysiological studies are in fact available. They show that in the acute stage the synchronous neocortical rhythms cannot be disrupted by olfactory stimuli (ARDUINI and MORUZZI, 1953;

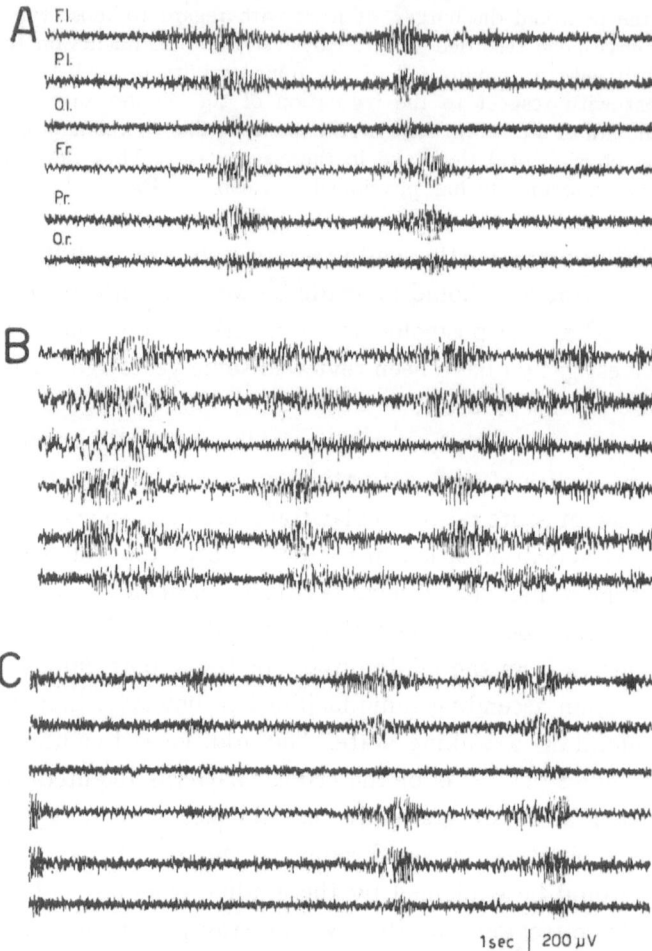

Fig. 5A–C. *Effects of acute visual deafferentiation on the EEG of the acute "cerveau isolé".* Records from frontal (*F*), parietal (*P*), occipital (*O*), left (*l*), and right (*r*) cortices in the acute low *cerveau isolé.* Striking enhancement of the background EEG synchronization (A) produced by increase of intraocular pressure (B): note increase in number or duration of spindles and appearence of slow waves during the interspindle lulls after 3 minutes of reversible retinal inactivation (B). There is a return to the original pattern (C) following decompression of the eyes. (From Bizzi and Spencer, 1962)

see Moruzzi, 1966). There are other differences between "low" and "high" versions of the *cerveau isolé* and between this preparation and the *encéphale isolé.* Their existence may be shown by the study of the arousal responses elicited by stimulation of the cerebral cortex (Bremer and Terzuolo, 1952, 1953, 1954; Mollica, 1958), of the hypothalamus (Rossi and Steffanon, 1953) and of the midline nuclei of the thalamus (Schlag and Chaillet, 1963).

2. The Chronic *Cerveau Isolé*

While studying the effects of partial sections of the midbrain, Genovesi et al. (1956) made an unexpected observation. When the final outcome of multistage lesions was a complete intercollicular interrruption of the brain stem, sparing only the pes peduncoli bilaterally, the cat could appear awake according to the usual ocular and EEG tests. These chronic animals corresponded, anatomically, to the "pyramidal" cat (WHITLOCK et al., 1953), which acute experiments had shown to behave as a *cerveau isolé*. Hence multistage lesions combined with a chronic study led to the conclusion that a state of alertness might be present in a cerebrum completely isolated from the ascending flow of impulses via the spinal cord and brain stem.

In the subsequent experiments we are now going to report, *complete* midbrain transection was carried out in a *single* operation, and the animals were followed for *longer* periods of time. Also in these chronic *cerveau isolé* preparations, behavioral and EEG signs clearly showed that after some time the animals were able to maintain a state of wakefulness.

BATSEL (1960) reported his studies on *cerveau isolé* dogs, which he had been able to follow for several weeks. The survival period lasted up to 73 days, a remarkable achievement which was mainly due to a new surgical technique of transection. He noted that as chronicity progressed there was a tendency for EEG spindles to diminish in number and frequency, until spontaneous desynchronization began to appear, followed by cycles of synchronization-desynchronization. His sections were practically *precollicular*, so that ocular behavior could not be controlled, and occasionally even the caudal diencephalon had been encroached upon. The spontaneous desynchronization occurred in these high *cerveau isolé* cats even after section of the optic nerves. Olfactory EEG arousal could not be produced when the EEG was spontaneously synchronized.

In a further group ef experiments, BATSEL (1964) studied the chronic behavior of low *cerveau isolé* cats. In three cases both olfactory bulbs were removed and the optic nerves were cut 2–3 weeks before the postcollicular transection of the brain stem. Even in these completely deafferented *cerveau isolé* preparations, spontaneous desynchronization tended to appear as chronicity progressed (Fig. 6). Although the caudal midbrain was damaged, EEG activation appeared in these "low" *cerveau isolé* after only one week; it was followed by the occurrence of the usual cycles of synchronization-desynchronization. The desynchronized EEG patterns appeared abruptly (Fig. 7) on a background of synchronization, while the reversal back to synchrony was slow and took place in several stages. When the preparation was not deafferented, slight pupillary dilation was occasionally noted during the period of EEG activation; and olfactory stimulation produced EEG activation, in confirmation of previous findings made in acute preparations (ARDUINI and MORUZZI, 1953). However olfactory arousal was unconstant. Surprisingly enough, spontaneous EEG activation eventually occurred notwithstanding the absence of olfactory arousal on a synchronized background. Tonic and phasic aspects

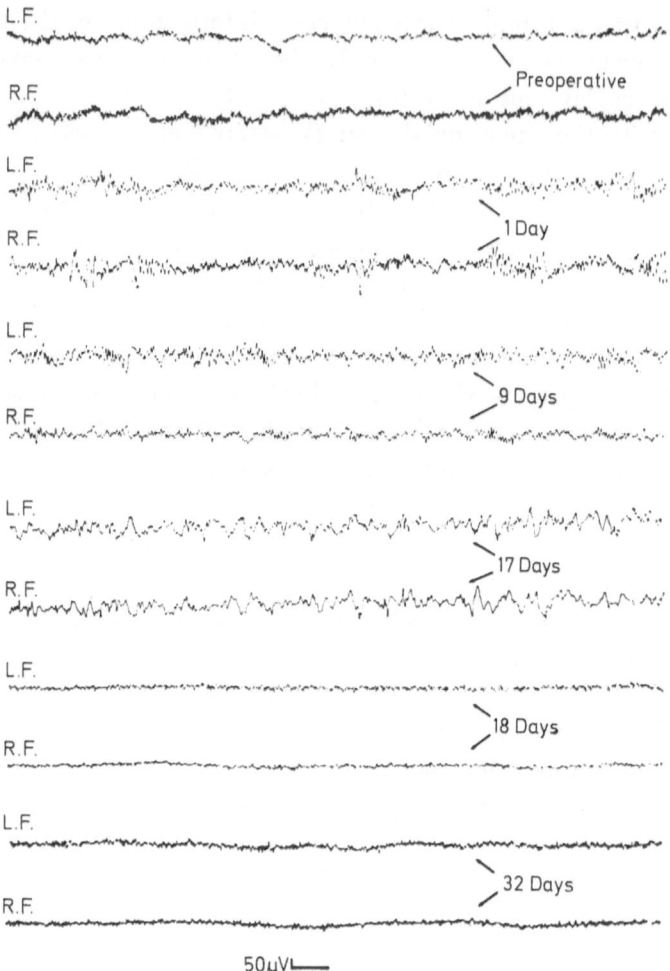

Fig. 6. *Appearance of desynchronized EEG patterns in the chronic, low "cerveau isolé" cat.* The upper pair of records shows desynchronization of the blind, awake animal, before the operation. Following low midbrain transection the EEG patterns remained synchronized, and the first period of desynchronization appeared at 18 days. Several long lasting periods of EEG desynchronization were observed in the later period of survival. Records are from left (*LF*) and right (*RT*) frontal leads. (From Batsel, 1964)

of EEG arousal could thus be dissociated. Adrian's induced waves (1950) were not recorded from the olfactory bulb, so that it would be impossible to dismiss the hypothesis that chronic or acute changes of the nasal mucous membranes might account for the negative results. That peripheral factors do not provide a satisfactory explanation is shown, however, by Villablanca's experiments (1965), to be reported below. Although cooling of the chronic preparation enhanced the tendency to EEG desynchronization, an observation later confirmed by Hobson (1965), spontaneous EEG activation occurred also at normal rectal temperatures. The effect of generalized cooling may be attributed to the activating influence of the thermoregulatory center, possibly exerted through the posterior hypothalamus.

The chronic *cerveau isolé* cats of VILLABLANCA (1962, 1965) had been transected at rostral midbrain (Fig. 8) or *precollicular* levels. In five animals the operation was combined with photocoagulation of the optic discs and ablation of the olfactory bulbs. Spontaneous periods of EEG desynchronization (Fig. 9), often lasting several hours, appeared from 7 to 10 days after the brain stem section, even in the completely deafferented cerebrum. The periods of EEG synchronization occupied from 35 to 50% of the total recording time.

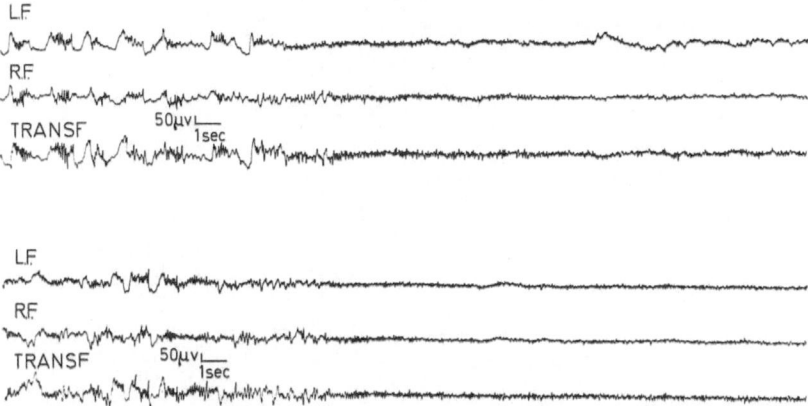

Fig. 7. *Sudden transition from synchronized to desynchronized EEG patterns in the chronic, low "cerveau isolé" cat. The isolation of the cerebrum had been completed by olfactory and visual deafferentiation. The synchronized EEG patterns change suddenly to low voltage fast activity. Records taken 9 (above) and 13 days (below) after operation, the desynchronized period lasting respectively 7 and 28 minutes.* (From BATSEL, 1964)

At their onset the synchronous rhythms could be easily disrupted by olfactory stimulation, but the effect was not seen when the period of synchronization had fully developed, an interesting finding, which possibly explains some of BATSEL's negative results (1964). Photic arousal was not obtained, in conformation of previous findings (ARDUINI and MORUZZI, 1953; BATSEL, 1960). No attempt was made to repeat BIZZI and SPENCER's experiments (1962) on the effect of reversible suppression of the retinal dark discharge on the synchronous rhythms.

These investigations were reported by VILLABLANCA in 1962 and 1965, and were completed (1966b) by an important study on pupillary behavior made in the *low, chronic cerveau isolé*. In order to eliminate the changes produced by photic reflexes, the optic discs were photocoagulated, thus reproducing the experimental situation of the sleep studies made by BERLUCCHI et al. (1964b) on the free moving cat. Extreme myosis was present in these visually de-afferented preparations during the first post-operative week, when EEG synchronization was complete, confirming the original findings of BREMER (1935, 1937, 1938). As lapses of EEG desynchronization began to appear, the myosis became less marked. The chronic *cerveau isolé* cats were followed for quite

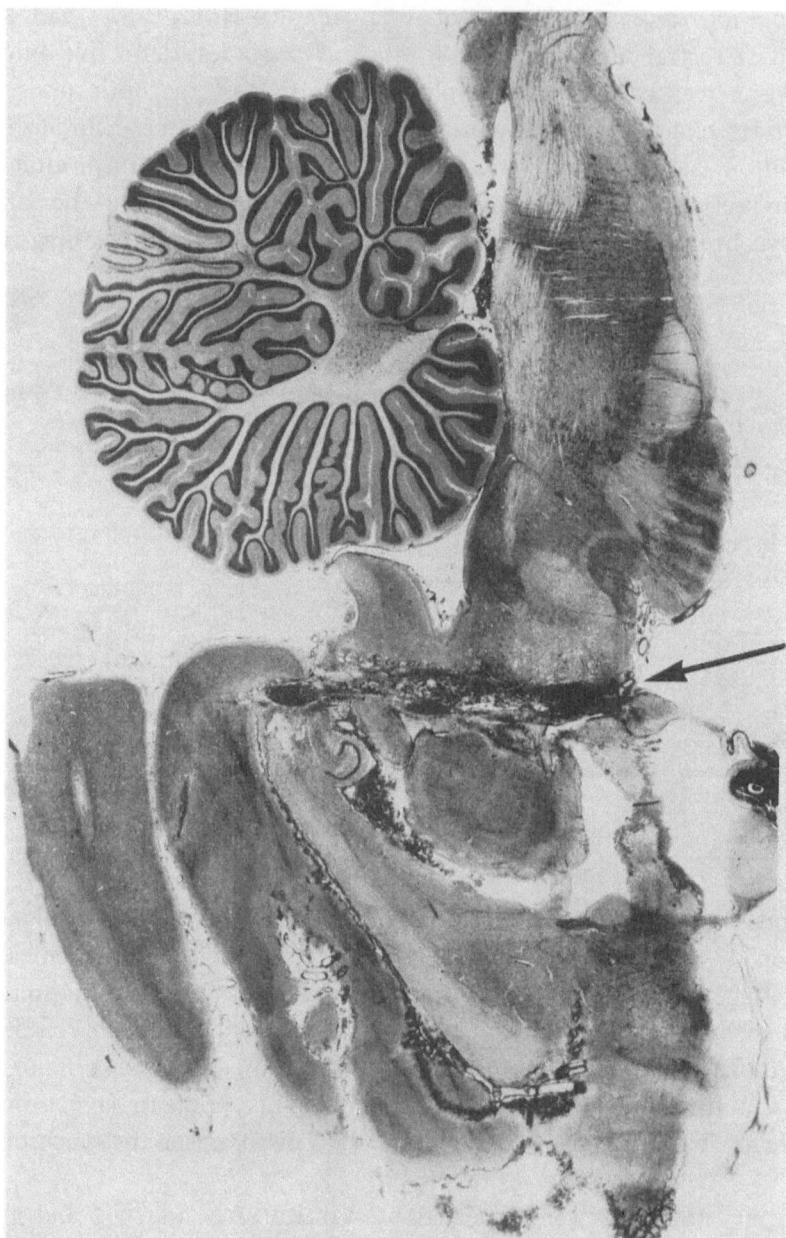

Fig. 8. *Rostral midbrain transection in chronic "cerveau isolé" cat.* Toluidine stained sagittal section. (From Villablanca, 1965)

long periods of time, from 15 to 72 days. As chronicity progressed the usual cycles of desynchronization-synchronization reappeared. They were accompanied by corresponding changes in the pupillary diameter, similar in sign to those observed during sleep and wakefulness in blinded cats by Berlucchi et al. (1964b). It turned out that, at the onset of the EEG synchronization, olfactory stimulation produced EEG arousal and pupillary dilation; however

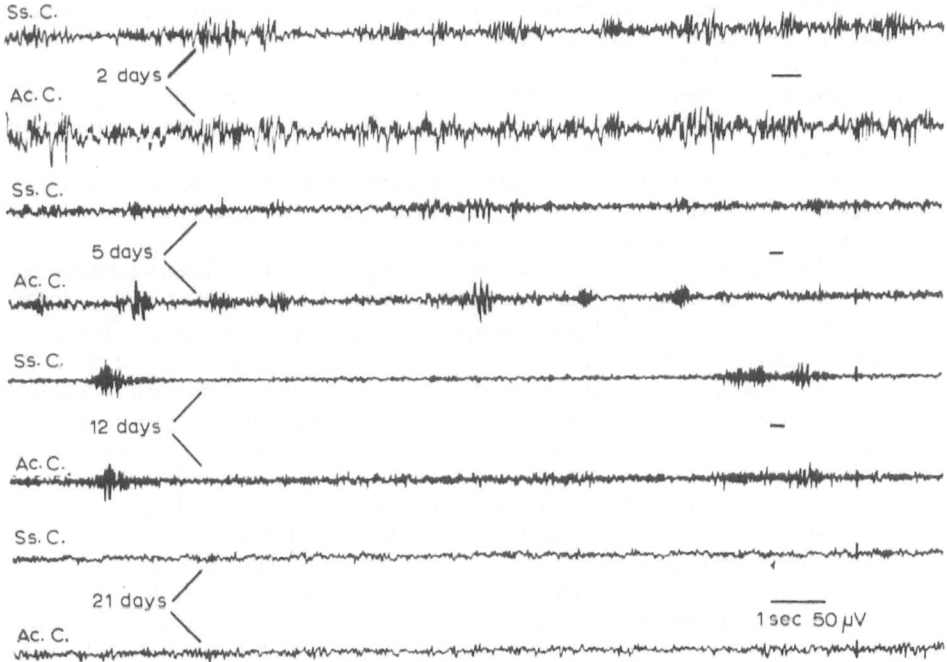

Fig. 9. *Progressive EEG activation in the chronic "cerveau isolé" cat.* Electrocorticograms from primary somatosensory cortex *(Ss.C)* and primary auditory cortex *(Ac.C)*, taken at 2, 5, 12, and 21 days after operation. Note progressive lengthening of the interspindle lulls, leading to full activation after three weeks. (From VILLABLANCA, 1965)

both effects disappeared, and the pupils became fissurated, as synchronization progressed. The pupillary dilation occurring during EEG arousal was less marked that in the free-moving, blinded cat (BERLUCCHI et al., 1964b), and the remarkable difference remained after sympathetic preganglionic denervation. The inhibition of the parasympathetic pupillary tone produced by arousal is thus apparently less strong after postcollicular section, an observation confirming ZBROZINA and BONVALLET's conclusions (1963) on the tonic inhibitory influence exerted by the medulla on the Edinger-Westphal nuclei.

SERKOV, MAKULKIN and TYCHINA (1966) confirmed VILLABLANCA's result on cats which survived up to 338 days to the intercollicular section. The EEG desynchronization appeared spontaneously after 7–9 days, but could be elicited by olfactory stimulation only at the 15th day.

In summary, chronic studies on cerveau isolé animals lead to the conclusion that a genuine state of wakefulness may be maintained in the isolated cerebrum. This statement is supported by the EEG records, and also by the ocular behavior, when the transection is carried out behind the third nerves.

The alternative hypothesis that the EEG desynchronization might be related with episodes of paradoxical sleep, which would be absent in the acute *cerveau isolé*, is refuted by the following data, reported and fully discussed by VILLABLANCA (1965, p. 583):

i) All EEG and behavioral signs of desynchronized sleep are related to, and probably produced by, neuronal discharges arising in the pons (see Jouvet, 1962, 1967a), therefore caudally to a postcollicular section;

ii) in the low *cerveau isolé* one observes only the rapid eye movements produced by the discharge of the VI cranial nerve neurons, which are caudal to the transection (Villablanca, 1966b).

iii) The episodes of spontaneous EEG desynchronization may continue for hours (Villablanca, 1965), while the episodes of paradoxical sleep occurring in the free moving cat last only 15–20 minutes (see Jouvet, 1967a).

iv) EEG and pupillary signs of activation may be produced by olfactory stimulation, which on the intact cat causes only arousal, never paradoxical sleep (Villablanca, 1965).

The objection can be raised that episodes of paradoxical sleep might reappear as chronicity progresses. In the absence of studies on the behavioral correlates of the EEG in the chronic *cerveau isolé*, it must be conceded that a distinction between two manifestations characterized by the same electrocortical patterns appears very difficult. According to this objection observations iii) and iv) might be concerned exclusively with genuine episodes of arousal, while the absence of complete ocular outbursts (ii) would be due to the fact that the third nerve motoneurons are separated by the postcollicular section from the triggering vestibular nuclei (see Pompeiano, 1967b).

However considerations reported under i) can be dismissed only with great difficulty. It might de maintained that, as chronicity progresses, there would be a recovery of the ability to promote episodes of paradoxical sleep, just as there is a recovery of the arousal capacity. However there is no proof whatsoever that neurons localized somewhere in the cerebrum may take over the triggering influence of the pontine center. It would be certainly interesting (see Berlucchi, 1970) to try to record from the pyramidal neurons of the motor cortex (see Evarts, 1964, 1965), in the chronic *cerveau isolé*. These experiments, however, have not yet been attempted.

The mechanisms underlying the comatose syndrome of the acute *cerveau isolé* and the recovery of the sleep-waking cycle in the chronic preparation may now be discussed. The two problems are obviously interrelated and should be reviewed together.

Two facts are undeniable: i) the structures responsible for the sleep-waking cycle in the chronic *cerveau isolé* are deeply disorganized immediately after the midbrain transection; ii) the disorder is reversible, as shown by the outstanding recovery that occurs after one week. For reasons stated in Berlucchi's review (1970), and in view of experimental results to be discussed later (see pp. 74–81), the structures responsible for the sleep-waking cycle are localized in the hypothalamus. We have the choice between two hypotheses: i) a purely *non specific damage of the hypothalamus*, and/or the midbrain overlying the section, might be produced by the surgical trauma; ii) a disorganization of hypothalamic centers connected with the waking behavior may result from a sudden withdrawal of the tonic influence of the ascending reticular system *(hypothesis of the reticular deafferentation)*. According to the first hypothesis, edema, local circulatory alterations, phenomena like spreading depression—in short any kind of local damage produced by the lesion of neighboring structures—might temporarily inactivate the structures lying rostrally to the midbrain transection, *independently of their specific function*. According to the second hypothesis the sudden withdrawal of the tonic influence exerted by the brain stem *on well defined hypothalamic nuclei* would be responsible for the disorder.

Let us start with the discussion of the *hypothesis of the unspecific damage*. Three lines of evidence suggest that such an assumption cannot explain the major findings of the acute syndrome.

i) Complete recovery from spreading depression—a phenomenon, incidentally, which is not easily elicited in the cat—takes from 30 minutes to a few hours (see BUREŠ and BUREŠOVÁ, 1960; BUREŠOVÁ and BUREŠ, 1969 for ref.). Thus the time factor is of an entirely different order of magnitude, since the ability to maintain wakefulness reappears, in the *low cerveau isolé*, after a minimum interval of one week.

ii) Convincing physiological evidence is available that the damage produced by the surgical trauma is amazingly limited, both rostrally and caudally to the midbrain transection. In fact neither precollicular (VILLABLANCA, 1966a; ZERNICKI et al., 1970) nor intercollicular (BREMER, 1935, 1937; ROSSI and STEFFANON, 1953) transections prevent the tonic activity of the neighboring Edinger-Westphal neurons, as shown by the presence of fissurated myosis. The experiments of VILLABLANCA (1966a) are particularly important in this respect, since the pupillary behavior of decerebrate (precollicular) cats was studied in chronic conditions by the same author who is responsible for the study of the chronic *cerveau isolé*. Incidentally, the fact that myosis was observed continuously for two weeks shows that we are concerned with release phenomena, a conclusion in line with the usual interpretation of decerebrate rigidity.

iii) EEG synchronization may be reversibly produced by cooling the midbrain in the intact cat (NAQUET et al., 1966). It might be objected that cold inactivation is likely to spread to the hypothalamus. However, cooling of the upper pons is followed by EEG synchronization and pupillary myosis (BERLUCCHI et al., 1964a) in both the *encéphale isolé* and the midpontine pretrigeminal preparation, therefore in experimental conditions where systemic circulatory changes and alterations of respiration are avoided. The behavioral and EEG signs of alertness of the midpontine pretrigeminal preparation are reversibly abolished, moreover, by cooling inactivation of the upper pons (l.c.). Finally the myosis shows that this cooling inactivation does not reach the upper midbrain, and therefore spares the hypothalamus.

Some local damage must obviously be produced by the section, and the electrolytic destruction is likely to be more dangerous in this respect (see SPRAGUE, 1967), but all the available evidence suggests that the lesion is probably limited to a thin strip of tissue lying immediately above and below the transection. These considerations by no means imply that hypothalamo-hypophyseal alterations are not present in the acute *cerveau isolé*. In fact neither i) midbrain transections of previously hypophysectomized animals nor ii) hypophysectomy of pretrigeminal preparations have been made so far. This line of work should not be neglected. However the core of the syndrome appears to be the consequence of the sudden withdrawal of the powerful influence exerted by the brain stem on the hypothalamus.

The *hypothesis of the reticular deafferentation* must now be discussed. Apparently some time after the midbrain transection there is a compensation for the sudden loss of the ascending influence of the brain stem, and the sleep-waking cycle reappears in the isolated cerebrum. The problem is to know how this compensation occurs.

Our interpretation starts from the following assumptions:

i) the acute effects of the midbrain transection are mainly due to the sudden withdrawal of an ascending barrage arising in the brain stem.

ii) The severity of the effects of the acute deafferentation of the brain should be in proportion to the intensity of the tonic impingement, in line with a principle usually accepted for the spinal shock.

iii) We are going to see later (see pp. 74–81) that there are in the hypothalamus and in the basal forebrain area two antagonistically oriented regions related respectively with waking and sleep behaviors.

iv) Since the phasic aspect of reticular activity is undoubtedly related to the arousal phenomenon it is a likely assumption that the tonic aspect is concerned with the maintenance of wakefulness. Hence the structures of the posterior hypothalamus related to the waking behavior are likely to be more heavily depressed by the acute deafferentation.

These four assumptions lead to the hypothesis that an imbalance between the antagonistic diencephalic structures related to sleep and waking behavior is the main cause of the acute syndrome of the *cerveau isolé*. According to this hypothesis, what chronicity provides is basically *compensation of an imbalance*. The sleep-waking cycle might in fact be present at any moment in the isolated cerebrum provided the imbalance between the antagonistic diencephalic structures concerned with sleep and wakefulness be not too strong. The imbalance, however, must be very strong if we accept the hypothesis that the midbrain transection is followed by a striking deactivation of the isolated cerebrum. The gradual recovery of the activating potentiality of the isolated cerebrum, during the days following the sudden withdrawal of the *ascending* influence of the brain stem, would follow a course similar to that of the recovery of the isolated segments of the spinal cord, after withdrawal of the *descending* influence of the diencephalon. In both cases denervation hypersensitivity (see STAVRAKY, 1961) might well play a role. When a critical level in the recovery of the activating mechanisms is reached in the isolated cerebrum, the ability to alternate between sleep and wakefulness suddenly reappears. Experiments on the midpontine pretrigeminal preparation, to be discussed in the next chapter, provide further evidence in support of the imbalance hypothesis.

3. The Pretrigeminal Preparation

When the brain stem is transected just in front of the origin of the trigeminal roots, i.e. a few millimeters behind the usual level of postcollicular sections, the preparation is characterized by EEG (Fig. 10) and ocular patterns entirely

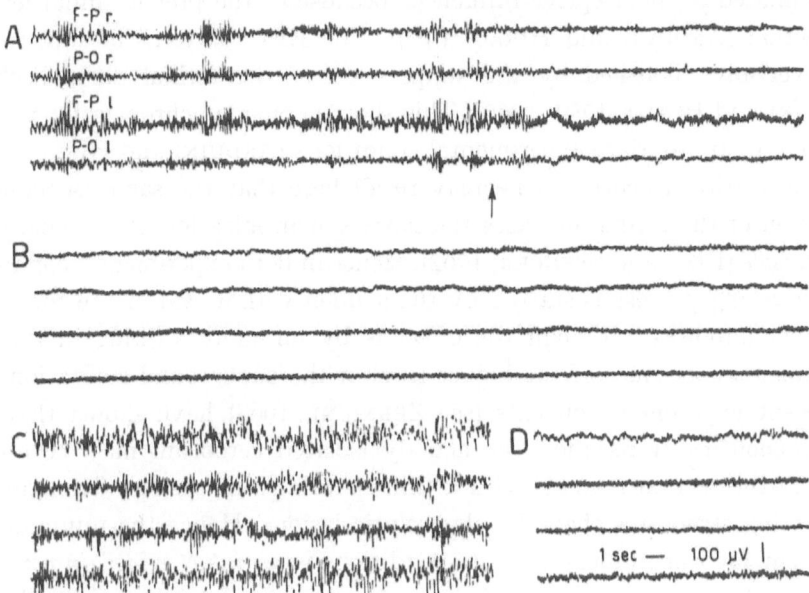

Fig. 10 A–D. *EEG patterns of an unanesthetized cat before and after midpontine pretrigeminal transection.* A, A normal unrestrained cat in a quiet environment displays synchronized EEG patterns and spindle bursts, easily interrupted by noise in the room (arrow). The same cat, following midpontine pretrigeminal transection, shows persistent EEG activation (B), which is reversibly replaced by synchronized rhythms (C) when sodium pentobarbital (10 mg/kg) is injected intravenously. D Recovery of activated patterns 4 hours following pentobarbital administration. Right (*r*) and left (*f*) fronto-parietal (*F–P*), and parieto-occipital (*P–O*) records. (From BATINI et al., 1959a)

different from those of BREMER's *cerveau isolé*. Even in the acute period, this midpontine pretrigeminal preparation presents clear-cut EEG and ocular signs of alertness (BATINI et al., 1958, 1959a). They are usually observed within two hours from the operation, occasionally even after only 20 minutes (ZERNICKI, 1968). Control experiments (l.c.) have shown that the shortest delay corresponds to the time required for the disappearence of the effects ot the ether.

The difference between these two acute preparations, the *cerveau isolé* and the midpontine pretrigeminal cat, appears particularly striking if one considers that in both cases only two afferent channels, olfactory and visual, are available to the isolated cerebrum. This fact has important theoretical implications, which will be developed in the next pages.

Anatomical deafferentation of both the olfactory and visual systems leads to an EEG synchronization which, however, disappears after 24 hours, thus showing that the EEG patterns of activation of the midpontine pretrigeminal preparation are not critically dependent on the tonic sensory inflow (Batini et al., 1959c). This flow, however, contributes to the activating influence, as shown by experiments of functional visual deafferentation made in the dark adapted preparation. Reversible abolition of the retinal discharge may be obtained by increasing bilaterally the intraocular pressure, an effect not contaminated by nociceptive influences because of the pretrigeminal level of the section (Arduini and Hirao, 1959). The EEG patterns of wakefulness are reversibly abolished by the suppression of the retinal dark discharge (Arduini and Hirao, 1959, 1960). This short lasting synchronization may be recorded, in appropriate experimental conditions (Arduini and Hirao, 1959), from the entire neocortex. One may recall here that the same ischemic in-activation of the retina increases the EEG synchronization of the acute, low *cerveau isolé* (Bizzi and Spencer, 1962). Hence in both experimental conditions a deactivating process is started by the sudden withdrawal of the tonic flow of retinal impulses, although the effect is by far more striking in the pre-trigeminal cat, because of the higher level of the background activation.

Recent experiments on cats (see Zernicki, 1968) have shown that also after prepontine transection the acutely isolated cerebrum may be awake, provided the section is made with a thin spatula, which is less damaging for neighboring structures than the electrolytic lesions. Hence the minimum re-quirement for wakefulness in acute conditions is *an intact midbrain connected with an intact cerebrum*, a conclusion fully confirmed by human pathology (Orthner, 1969). Obviously when also the upper pontine nuclei are connected to the cerebrum, as in the midpontine preparation, the maintenance of wake-fulness is likely to be easier. A study of the differences between prepontine and midpontine cats has not been made. At the present state of our knowledge it is perhaps advisable to follow Zernicki's suggestion (1968), and to refer simply to the pretrigeminal preparation.

The sleep-waking cycle is severely imbalanced, but never abolished, in the pretrigeminal preparation. This is an important difference with respect to the acute *cerveau isolé*, where any cyclic change is entirely absent. It has been maintained that true sleep—that is, a reversible state characterized by EEG synchronization and behavioral unresponsiveness—would appear only at the third day (Slósarska and Zernicki, 1969). During the first two days the synchronous EEG patterns would be related to a state of drowsiness, rather than of true sleep, as shown by the vigorous ocular responses to visual stimuli (l.c.). However recent experiments by the same authors (1971) have shown that genuine episodes of synchronized sleep appear already during the first day, when the preparation is kept comfortably in the incubator and fed with glucose. They may be interrupted by olfactory stimulation. In summary, with the midpontine pretrigeminal preparation a sleep-waking cycle may undoubtedly be present even during the first day, i.e. in acute conditions, although it is greatly distorted by a remarkable tendency to insomnia, as will be pointed out later (see pp. 31–36). The tendency to insomnia may increase, thus giving the impression of the abolition of the sleep-waking cycle, as a consequence of fall of temperature or of hypoglycemia.

Five years after the first preliminary report presented in 1958 by BATINI et al. (l.c.), the progress of our knowledge on the pretrigeminal preparation had been extensive and several lines of convergent evidence could be quoted (MORUZZI, 1963 b) in support of the statement that a genuine state of alertness was present when the EEG was desynchronized.

i) The pretrigeminal cat follows with vertical eye movements any object passing across its visual field (BATINI et al., 1958, 1959; KING and MARCHIAFAVA, 1963).

ii) A pupillary dilation occurs, occasionally, when a visual stimulus is suddenly presented (BATINI et al., 1959a; AFFANNI et al., 1962a).

iii) Visual accomodation to a near object is present (ELUL and MARCHIAFAVA, 1964).

iv) The ocular symptoms described from i) to iii) are absent when the EEG is synchronized (BATINI et al., 1959a; AFFANNI et al., 1962a; ELUL and MARCHIAFAVA, 1964).

v) The tracking eye movements and the mydriatic response to sudden illumination are easily extinguished upon repetition, an observation suggesting that they are likely to represent the fragmentary expression of the orienting reflex (AFFANNI et al., 1962a).

vi) When trains of flashes are repeatedly followed by hypothalamic stimulations producing pupillary dilation, the photic stimulus begins, after a number of trials, to elicit the mydriatic response (AFFANNI et al., 1962b). Control experiments (l.c.) show that true conditioning may be obtained in the isolated cerebrum.

As will be pointed out later, the behavioral signs of alertness may be absent notwithstanding the presence of a desynchronized EEG. However the significance of the positive behavioral findings is proved by their reversible disappearence when the EEG becomes synchronized. These conclusions are supported also by the results of experiments on the lateral geniculate units which we are now going to review. MAFFEI et al. (1965) reported that the modulation of the firing of lateral geniculate units by sinusoidal light was present, in their acute midpontine cats, on a background of behavioral and EEG wakefulness (Fig. 11). The best results (less distortion in the sinewave shape) were obtained when the EEG was fully desynchronized and the animal showed tracking ocular movements. This effect disappeared as soon as the

EEG became synchronized (Fig. 11), and was never observed on retinal units (Fig. 12). Maffei and Rizzolatti (1965) found, moreover, a striking enhancement of the responses to flashes of lateral geniculate units when the midpontine preparation presented the typical waking patterns (Fig. 13). Hence three lines of evidence—behavioral, EEG, and unit recording—prove that the acute midpontine cat may be awake.

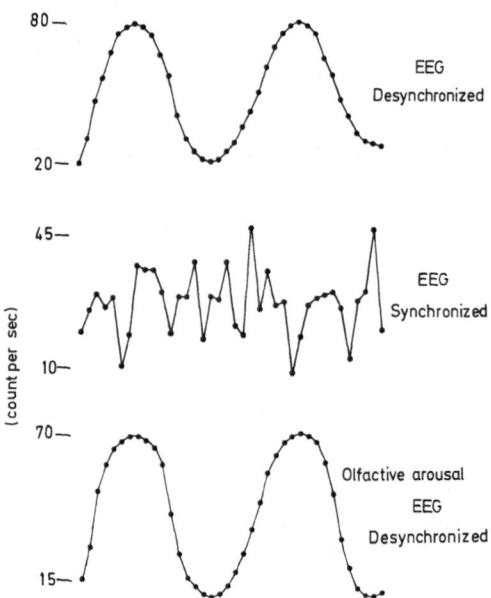

Fig. 11. *Average firing of the same LGB unit in response to sinewave photic stimulation during spontaneous episode of synchronized sleep and during wakefulness induced by olfactory stimulation.* The linear modulation by sinewave light of the firing rate of the lateral geniculate unit is present, in the acute midpontine cat, during spontaneous behavioral and EEG wakefulness (above) and during arousal produced by olfactory stimulation (below), while it is absent altogether during the episode of EEG synchronization (middle). Dots give the frequency of the unit averaged over 10 cycles by the computer. The frequency of the sinewave modulation of light intensity was 0.1 cps. (From Maffei et al., 1965)

In summary, there is little doubt that an acute pretrigeminal preparation can be awake, and that it is in fact frequently awake, although this obviously does not imply that the cat is behaviorally awake whenever the EEG is desynchronized.

During the following years important experiments leading to the same general conclusions were made by Zernicki, who devoted two extensive reviews to the pretrigeminal preparation (1964, 1968). Reference is made to the detailed analysis of the fixation reflex and of the tracking eye movements (Dreher and Zernicki, 1969); to the patterns of visual stimulation leading to the mydriatic response (Zernicki and Dreher, 1965; Zernicki et al., 1967); to the arousing effect of olfactory stimulation (see Zernicki, 1964, 1968); to the studies of habituation (l.c.); and, finally, to the conditioning experiments on the pupillary response (Zernicki and Osetowska, 1963). The wide litera-

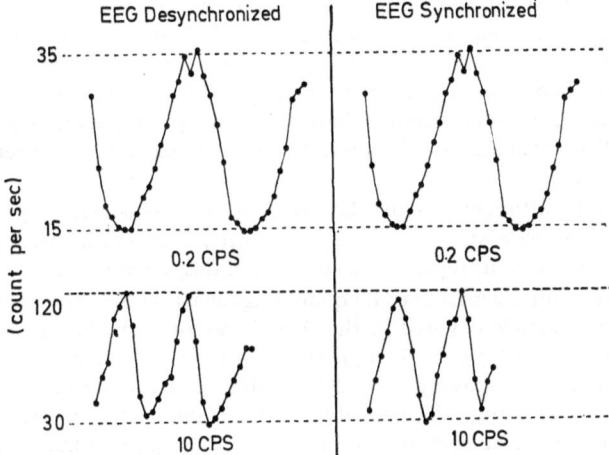

Fig. 12. *Firing rate of retinal ganglion cell in response to low and fast sinewave variations of light intensity unaffected by sleep and wakefulness.* The numbers below each diagram represent the frequency of the sine-wave stimulation. No clear-cut difference between states characterized by EEG desynchronization and synchronization (obtained by cooling the upper part of the pons). Midpontine pretrigeminal cat. (From MAFFEI et al., 1965)

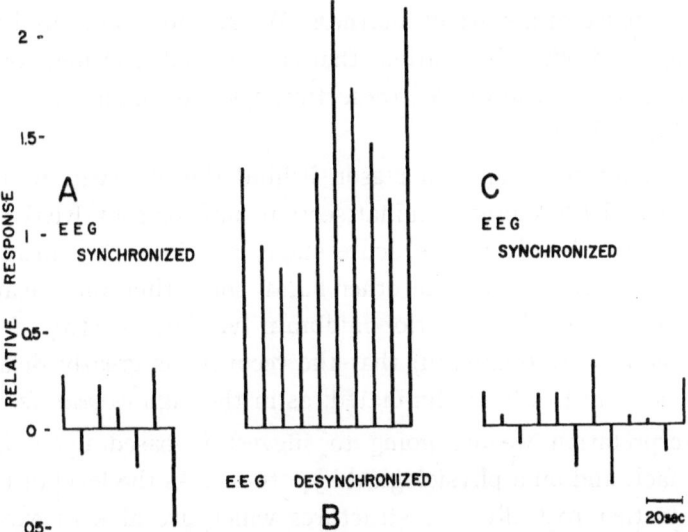

Fig. 13 A–C. *Responses of LGB unit to flashes of light of weak intensity during synchronized sleep and wakefulness.* Flashes of light 300 msec in duration of constant intensity (1 Lux) were applied every ten seconds. Relative rate response of lateral geniculate cell (vertical scale) is given by $\frac{X-\bar{X}}{\bar{X}}$, where X is cell firing of LGB cell starting with the time of light pulse (the time of measure is 1 sec) and $\bar{X}$ is the average rate during 6 seconds prior to pulse. The intensity of the flashes of light was decreased in such a way that no response was detected during synchronized sleep (A, C). A stimulus that was ignored during sleep gave a clear-cut response during wakefulness (B). Both sleep and wakefulness alternated spontaneously in this acute midpontine cat. (From MAFFEI and RIZZOLATTI, 1965)

ture will not be reviewed in detail, since it is not exclusively concerned with sleep physiology.

An exception will be made for the puzzling observation of EEG desynchronization without signs of behavioral alertness. This absence of behavioral signs of wakefulness is of little physiological interest when it is permanent or long lasting. Coma without slow EEG activity following pontine lesions is well known in the clinical literature (Loeb and Poggio, 1953; Chatrian et al., 1964), although signs of consciousness may be present in man after complete destruction of the pontine reticular formation (Orthner, 1969), in confirmation of experimental findings on the pretrigeminal cat (see Moruzzi, 1963 b, p. 240). The pathology of the pretrigeminal preparation has been discussed by Zernicki (1964, 1968).

It is an entirely different matter, however, when the ocular responses disappear for a short time, while the EEG remains desynchronized. These short periods of behavioral unresponsiveness might be intrepreted as episodes of desynchronized sleep.

The lines of reasoning which may be quoted against this hypothesis are summarized in the section of this article devoted to the chronic *cerveau isolé* (p. 21). While one must agree with Berlucchi (1970) that the hypothesis of the reappearence of the paradoxical episodes by a process of recovery should be tested in a *chronic* isolated cerebrum, it appears justified to state that the available evidence does not support the assumption that episodes of paradoxical sleep may occur in the *acute* pretrigeminal cat. While phenomena of adaptive compensation cannot be formally denied until experimentally disproved, the assumption that paradoxical episodes may appear within a few hours after a prepontine section obviously conflicts with the present views on the pontine localization of the paradoxical episode (see Jouvet, 1962, 1967).

Our attention will be concentrated upon the positive cases leaving no doubt about the existence of a state of alertness. We are not confronted with occasional findings, but with observations that can be made routinely on any good preparation. The implications of these findings are important and may be summarized as follows:

a) There must be a common reason behind the observations i) that the cerebrum isolated by a pretrigeminal section may present EEG and ocular signs of alertness already in the acute stage, provided the entire midbrain has been spared; ii) that, on the other hand, some time must elapse before recovery occurs when the posterior midbrain has been destroyed, as in the low *cerveau isolé*; and, finally, iii) that the recovery is greatly delayed when the entire midbrain has been eliminated, as in the high *cerveau isolé*.

The interpretation we are going to suggest is based upon an obvious anatomical fact, and on a physiological hypothesis. As the level of transection is brought further rostrally, the structures which are able to exert a tonic activating influence on the brain are reduced in extent. The physiological hypothesis may be made that this anatomical reduction is compensated, as chronicity progresses, by recovery processes, and therefore by an increased efficiency of the remaining activating mechanisms. A prepontine section with full integrity of the midbrain is the most rostral level of cerebral deafferentation which still permits the maintenance of wakefulness in the acute stage. This is likely to be easier when the upper pons is still connected to the cerebrum, as in the midpontine pretrigeminal cat. Only chronic experimentation, however, will permit the *cerveau isolé* preparation to reach the critical level of activation required for alertness, when the section is made at post-, inter- or precollicular levels. The recovery is eventually never complete, particularly when the level

of the section is high, as shown by the fact that the insomnia of the pretrigeminal cat (see p. 32) is never observed in the chronic *cerveau isolé*.

   b) The remarkable differences between the *cerveau isolé* and the pretrigeminal preparations (see MORUZZI, 1964), a striking phenomenon when
EEG and ocular behavior are studied in the acute stage, is a new piece of
evidence (BATINI et al., 1959a) in favor of the hypothesis (MORUZZI and
MAGOUN, 1949, p. 470) that withdrawal of the tonic influence of the ascending
reticular system is responsible for the coma of the acute *cerveau isolé*. In both
preparations the sensory deafferentation is in fact complete, with the exception
of the olfactory and visual systems. Hence only the neurons lying between
the two sections are the likely candidates for explaining the critical differences
in both the EEG and ocular behavior.

   The alternative hypothesis (see BERLUCCHI, 1970) that some kind of unspecific, reversible damage may occur in the activating structures of the
cerveau isolé lying above the brain stem transection, one that would be absent
or critically unimportant after pre- or midpontine transections, cannot be
dismissed altogether, although it appears on the whole less convincing. We
have seen (p. 23) that some degree of reversible alteration must occur immediately above and below any brain stem transection, but we have also
stated at length (pp. 23–24) the reasons for our belief that these minor and
inconstant alterations do not provide a sufficient challenge to the classical
interpretation.

   In our opinion, a *unitary explanation* of the behavior of both the acute
and the *chronic cerveau isolé*, and of the pretrigeminal preparation, is given
by the hypothesis that an *imbalance* of the sleep and waking centers of the
brain is the major consequence of any brain stem transection. We are going
to see that in the pretrigeminal cat the imbalance is in fact just opposite,
and probably also less severe, than that of the acute *cerveau isolé*.

   The study of the pretrigeminal preparation has led also to another unexpected conclusion, namely that in the lower brain stem there are structures
oriented antagonistically with respect to the activating reticular system.
BATINI et al. (1959a) reported that EEG activation patterns were present
one day after olfactory and visual deafferentation, when the cerebrum is
isolated by a midpontine pretrigeminal section. In this experimental situation
the isolated cerebrum was completely deafferented. These results obviously
conflicted with the generalized EEG synchronization elicited in the *encéphale
isolé* cat by bilateral gasserectomy (ROGER et al., 1956), as well as with
the older observation that extensive sensory deafferentation may produce
a strong tendency to sleep, in the otherwise intact cat or dog (see ZERNICKI,
1964, p. 257, for the literature). The great tendency to sleep after sensory
deafferentation and the complete absence of this phenomenon in the pre-

trigeminal cat can be explained only by postulating a deactivating influence arises below the pretrigeminal transection.

This conclusion is supported by an entirely different line of evidence. In the midpontine pretrigeminal preparation there is a striking tendency to remain awake, a true experimental insomnia. To quote from the first preliminary note of BATINI et al. (1958):

"Hourly electroencephalographic examination, for not less than 24 consecutive hours, of the same cats before and after midpontine transection brought out some striking differences. While the normal intact cats, when isolated in a quiet environment, would display low-voltage, fast electroencephalographic activity for not more than 20 to 50 percent of the total recording time, a waking electroencephalographic pattern persisted in the midpontine preparations for at least 70, and frequently 90, percent of the total time of recording, in spite of the complete absence of any intentional stimulation" (l.c., p. 31). Therefore "a synchronizing, or possibly sleep-inducing, influence exerted by some structure in the caudal brain stem can be tentatively envisaged" (l.c., p. 32).

We have seen above (p. 29), that the pretrigeminal preparation is not necessarily behaviorally awake when the EEG is desynchronized. The periods of behavioral inactivity will reduce the percentage of time which may be classified as wakefulness. But the tendency to insomnia is too striking to be denied, even if one leaves aside the fact that the release hypothesis is fully supported by other lines of evidence.

The only alternative explanation is the irritative hypothesis, which therefore will be discussed at length.

Clinical neurologists who are so frequently concerned with processes of local irritation lasting for years, may object to the insistence of the physiologist on the temporary character of the irritative phenomena. The reasons of this attitude should be justified. When a given phenomenon is enhanced as a consequence of a lesion, the effect may be due either to *release* or to *irritation*. A release phenomenon, such as decerebrate rigidity, is easily recognized whenever it is clearly due to the *stabilized deficiency of a tonic inhibitory influence*. The irritation, on the other hand, may be due to injury discharges, to slight local circulatory changes giving rise to ischemic excitation of neurons, and of course to several other comparatively short lasting factors. These are generally transient phenomena, since it is extremely unlikely that an irritative situation remains stabilized in both space and time. Irritation either disappears or spreads to neighboring regions. A prolonged ischemia, moreover, is likely to give rise to deficiency symptoms, hence to a reversal of the syndrome. The pathological situations causing convulsive fits, as focal epilepsy, are rare outside the neocortex and the hippocampus.

A small degree of irritation may be present immediately after the transection and this phenomenon could actually explain some findings observed immediately after the operation. However three lines of evidence show that the striking imbalance in favor of activation EEG patterns cannot be due to irritation.

a) The low voltage fast EEG activity persisted unmodified throughout the survival time, up to 9 days, in the experiments of BATINI et al. (1959a). The midpontine and prepontine pretrigeminal cats of SLÓSARSKA and ZERNICKI (1969) survived up to 22 days, and the amount of sleep with EEG synchronization was always less than in the normal animals.

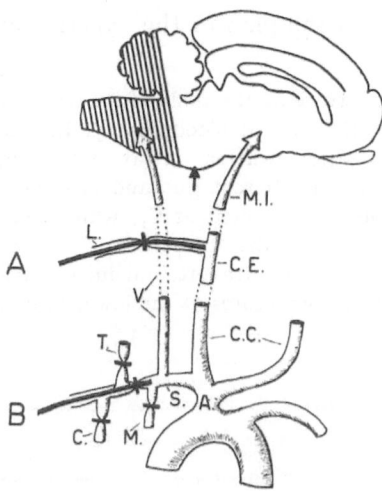

Fig. 14. *Schematic drawing showing procedure for separate perfusion of either medulla and caudal pons, or rostral pons, midbrain, and upper cerebrum.* Upper part: outline of cat's brain. Small arrow indicates level of clamping of the basilar artery, cross hatched area region supplied by vertebral circulation. Lower part: schema of the vascular circuits. Anatomical abbreviations: *A* arteria anonyma (truncus brachiocephalicus); *C* truncus costocervicalis; *C.C.* arteria carotis communis; *C.E.* arteria carotis externa; *L* arteria lingualis; *M* arteria mammaria interna; *M.I.* arteria maxillaris interna; *S.* arteria subclavia; *T* truncus thyreocervicalis; *V.* arteria vertebralis. (From MAGNI et al., 1959)

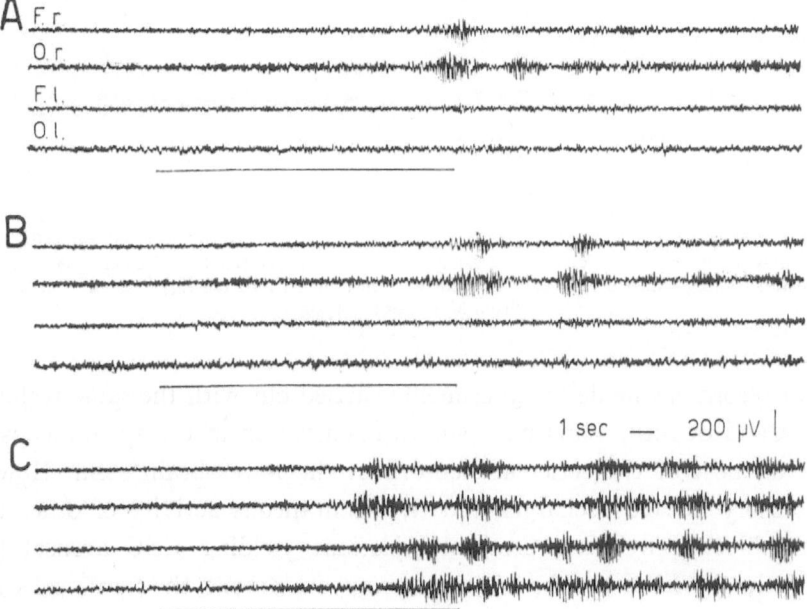

Fig. 15 A–C. *Unilateral and bilateral EEG effects of different intracarotid doses of thiopental.* "Encéphale isolé" animal. EEG recordings from right (*r.*) and left (*l.*), frontal (*F.*) and occipital (*O.*) areas. The black signals indicate introduction of 0.1 mg (A), 0.3 mg (B) and 0.6 mg (C) of thiopental into the right carotid circulation. Only the last dose has a definite bilateral effect. (From MAGNI et al., 1959)

b) MAGNI et al. (1959) produced EEG arousal by pharmacological inactivation of the lower brain stem in the *encéphale isolé* cat. Figs. 14–16 are reproduced from their work, and the plan of the experiment may be summarized as follows:

"Clamping of the basilar artery at the midpontine level shifted the rostral pontine region from the vertebral to the carotid blood supply. Intracarotid injections of thiopental led to EEG synchronization, whereas intravertebral injections had exactly the opposite effects. Changes of systemic circulation and respiration were lacking since the spinal cord had been previously transected at C_1, while several controls showed that the EEG effects were really due to the thiopental *per se*. Several considerations led to the conclusion that the effects of intravertebral injection were due to short-lasting inactivation of the synchronizing structures of the lower brain stem" (MORUZZI, 1963b, p. 243).

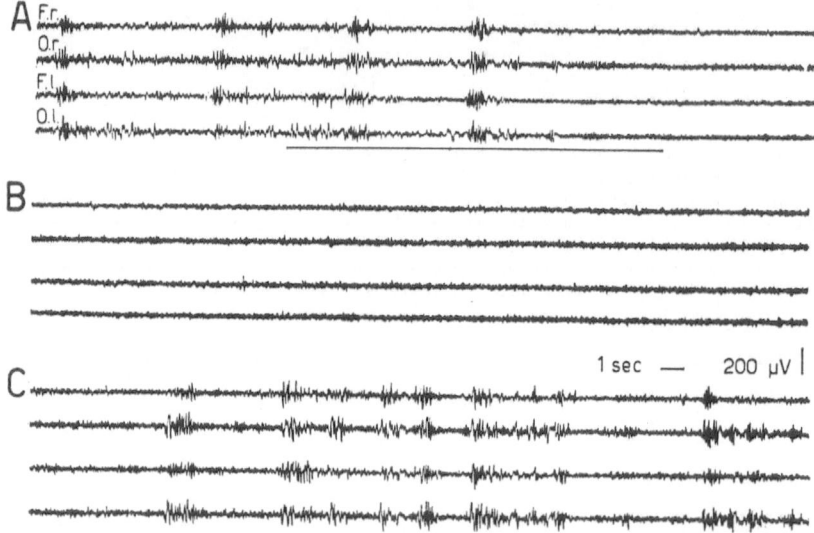

Fig. 16A–C. *EEG arousal following intravertebral injection of thiopental.* "Encéphale isolé" cat. EEG recordings as in previous figure. 5 sec intervals between A and B, and between B and C. The black signal indicates the time during which 0.3 mg of thiopental were introduced into the right vertebral artery.
(From MAGNI et al., 1959)

Unit recordings made in experiments carried out with the same technique (ROSINA and SOTGIU, 1967) have shown inactivation of the spontaneous discharge of most of the reticular neurons of the lower brain stem. However their sensory driven discharges and the spontaneous activity of other units was not blocked by intravertebral injections yielding EEG arousal (l.c.). Apparently the EEG synchronizing neurons are among the nerve cells most severely affected by the drug.

The EEG effects have been confirmed by intravertebral injections of fast acting depressing drugs in the cat (ROSINA and MANCIA, 1966), the rabbit (ROSADINI et al., 1964) and man (GENTILOMO et al., 1964; ALEMÀ et al., 1966; ZATTONI and ROSSI, 1967). In chronic experiments made on cats, ligature of the basilar artery was made at different levels, and it was concluded that the lower brain stem structures have a depressing influence both on the arousal system and on the pontine center responsible for paradoxical sleep (ROSINA and MANCIA, 1966). In all these experiments the medulla was connected with the spinal cord, so that effects produced by respiratory and circulatory changes (GENTILOMO et al., 1964) could not be avoided. Although there are reasons (ROSINA and MANCIA, 1966) to believe that these systemic contaminations did not represent a critical factor, it is apparent that they might be eliminated only by a cervical section.

Although not strictly related to the problem of sleep, the work of ALEMÀ et al. (1966) deserves separate discussion, because of the importance of its theoretical implications. These authors have found EEG arousal following intravertebral injections of amobarbital in man, using doses which produce temporary abolition of the corneal reflex. When the basilar artery is not ligated, it is of course very difficult to determine the extent of the brain stem which has been pharmacologically inactivated. Even a complete inactivation of both medulla and pons, leading to a temporary prepontine preparation, should be expected to be compatible with consciousness, as proved by the effect of lesions in both cat (ZERNICKI et al., 1967) and man (ORTHNER, 1969). The inactivation, moreover, is never complete, as shown i) by unit recording (ROSINA and SOTGIU, 1967); ii) by the fact that respiration is affected, but not abolished (GENTILOMO et al., 1964), and iii) by the observation that hearing is apparently not depressed in man (ALEMÀ et al., 1966). Hence the real problem is to know whether the mesencephalic portion of the ascending reticular system was inactivated in the experiments of ALEMÀ et al. (1966). According to these authors, the proof that the midbrain was affected is given by the bilateral ptosis and by the mydriasis produced by the injection. In fact only the ptosis may imply a depression of some oculomotor neurons; however a complete study of the eye movements has not yet been made. Pupillary dilation does not necessarily imply midbrain inactivation. A release of hypothalamic and midbrain structures from the inhibitory influence of the medulla (ZBROZINA and BONVALLET, 1963) may also produce mydriasis. The inhibition of the Edinger-Westphal neurons and an enhancement of the sympathetic discharge have a powerful pupillo-dilator effect. We shall see that bulbar cooling produces EEG arousal and pupillary dilatation, which are undoubtedly release phenomena (BERLUCCHI et al., 1964a, 1965).

c) BERLUCCHI et al. (1964a, 1965) have shown that cooling of the medullary floor of the fourth ventricle, carried out in the *encéphale isolé* cat on a background of drowsiness or of synchronized sleep, is followed by EEG and behavioral (Fig. 17) signs of arousal. These effects were attributed to the release from EEG synchronizing influences of the lower brain stem, since opposite results were obtained when the pontine floor of the fourth ventricle was cooled (l.c.). That the effects of bulbar cooling were due to functional inactivation of the underlying medulla was shown by the increase of heart rate. Cardiac acceleration occurred even after prepontine section (l.c.). These data, and other controls reported by the authors (l.c.), suggest that cooling of the bulbar surface of the fourth ventricle acted by abolishing the tonic discharge of underlying neurons. These results have been confirmed by NAQUET et al. (1965, 1966, p. 162) in the *intact* cat.

The demonstration of the existence within the lower brain stem of neurons oriented antagonistically with respect to the ascending reticular system has given rise to several experiments and interpretations, which may be grouped along the following lines.

a) Stimulation (Magnes et al., 1961 a, b) and lesion (Bonvallet and Bloch, 1961; Bonvallet and Allen, 1963) experiments would suggest that neurons endowed with a deactivating influence are localized in the region of the solitary tract. This approach has been fully reviewed by Moruzzi (1963 b) and Bon-vallet (1966), so other details will not be given here. These conclusions by

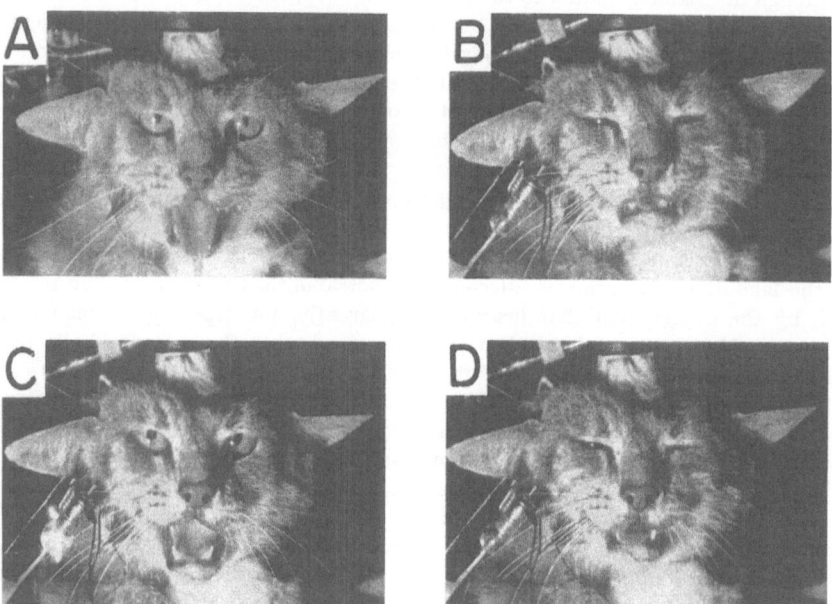

Fig. 17 A–D. *Behavioral effects of bulbar cooling.* A "Encéphale isolé" cat during wakefulness. B Spontaneous sleep. C Arousal elicited by bulbar cooling. D Reappearance of sleep behavior after the end of bulbar cooling. (From Berlucchi et al., 1964)

no means imply that this is the only localization of neurons antagonistically oriented with respect to the ascending reticular system. Insomnia has been reported following lesions of the raphe nuclei or sagittal sections of the medulla. This line of investigations, to be reviewed later (p. 62), has interesting neuro-chemical implications, which will be discussed by Jouvet in these reviews.

b) The problem of the mechanism of the EEG synchronizing and possibly sleep inducing influence of the lower brain stem has been the theme of a review (Moruzzi, 1960) and will again be discussed in this article (see p. 99). The deactivating structures of the lower brain stem might act by inhibiting the ascending reticular system or by counteracting its influence at the level of the diencephalon.

c) A third problem concerns the sensory (reflex) control of the synchronizing structures of the lower brain stem. It will be reviewed in another section (pp. 84–89), devoted to the production of sleep by sensory stimulations.

4. The Chronically Decerebrate Preparation

The first study on chronically decerebrate cats was made by BAZETT and PENFIELD (1922), who showed that occasionally the animals quieted down and their extensor rigidity decreased. They stated (l.c., p. 200) that "passage from a flexed to an extended posture seemed to be induced by mere auditory stimuli". Also HEAD (1923) reported that the decerebrate preparation may occasionally be found with its limbs in a state of flexion, "as if asleep", and confirmed that extensor rigidity reappeared following stimuli exerting an arousing influence in the normal cat. These cyclic changes were compared by PENFIELD (1954) to those occurring in the intact animal during the alternation between sleep and wakefulness.

The main problem arising from these early observations is not whether real sleep can be obtained in the decerebrate animal—this would be obviously impossible—but whether the *motor and postural aspects of sleep and waking behavior* may still occur in this simplified preparation. These behavioral signs might in fact be utilized for the study of the cycles of activation and deactivation in the brain stem, after its isolation from the cerebrum.

There are semantic difficulties in this approach to hypnic physiology, as shown by a lively discussion at the St. Marguerite Symposium (ADRIAN et al., 1954, see pp. 133–135). While it may be misleading to call "sleep" the cataplexic episodes of the decerebrate cat, to be described below, it would not be advisable to restrict the sphere of sleep studies only to conditions involving changes of consciousness. This limitation would, in fact, prevent—at least for the lower vertebrates and certainly for the insects—any comparative approach to the study of the functional significance of the cycles of activity and rest. These studies should be made, in our opinion, quite independently of any definition of sleep. There is no reason to disregard behavioral investigations on mammalian preparations certainly devoid of consciousness, as the decerebrate cat, although one has to be aware of the difficulties involved in any attempt to extrapolate these findings to genuine physiological sleep. The studies we are going to report were made in the great majority of cases on chronically decerebrate cats.

RIOCH (1954) briefly related, in a discussion at the St. Marguerite Symposium, that he had been able to observe cyclic changes of posture recalling sleep, with increased threshold to external stimuli, in his chronically decerebrated cats and dogs. He observed these phenomena starting from 10 days after precollicular decerebration, but never after postcollicular transection, where only slight changes of decerebrate rigidity could be detected.

BARD and MACHT (1958) were able to study their decerebrate cats for much longer periods of time, an achievement probably due to the fact that they left an isolated hypothalamic island in connection with the hypophysis. Two precollicular cats, which survived for 31 and 154 days, "slept" in a crouched position, marked by loss of rigidity and drooping of the head. They responded to slight auditory stimuli by raising the head, standing, and even walking. After pontine decerebration the cat did not change its position, but a marked fall and even complete absence of rigidity revealed the existence of a state of "sleep" of the brain stem; while "awakening" was marked by resumption of rigidity and raising of the head. Hence even after pontine section cyclic changes could be detected in the lower brain stem, although the manifestations were limited to reversible changes of extensor rigidity.

JOUVET (1962, see 1967a) made an important step forward with his studies on the rhythmic activities occurring both below and above brain stem transection. In the unanesthetized, free moving cat he had noted that the onset of the paradoxical episodes of sleep was characterized by a sudden fall of the antigravity tonus, as shown by the EMG silence of the cervical muscles. He rightly pointed out that the collapse of extensor rigidity occurring in the chronic preparation after *rostropontine* transection corresponds to the postural changes observed in the intact animal during the paradoxical phase of sleep. He reported that these *cataplexic episodes* were present after *cerebellectomy*, but disappeared after *caudopontine transection*. He thus concluded that the rhythmically active structures were localized in the pons, in confirmation of his electrophysiologically findings (see JOUVET, 1967a), and suggested that the same triggering neurons were responsible for both the cataplexic episodes after chronic decerebration and the periods of paradoxical sleep of the intact animal. Of course the pontine volleys could no longer reach the cerebrum after brain stem transection and the dissociation between EEG and postural changes was complete (see below, pp. 42–43).

The survival of the first group of cats studied by JOUVET had not been longer than 10 days. In a second group of experiments reported at the Lyon Symposium (JOUVET, 1965c), he was able to follow his rostropontine cats for two months, by utilizing BARD and MACHT's technique (1958) of the hypothalamo-hypophyseal island. At these later stages of chronicity the animal was able to respond with a rotation of its head to strong auditory stimuli. The average duration of single cataplexic episodes was 6 minutes, i.e. about the same as that of the paradoxical episodes of the intact animal; the circadian percentage was slightly less, 10% instead of 15%. Moreover the cataplexic phenomena were distributed along the nychthemeron, while in the intact animal the paradoxical episodes occurred only during behavioral sleep, a condition taking about 70% of the time in a cat "habituated" to its sound-proof cage (see BATINI et al., 1959a). In view of the fact that ocular movements due to sixth nerve innervation, as well as respiratory and heart acceleration, occurred during the cataplexic episodes (VILLABLANCA, 1966a), the hypothesis of their relations with paradoxical sleep appears a very likely one.

JOUVET stated that no behavioral signs corresponding to sleep with EEG synchronization could be found in his decerebrated, pontine cats. He rightly emphasized (JOUVET, 1965a, p. 617) that the absence in the slow phase of clear-cut behavioral signs makes it difficult to identify this stage after decerebration. The paradoxical phase is easily recognizable because of the sudden collapse of extensor tonus, the onset of rapid eye movements and the clonic twitches. Precollicular cats might be thought to provide a more convenient experimental subject, since righting reflexes and pupils could be studied as well. This advantage has been exploited by VILLABLANCA (1966a), with chronic experiments we are going to report.

HOBSON (1965) made a parallel study of the neural structures lying above and below the chronic brain stem transection. After prepontine section short periods of atonia appeared without any relation with the EEG, which remained fully synchronized. Hence after neural disconnection the isolated cerebrum and the bulbo-pontine structures appeared to work independently, an observation that migth be expected, yet one which suggests added complications for the humoral theories of sleep. Periodic atonia was present even when decerebration was combined with anterior lobe ablation, an observation suggesting that the inhibitory reticular formation of the medulla might be responsible for periodic atonia. The dissociation between postural behavior and EEG was complete even after a subthalamic section sparing the midbrain entirely.

VILLABLANCA (1966a) published in 1966 an important study on the behavior of chronically decerebrate cats, which he followed for a mean time of 485 days. Three levels of decerebration were utilized: i) precollicular sections, rostral to the third nerve nuclei (high mesencephalic cats); ii) rostral mesencephalic section, damaging or destroying these nuclei; and finally, iii) postcollicular sections. The long survival was made possible by the technique of the diencephalo-hypophyseal island (BARD and MACHT, 1958). The observations made after 15–20 days in the high mesencephalic cat are particularly important for sleep physiology.

The waking behavior was described as follows:

"In prolonged observation a high mesencephalic cat could be found crouching, sitting, attempting to walk or engaged in iterative climbing if suddenly stopped by a wall. These postures included a variable degree of hyperextension of the neck as a residual manifestation of decerebrate rigidity. The eyelids were open, the nictating membranes retracted, and pupils were widely dilated so that, at times, the iris was hardly visible. The eyeballs either were motionless in the center of the orbit or exhibited slow conjugated movements. At this stage the animal would respond to sudden noises with rotation of the head toward the source of noise and was able to display defense reactions to nociceptive stimuli" (VILLABLANCA, 1966a, p. 563).

The "waking" behavior of the pupils is shown in Fig. 18.

VILLABLANCA (1966a, p. 573) rightly emphasized that the mydriasis observed in a state of "wakefulness" in the chronic precollicular cat was surprisingly similar to that of the free moving, unanesthetized animal at the

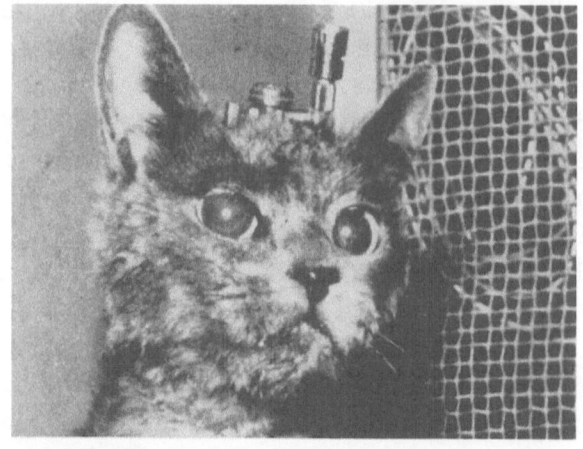

A

B

Fig. 18 A–D. *Ocular behavior in the chronic precollicular cat*. A Widely opened eye lids and strongly dilated pupils in the "waking" preparation. B Medium sized myotic pupil and slight inward rotation during light "sleep" (*VMS*). C, D Stage of extreme myosis, corresponding to the paradoxical phase of the normal cat (*EMS*). In C the pupils are covered by the nictitating membrane, but the downward and inward rotation of the eyeballs is clearly visible. In D the nictitating membranes were pulled out of the plastic cylinder to show extreme myosis. Detailed description and terminology in the text. (From Villablanca, 1966a)

moment of the arousal from sleep or during the waking state. Pupillary dilation, in fact, i) occurred as integrated patterns of ocular signs of wakefulness; ii) could be evoked by auditory stimuli; and finally iii) was present after preganglionic sympathectomy. Hence the effect was mainly due to inhibition of the Edinger-Westphal nucleus.

Villablanca (l.c.) was also able to recognize in the precollicular cat a pupillary behavior similar to that of the normal animal during sleep with EEG synchronization. He called this period "variable myosis stage" (VMS). To quote:

"If a chronic, 'waking' high decerebrate cat was left undisturbed, its motor activity decreased; the waking posture tended to collapse and the animal lay down in a random

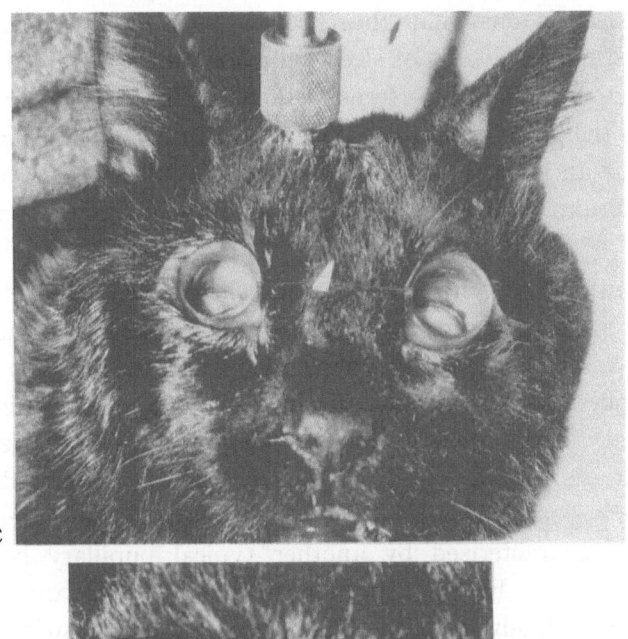

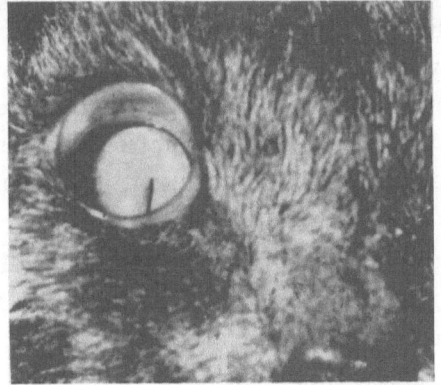

Figs. 18C and D.

position. The eyelids and nictating membranes closed and the pupils exhibited fluctuating myosis. The eyeballs exhibited slow and often dissociated movements tending toward an inward and downward rotation. At this stage of sleep the brainstem electrical patterns did not change significantly. Spontaneously or following exteroceptive stimuli the pupils would widen, the nictitating membranes and eyelids retract and the eyeballs would quickly revert to a symmetrical central position. If sufficiently intense, the stimuli would result in postural and EMG arousal. Auditory stimuli were very effective in causing arousal in the chronic animal, with weak sounds evoking marked, although usually short lasting, pupillary dilation. The stimulus intensity required to produce the reversal of the ocular sleep pattern was higher when the pupillary diameter was small and had been maintained for a longer time" (VILLABLANCA, 1966a, pp. 565–566).

The postural and ocular (Fig. 18A, B) behavioral patterns of the chronic precollicular cat during VMS fit very well those of the unanesthetized, free moving cat, during the episodes of sleep with EEG synchronization (see BER-LUCCHI et al., 1964b). Obviously the typical behavior, clearly instinctive in nature, that occurs in the normal cat during the phase immediately preceding the onset of sleep is missed in the chronic precollicular preparation. Only ocular and postural fragments of sleep behavior can be obtained. This result could be expected. Although patterns of sexual and of fear behavior can still be observed in the precollicular cat (BARD and MACHT, 1958), they are always fragmentary or at least incomplete. Instinctive behavior is observed only when the cerebrum, or at least the hypothalamus, is connected to the brain stem.

The relationship between the VMS of the chronic precollicular preparation and the ocular and somatic manifestations of sleep with EEG synchronization in the normal cat is suggested by the sequence of events in the decerebrate preparation, more than by behavioral similarities, whose significance might be controversial. Periods characterized by pupillary dilatation, clearly related with the mydriasis of the waking cat, were followed by episodes of VMS, *in the absence of any fall of extensor tonus.* Since the cataplexic episodes occurred later and were characterized by another typical pupillary behavior (the extreme myosis stage, EMS, to be described below) VILLABLANCA's hypothesis appears as a likely one. The VMS is related neither with arousal nor with paradoxical sleep, but rather with an intermediate stage in the temporal sequence of events. In the normal cat this intermediate stage is actually sleep with EEG synchronization. This conclusion is of course strengthened by the behavioral similarities described above.

Such a hypothesis is supported by the fact that the VMS is followed by the *extreme myosis stage* (EMS), which clearly corresponds to the *cataplexic episodes* of the chronic rostropontine preparation. To quote, once more:

"If the cat was not disturbed during VMS both ocular and body sleep patterns tended to become accentuated. At any moment the myosis could become extreme (*"extreme myosis stage"*; EMS) and the muscular tonus would decrease (drooping of the head, relaxation of the perioral muscles and lowering of the whiskers), down to complete limb and body atonia. Neck muscles EMG became silent as soon as pupils reached maximum constriction. Simultaneously brief bursts of clonic muscular activity appeared in any part of the body, but particularly in the face where contraction of the extrinsic muscles of the eyes gave rise to rapid eye movements (REMs)" (VILLABLANCA, 1966a, p. 566).

These postural and ocular behavioral patterns (Fig. 18C, D) recall those of the normal cat during sleep with EEG desynchronization (see BERLUCCHI et al., 1964b). They are in fact associated with the electrophysiological pontine manifestations, which are always present when the phase of sleep occurs in the normal animal (see JOUVET, 1967a). The duration of the EMS episodes might be as long as 20–25 minutes, and could be interrupted by natural or reticular stimulations. The threshold of arousal was much higher during EMS than during VMS, an observation recalling similar differences between the episodes of sleep characterized, respectively, by synchronized and desynchronized EEG (see l.c.).

We have seen (p. 7) that spinal reflexes are severely depressed during paradoxical episodes of sleep in the normal cat (see POMPEIANO, 1965, 1966, 1967b). As it might be expected these results were duplicated during the cataplexic episodes of the chronic pontine cat. However IWAMURA, STERMAN, and McGINTY (1969) have come to the surprising conclusion that mono- and polysynaptic reflexes are enhanced during the atonic phase of the chronic mesencephalic or diencephalic preparation.

WOODS (1964) studied nine rats which survived up to 98 days after precollicular decerebration. Recovery of motor and postural activities was by far faster and more complete than in the corresponding feline preparations, and well integrated forms of behavior were observed. These motor manifestations

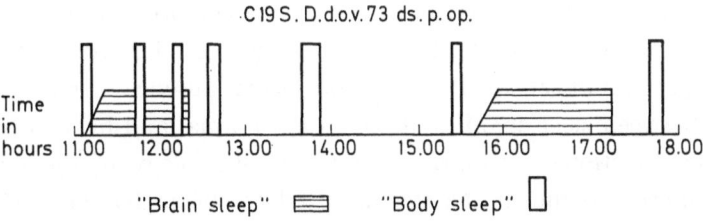

Fig. 19. *Independence of sleep episodes of the isolated cerebrum and of "sleep" episodes of the brain stem.* Chronic (79 days) "cerveau isolé" cat, after olfactory and visual deafferentation. Schematic drawings represent synchronization periods of the cerebral cortex ("brain sleep") and cataplexic episodes arising in the truncated brain stem ("body sleep"). The 7 hour recording session shows complete independence of the phenomena. (From VILLABLANCA, 1966)

disappeared during periods in which the animal presented the physiological sleep posture, while auditory and tactile stimulation gave rise to typical manifestations of arousal. The periods of sleeping were not related to the daylight hours, as in the normal rat. WOODS made no attempt to distinguish between behaviors related respectively to slow and paradoxical sleep.

In summary, cyclic changes of behavior, clearly related with the physiological alternation of sleep and wakefulness of the normal animal, may be found in the chronic decerebrate preparation, a conclusion fitting the observations made on human anencephalic monsters (GAMPER, 1926, see JOVANOVIĆ, 1969, p. 67, for the literature). However these rhythmic changes of brain stem activities are affected by the chronic decerebration in at least three ways.

a) We have seen that the cataplexic episodes of the decerebrate preparation occur throughout the nychthemeron (JOUVET, 1965c), while the paradoxical sleep episodes occur, in the *normal animal,* only when it is behaviorally asleep. A likely explanation is that the outbursts of pontine activity responsible for the paradoxical episodes (see JOUVET, 1967a) are blocked by the cerebrum during the waking state, although their circadian percentage, probably related to local neurochemical factors, is approximately the same (l.c.). In the chronic *cerveau isolé* the EEG signs of synchronization ("brain sleep") and the cataplexic episodes ("body sleep") occur independently (see Villablanca, 1966a; Fig. 19).

b) The extreme myosis stage, corresponding to the paradoxical episodes of the normal cat, appears during the first ten days after the decerebration, while the variable myosis stage, corresponding to the sleep with EEG synchronization, appears only after 10–15 days. Hence the brain stem mechanisms underlying the paradoxical episodes are less severely affected by decerebration.

c) Even with experiments characterized by long term chronicity, slow synchronous waves ("spindles") never appear in the truncated brain stem, while they are constantly present in the normal animal (see Jouvet, 1967a). These electrical manifestations are driven by the cerebrum, probably through pyramidal (Adrian and Moruzzi, 1939) or corticoreticular (von Baumgarten et al., 1954; see Moruzzi, 1954) connections. Their absence represents a major difference between sleep manifestations in the normal and in the decerebrate animal. The functional significance for sleep physiology of the descending flow of impulses arising in the synchronized cerebrum has been the theme of a lively discussion at the Lyon Symposium (Jouvet, 1965a; p. 614–618).

5. The Isolated Midbrain

Zernicki et al. (1970) obtained an isolated midbrain preparation by combining a pretrigeminal transection with either i) a premesencephalic section or ii) a cut starting 3–4 mm anterior to the superior colliculus and ending in the preoptic area. The preparations were studied in both acute and chronic conditions, the latter lasting as long as 2 weeks.

In the *pretrigeminal-preoptic cat* the midbrain was connected with the retina, as shown by the appearence of the photic reflex and by the flash-evoked responses of the superior colliculi. However consistent signs of tracking ocular movements were never observed. Spontaneous electrical activity was absent from several areas of the isolated midbrain reticular formation, and even in chronic conditions there was no remarkable recovery. This might be the cause of the absence of the fixation reflex, which eventually reappeared, but only for a few seconds, after electrical stimulation of the midbrain reticular formation. The pupils remained myotic even in darkness.

In the pretrigeminal-preoptic cat the hypothalamus was still connect to the mesencephalon. The midbrain was really isolated only in the *pretrigeminal-premesencephalic cat*. The symptoms were, however, essentially the same in this visually deafferented preparation. There was also fissurated myosis, in confirmation of Keller's observations (1946), which may be attributed to an autochthonous tonus of the Edinger-Westphal nucleus. Pupillary dilation was produced by electrical stimulation of the midbrain reticular formation, an important finding suggesting that the mydriatic response produced by hypothalamic stimulation might be mediated by short-axoned inhibitory interneurons localized in the mesencephalon.

The authors have pointed out that the striking depression of the isolated mesencephalic reticular formation is likely to be due to withdrawal of a descending influence exerted by the cerebrum on the midbrain reticular formation. A non-specific postsurgical damage should not be a major factor, as shown by the fissurated myosis. The descending influences should be mainly ipsilateral, since striking asymmetries of midbrain reticular activity were observed when the premesencephalic section had been made on one side only. It will be shown later that also the tonic ascending influence arising from the midbrain is mainly ipsilateral (see p. 51).

Unfortunately only macroelectrode recordings were made, but it is likely that also unit activity was severely depressed, at least when gross recording showed electrical silence (ZERNICKI et al., 1970; see Fig. 6B). This conclusion is in sharp contrast with the presence of marked spontaneous and driven activity of midbrain reticular units, following precollicular decerebration (MANCIA et al., 1957). Any deterioration of midbrain is indicated at once, in this preparation, by the appearence of mydriasis and by the absence of spontaneous activity of midbrain units (l.c.). Even reticular units of the lower brain stem are spontaneously quite active after precollicular (GAUTHIER et al., 1956) and either inter- or postcollicular (MOLLICA et al., 1953) decerebration. Thus low reticular activity with myosis would imply that most of the spontaneous activity of the midbrain reticular formation is not autochthonous in nature, but that adequate support is required by either ascending or descending influences.

A second, important consequence would be that the strong activity of the reticular formation lying above the pretrigeminal transection is in fact driven by a descending barrage. We have seen that a release enhancement of this activity is likely to appear as a consequence of the withdrawal of an inhibitory influence arising from the lower brain stem (see pp. 31–36). Thus the striking syndrome of alertness of the pretrigeminal preparation might be due to release of cortico-reticulo-cortical loops or, more generally, of the driven activity of the mescencephalon. The midbrain would have a critical importance in these physiological circuits, as shown by the striking differences existing, in the acute stage, between midpontine or prepontine cats and the *cerveau isolé*.

6. Conclusions

BERLUCCHI (1970) has rightly pointed out that in the study of the symptoms elicited by experimental or clinical lesions of the central nervous system, a distinction should be made between *abolition* and *imbalance* of the sleep-waking cycle.

It is usually stated that the acute *cerveau isolé* is characterized by a state of coma, rather than by one of sleep, since arousal can no longer be obtained with any of the sensory stimulations which are still available. The significance of this observation for the physiologist may be better expressed by stating that the sleep-waking cycle is abolished altogether. In fact the animal is neither able to remain awake nor to sleep, it is comatous. The difference from the acute pretrigeminal cat deserves to be stressed, since in the latter preparation the sleep-waking cycle is not abolished (although it is seriously imbalanced), as shown by the presence of episodes of true, reversible sleep occurring on a background of prolonged behavioral and EEG wakefulness.

The major finding of the experiments on the chronic *cerveau isolé* is therefore not only i) the demonstration that the ability to remain awake may eventually be recovered, but also ii) the discovery that *structures lying within the isolated cerebrum may account for the reappearence of a sleep-waking cycle*. The sleep of the chronic preparation is of course different from that of the normal animal, if we assume that the paradoxical phase is permanently abolished after a prepontine and possibly also following a midpontine pre-trigeminal section. But the point to be stressed is that the sleep-waking cycle, which is present in the acute stage of the pretrigeminal preparation, reappears in the *cerveau isolé* only after several days. Since the recovery process may be observed even in the high *cerveau isolé*, i.e. after precollicular transection, the conclusion is drawn that *the sleep-waking rhythm may appear in the absence of any brain stem control*.

It would be a serious error of methodology to start our discussion on the effects of total brain stem transections by assuming that a choice should be made between the conflicting results of acute and chronic experimentation. Opposite observations are in fact conflicting only when they are obtained in the same experimental conditions. The extensor rigidity occurring in the carnivores immediately after cerebellectomy is an important physiological phenomenon, just as is the disappearance of the hypertonus one week later. We are going to show that both the coma of the acute *cerveau isolé* and the recovery of the sleep-wakefulness cycle occurring in the chronic preparation are equally important for unravelling the neural mechanisms underlying the rhythmic activity of the diencephalon, and the tonic and phasic influences exerted thereon by the brain stem.

Several lines of convergent evidence, which we have already discussed, lead to the conclusion that the acute syndrome of the *cerveau isolé* (Bremer, 1935, 1937, 1938) is basically due to sudden withdrawal of an ascending tonic influence arising in the regions of the brain stem underlying the section. This conclusion alone, combined with the observations made in the midpontine pretrigeminal cat (BATINI et al., 1958, 1959a), fully justified the study of the acute preparation. To dismiss this approach, because of the undeniable fact that a state of coma, not one of true physiological sleep, is observed in the acute *cerveau isolé*, would be tantamount to denying that decerebrate rigidity, which is also a pathological phenomenon, is important for the study of the neural mechanisms underlying postural tonus.

A similar misinterpretation, one leading however to opposite conclusions, would be made by attributing the amazing recovery of a sleep-waking cycle exclusively to denervation hypersensitivity, an ill defined phenomenon in the central nervous system (see SHARPLESS, 1964). Needless to say, we are interested in knowing the functions of the cerebrum in normal conditions. The physiological significance of the amazing recovery of the chronic *cerveau isolé* is therefore contingent upon the demonstration that *the sleep-waking cycle arises within the cerebrum not only after chronic midbrain transection, an unescapable conclusion, but also in the normal animal, when cerebral activities are tonically and phasically modulated by the brain stem.*

The recovery of the ability to maintain wakefulness may be influenced by denervation hypersensitivity, but it is difficult to believe that the reappearence of the sleep-waking cycle may be explained exclusively by CANNON's law of denervation (see STAVRAKY, 1961). The acute syndrome of the *cerveau isolé* may be contaminated by aspecific phenomena, just as the chronic syndrome may be modified by denervation hypersensitivity. However the basic mechanisms cannot be entirely explained by these hypotheses.

We believe that an attempt at a physiological interpretation must start from the observations made on both acute and chronic experiments. The only way to find an unitary explanation of both the far reaching disorganization of the brain in the acute stage, and of the slow recovery that occurs as chronicity progresses, lies in the assumption that we are confronted with mechanisms basically similar to those underlying the spinal shock and its disappearence in the chronic state.

We are led, therefore, to the conclusion that the neurons responsible for the sleep-waking cycle are present and fully active even in the normal cerebrum, just as the neural mechanisms responsible e.g. for the knee jerk are present, and fully active, in the spinal segments of the normal animal. Hence *the sleep-waking cycle would arise within the cerebrum, even in the normal animal, and would simply be controlled by the ascending flow of brain stem impulses.* The ascending influence exerted by the brain stem on waking structures of the diencephalon would be both tonic and phasic in nature, thus explaining the maintenance of wakefulness and the onset of arousal on a background of sleep or drowsiness. The sudden withdrawal of the ascending influence of the brain stem would be the cause of the coma of the acute *cerveau isolé*, just as the sudden fall of its descending influence leads to the deep disorganization of segmental activities which we call spinal shock.

The comparison between the coma of the acute *cerveau isolé* and the *spinal shock* is justified by the fact that in both cases we are concerned with the effects of the *sudden withdrawal of a tonic influence of the brain stem*. However the consequences of these acute defacilitations are entirely different, since spontaneous activity is still going or in the acutely isolated cerebrum (Whitlock et al., 1953). The firing of the cerebral neurons is severely disorganized, but by no means suppressed, while the firing of most of the segmental motoneurons is abolished during spinal shock. Thus the old concept of cerebral tone is entirely different from that of postural tonus. These ideas have been developed also elsewhere in this article (pp. 115–117).

If this assumption is correct—and we shall see (see pp. 115–117) later that entirely different lines of evidence lead to the same conclusion—*the brain stem would be only indirectly related to the sleep-waking cycle*. In fact the ascending reticular system and the antagonistically oriented structures of the lower brain stem are both likely to exert an influence on the cerebrum which is much wider in scope. The regulation of the sleep-waking cycle would symply be one aspect of the control of all cerebral activities, during conscious and unconscious life. These concepts will be developed in the final chapter of our review (see p. 142).

We are now going to discuss the experiments of *chronic decerebration*. The fact that postural changes related to sleep behavior may still be observed in the chronically decerebrate cat is an interesting finding. However it is not an isolated observation, since fragments of sexual behavior may be observed in the chronically decerebrate female cat, following injection of sexual hormones (Bard and Macht, 1958). Similar observations have been made for the "rage" responses to nociceptive stimulation (l.c.) and for the thermoregulatory responses which occur in the absence of hypothalamic control (Thauer, 1939, 1967). Much more important is the demonstration that the brain stem structures underlying the somatic aspects of sleep and waking behavior have some kind of reciprocal organization, which leads to the reappearence of rhythms recalling the sleep-waking cycle of the normal animal.

These considerations may serve as an introduction to the wider problem of the origin of the *physiological rhythms*. We have seen that both the structures lying above and those localized below the brain stem transection may recover the ability to maintain a rhythm, which appears closely related to the sleep-waking cycle of the normal animal. The rhythms occurring in a cerebrum chronically isolated from the brain stem are of course quite different from those of a brain stem chronically deprived of the cerebral influences. And yet both preparations share two general properties: i) the existence in each of them of functionally antagonistic structures, and ii) the imbalance produced in both cases by the midbrain transection. We shall see later (p. 83) that two antagonistic neuronal mechanisms, identified as the sleep and the waking centers, are present in the diencephalon; while the study of the *cerveau isolé* and of the pretrigeminal preparation shows that two other antagonistic neuronal populations are localized in the brain stem and control the activity of the cerebrum. A striking imbalance is produced in the cerebrum, as in the brain stem and the spinal cord, by a midbrain section. The cerebral manifestation of this imbalance is the coma of the *cerveau isolé*, while extensor rigidity is the major sign of an imbalance in the brain stem and the spinal cord. A rhythm eventually reappears on both sides of the transection as chronicity progresses, an observation suggesting i) the existence of reciprocal connections between the antagonistic structures and ii) a reduction of the imbalance produced by some process of recovery, the nature of which is still unknown. It is likely that rhythmicity becomes again possible when the levels of activity of the antagonistic structures oscillate within critical ranges, not too far from a state of equilibrium.

That a rhythm may be disrupted by a severe imbalance is a classical notion in physiology. Respiratory rhythmicity is greatly impaired when a midpontine section combined with bilateral vagotomy gives rise to apneustic breathing (see WYSS, 1964, p. 17–49, for ref.). The respiratory centers are severely imbalanced by the suppression of the central (pneumotaxic center) and peripheral mechanisms (vagal reflexes) checking the intensity and duration of the inspiratory discharge.

The decerebrate preparation is characterized by a striking imbalance in the brain stem structures controlling the spinal segments (WARD, 1947), one that has been attributed to deprivation paralysis (l.c.) of MAGOUN's bulbo-inhibitory center (see MAGOUN and RHINES, 1947), i.e. to withdrawal of a facilitatory influence exerted upon it. When the balance between inhibitory and facilitatory influences of the brain stem is upset, the overactivity of the extensor mechanisms of the spinal segments leads to decerebrate rigidity. This imbalance between extensor and flexor "half-centers" is reversed when the spinal cord is severed in a decerebrate preparation (SHERRINGTON, 1910; SHERRINGTON and SOWTON, 1915). The extensor mechanisms are predominantly depressed during the spinal shock, an effect attributed to the withdrawal of a facilitatory influence (LIDDEL, 1934), which in the carnivores arises mainly in the vestibular nuclei and in the reticular formation (see MAGOUN and RHINES, 1947).

The classical neurophysiological literature shows that the rhythmicity leading to an alternation between antagonistic activities is critically affected by any imbalance. It has been a widely accepted notion since SHERRINGTON's time that the rhythmic activity of the spinal cord requires some sort of equilibrium between two antagonistic "half-centers" (GRAHAM BROWN, 1914) controlling, respectively, the extensor and flexor muscles. The reflex stepping and the scratch reflex, two phenomena implying a rhythmic activity within spinal segments, are less easily obtained in the decerebrate carnivores, where the imbalance is extreme, than in the decapitate or spinal preparation (SHERRINGTON, 1906a; SHERRINGTON and SOWTON, 1915). SHERRINGTON noted (1910, p. 113), moreover, that "in those exceptional cases in which in the decerebrate preparation the rigidity does not develop or lapses, the proneness to alternating reflexes is greater than when rigidity is present". BREMER (1922, p. 209–212) stressed the importance of an optimal tonus for the appearence of rhythmic movements in the decerebrate preparation. Finally in his studies on the stepping movements produced by spinal section in the decerebrate preparation, GRAHAM BROWN emphasized (1911, 1916) that a critical equilibrium between antagonistic half-centers is required in order to obtain a transient spontaneous alternation between extensor and flexor movements (1916, p. 647). Undoubtedly the imbalance that counteracts rhythmicity is a central phenomenon, one that is likely to be gradually reduced during recovery from spinal shock. A penetrating observation made by Sherrington (1906b) shows that the rhythmic activities are critically affected by the imbalance of the spinal shock. He stated (l.c., p. 42) that the spinal shock "affects the scratch reflex more than the knee jerk, the flexion reflex, or even the 'extensor thrust'. From it the scratch reflex emerges slowly in the course of weeks or months". Apparently the critical equilibrium between antagonistic centers, which is required for the onset of rhythmicity, is a result that is attained only slowly during the recovery processes following the spinal shock.

The time course of the sleep-waking cycle is of course slower by far than that of the neural rhythms we have considered above. Moreover, endogenous factors, probably neurochemical in nature, influence the brain stem and possibly also the diencephalic structures concerned with sleep and wakefulness. One is left, nonetheless, with the impression that even for these slow rhythms the imbalance produced by the midbrain section is the main cause of the loss, or of the impairment, of the rhythmic alternation in both cerebral and spinal activities. Both cerebrum and spinal segments are in fact controlled by the brain stem through ascending and descending paths (see Brodal and Rossi, 1955; Rossi and Zanchetti, 1957). Moreover, histological (Scheibel and Scheibel, 1958) and physiological (Magni and Willis, 1963) evidence suggests that the axons of some reticular neurons project both rostrally and caudally. This is an additional justification for a parallel study of the effects produced by the sudden withdrawal of both the ascending and descending influences of the brain stem.

We have already stated that the acute coma and its extremely slow recovery following midbrain transection present undeniable similarities with the spinal shock and its recovery. Two imbalances opposite in sign are present, respectively, in the *cerveau isolé* and in the pretrigeminal preparation; just as two opposite imbalances of spinal activity are present, respectively, in the decerebrate and in the decapitate animal. It may therefore be convenient to try to regard the recovery of a sleep-waking cycle, either in the isolated cerebrum or in the truncated brain stem of the chronically decerebrate cat, as exemples of recovery from an imbalance.

The sleep-waking rhythm would then be thought to reappear in the *cerveau isolé* when the imbalance is suitably reduced. As soon as some degree of wakefulness becomes again possible in the isolated cerebrum one observes a tendency to an alternation between sleep and waking states. Similarly, the first sign of recovery in the decerebrate preparation is the reappearence of the inherent ability to inhibit decerebrate rigidity. The "cataplexic episodes" provide of course only the postural aspect of sleep behavior, but their importance lies mainly in the demonstration that some kind of sleep-waking cycle has again become possible. More elaborate postural and motor manifestations of sleep and wakefulness will appear later, as chronicity progresses.

The idea that some kind of balance between antagonistically oriented structures is required for the sleep-waking cycle, that this balance is temporarily disrupted by the midbrain transection, and that it eventually reappears as chronicity progresses, does not conflict with the hypothesis that special pacemaker neurons, endowed with endogenous mechanisms, are responsible for the tendency to circadian rhythmicity. The problem has been dealt with at length in BERLUCCHI's review (1970), and the accumulation of neurochemically active substances in well defined groups of brain stem neurons has been shown by JOUVET with investigations to be reviewed in *Ergebnisse der Physiologie*. We simply want to emphasize that whenever a rhythm is due to the alternation of activities of two antagonistically oriented structures, the background of these activities must also be considered. When the imbalance is too marked, as in the acute *cerveau isolé* following the sudden fall of the tone of the activating structures of the brain, rhythmicity is no longer possible.

The damaging effects of vague terminology cannot be overemphasized. VON MONAKOW's diaschisis (1902) has been loosely utilized to cover all kinds of functional alterations of a group of neurons produced by the lesion of another group of nerve cells. The causes of these distant actions may of course be entirely different and should not be covered by the same terminology. The alterations produced by a section or an ablation are either non-specific, or specifically produced by the withdrawal of a tonic flow of impulses. Nonspecific effects are of course devoid of any physiological interest: they are merely causes of error, which should be carefully considered by the experimenters. We have shown that the brain stem transection acts mainly by withdrawing a tonic influence normally exerted on the cerebrum. Connections between sleep and waking centers, or between diencephalon and cerebral cortex are of course always likely to be present, even immediately after midbrain transection. In the chapter dealing with the acute *cerveau isolé* we have reviewed the evidence showing that typical responses may be produced by electrical stimulation of hypothalamic structures in this experimental situation. Neurons are excitable, pathways are available; a situation that may be hardly reconciled with VON MONAKOW's diaschisis. What is present in the acute stage is basically a marked alteration in the tonic activities, one leading to a severe imbalance between sleep and wakefulness structures. The recovery process can be best seen as essentially a compensation of this imbalance.

B. Unilateral Brain Stem Transections
1. Midbrain Hemisections

KNOTT, INGRAM, and CHILES (1955) reported that the EEG was synchronized ipsilaterally to a chronic lesion of the midbrain tegmentum. Their results were confirmed by ROTHBALLER (1956). The conclusion of these experiments, one which was in full agreement with anatomical data (BRODAL and ROSSI, 1955) and with occasional electrophysiological observations (MORUZZI and

4*

MAGOUN, 1949; see l.c. Fig. 1 D), was that the pathways of the ascending reticular system are predominantly ipsilateral, at least at the level of the mesencephalon.

CORDEAU and MANCIA (1958) devoted a detailed study to the temporal course of the EEG asymmetry and to the mechanisms of its compensation. The EEG asymmetry elicited in the cat by hemisection at precollicular level appeared in fact as a striking phenomenon only during the first day after the operation. Two or three days later an EEG arousal could be elicited on the lesion side by auditory stimulation, even though this effect had been absent immediately after the operation; some asymmetry persisted, however, since the EEG synchronization occurred earlier on the side ipsilateral to the lesion. Full compensation, as shown by the disappearence of any trace of tonic and phasic EEG asymmetries, occurred occasionally by the 4th day, more often between the 5th and 8th days.

When full compensation had been attained, the opposite side of the midbrain was also transected, thus producing, by a two stage procedure, a half-acute, half-chronic, high *cerveau isolé* preparation. The second operation precipitated the EEG synchronization on both sides, showing that the tonic flow of activating impulses, which after the first operation could reach the brain only through the midbrain gate that had remained open, influenced the brain also on the same side of the lesion, obviously through crossed intracerebral pathways. An important observation made after the second operation was the reversal of the EEG asymmetry, as shown by the fact that the hemisphere ipsilateral to the first hemisection showed a greater tendency to remain desynchronized. Soon after the midbrain transection had been completed, a clear-cut cycle of synchronization-desynchronization could be detected only on the side of the first lesion. When both hemispheres were synchronized, the arousing effect of an olfactory stimulation appeared first on this side. Thus rostrally and ipsilaterally to the first midbrain hemisection some kind of local recovery process had occurred. A phenomenon of this kind had been previously observed in the recovery from hyporeflexia after spinal hemisection (FULTON and McCOUCH, 1937). These similarities support the view that recovery processes from coma and from spinal shock are related phenomena (see p. 48).

The main conclusion to be drawn from these experiments is that the reappearence of the ability to remain awake in the chronic high *cerveau isolé* is due to a local process of recovery occurring within the isolated cerebrum, rostrally to the precollicular transection. The recovery process occurs asymmetrically when the midbrain hemisection is made in two stages. It is likely, moreover, that each midbrain may exert also an effect on the opposite cerebral hemisphere, through crossed pathways. The physiological importance of these cross connections, probably of little significance in normal conditions, becomes detectable only when the excitability of the neurons overlying the first hemi-section has been enhanced by the local process of recovery; or when the flow of impulses reaching the cerebrum through the open gate is particularly strong, as we shall see below.

When the same midbrain hemisection was made in the *midpontine pre-trigeminal cat*, the EEG asymmetry again appeared, showing that the uni-lateral withdrawal of a tonic influence arising from neurons lying between the two sections was fully adequate to precipitate synchronization on the same side. A unilateral interruption of the flow of impulses ascending through the lemniscal pathways was obviously out of the question in a midpontine pretrigeminal preparation. An important observation, one usually neglected, was made on the pretrigeminal cat: *the recovery* of the ability to maintain EEG patterns of wakefulness, ipsilaterally to the precollicular hemisection, *was an extremely fast process.* In a matter of a few hours the EEG could again appear desynchronized, and at the same time the cycle of synchronization-desynchronization started again. This observation can hardly be reconciled with the hypothesis that the recovery of the deactivated cerebral hemisphere was due to CANNON's sensitization of some structures lying above the first hemisection, possibly in the hypothalamus. There is of course no reason why denervation hypersensitivity should occur so early, in the acute pretrigeminal cat. Apparently the marked tendency to EEG desynchronization which charac-terizes this preparation is due to an enhanced flow of activating impulses arising in the upper brain stem and reaching the cerebrum through the mid-brain gate that has remained open. The imbalance produced by the sudden withdrawal of the tonic activating influence of the brain stem on one side is thus more easily compensated by the intense flow of ponto-midbrain activat-ing impulses reaching the cerebrum through the gate which has remained opened on the opposite side. This interpretation would support a hypothesis we have previously put forward (pp. 22–24), namely that the sudden fall of the brain stem activating influence produces an *imbalance, which is the im-mediate cause of the disappearance of the sleep-waking cycle in the acute cerveau isolé.* The recovery occurring in the chronic stage mainly results from a cor-rection of this imbalance. This imbalance is corrected much earlier in the pretrigeminal preparation, because of the enhanced background of activation (see p. 50).

The complete midbrain hemisections of CORDEAU and MANCIA (1958) were produced stereotaxically, through electrodes introduced vertically into the brain. Obviously the ipsilateral hemisphere was slightly damaged with this procedure. The demonstration by TIBERIN et al. (1961) that EEG synchroniza-tion may be observed for a few hours after ipsilateral craniotomy, particularly when the dura is open, suggests that the EEG asymmetries observed by CORDEAU and MANCIA might not be entirely due to the midbrain hemisection.

ROSSI et al. (1963) obtained, however, qualitatively the same results by introducing the electrode into the brain stem through the cerebellum, thus sparing entirely the cerebral hemisphere. Even two days after the operation the EEG of the ipsilateral hemisphere was characterized by a greater tendency to EEG synchronization at the beginning of sleep, and by its delayed disappearence during arousal. However a spontaneous ipsilateral synchronization during wakefulness was observed inconsistently and only in the first recording sessions. Thus a contribution by the slight damage of the cerebral cortex to the ipsilateral synchronization described by CORDEAU and MANCIA (1958) deserves consideration.

The recent work by ZERNICKI et al. (1970), by showing that the hemi-cerebrum has a facilitating influence on the ipsilateral midbrain reticular formation (see p. 45) suggests that the EEG asymmetries observed by TIBERIN et al. (l.c.), though less intense and fully reversible, might well have a neural mechanism similar to that involved in the ipsilateral synchronization produced by midbrain hemisection. In other words, opening of the skull and the dura would be followed by temporary slackening of the tonic cortico-reticular barrage and therefore of the spontaneous activity of the mesencephalic reticular neurons.

ROSSI et al. (1963) showed, moreover, that the EEG desynchronization characterizing the paradoxical phase of sleep was strikingly reduced ipsi-laterally to the lesion. Their interpretation of these results involves the assumption that the hemisection interrupts not only i) the tonic flow of ascending impulses from the activating reticular system, but also ii) the flow of impulses ascending from pontine structures which are responsible for para-doxical sleep. Both connections between brain stem and cerebrum would be mainly ipsilateral in character.

2. Pontine Hemisections

The interaction of the activating and deactivating influences of the brain stem is shown by another group of experiments made by CORDEAU and MANCIA (1959).

When the hemisection is carried out at midpontine pretrigeminal level, the acute preparation presents an EEG asymmetry, which is opposite in sign to that elicited by a midbrain hemisection, as shown by the fact that the episodes of synchronization usually start contralaterally to the lesion (l.c.). The EEG asymmetry is present in the *encéphale isolé*, is not abolished by baroceptive denervation, and is reversed when a second hemisection is made at midbrain level. We have seen that a tonic EEG synchronizing influence is exerted on the cerebrum by the lower brain stem (BATINI et al., 1958, 1959a). This effect is present also when the deactivating influence of the baroceptive

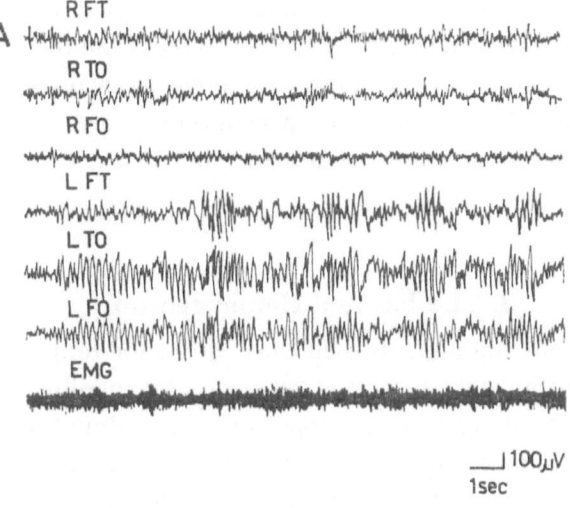

R FT

R TO

R FO

L FT

L TO

L FO

EMG

100μV
1sec

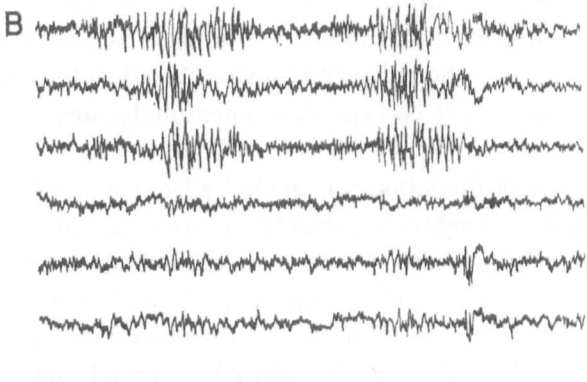

B

Fig. 20 A and B. *Effects of right midpontine hemisection on the EEG patterns of slow wave and paradoxical sleep.* Bipolar EEG records from right (*R*) and left (*L*) fronto-temporal (*FT*), temporo-occipital (*TO*) and fronto-occipital (*FO*) regions, and bipolar EMG records from posterior cervical muscles (*EMG*). Records taken two days after operation. A Beginning of sleep. No synchronized rhythms of slow wave sleep are present on the right side. B Paradoxical sleep. No desynchronized patterns of paradoxical sleep are present on the right side. (From ROSSI et al., 1963)

barrage (BONVALLET et al., 1954; see MORUZZI, 1963 b, p. 259, for the literature) has been abolished. It is probably mediated by ipsilateral ascending pathways.

ROSSI et al. (1963) confirmed these results. They showed, moreover, that the EEG asymmetries produced by the midpontine hemisection were opposite in sign during the slow phase and the paradoxical episodes of sleep (Fig. 20). Thus while the tendency toward sleep synchronization was enhanced by midbrain hemisection and decreased by midpontine hemisection, the paradoxical phase was reduced by both types of hemisection. It is likely that the midpontine hemisection interrupted, or preferentially decreased, the tonic flow of impulses i) from the lower brain stem structures, antagonistically oriented with respect to the activating reticular system, and ii) from the pontine neurons responsible for paradoxical sleep. Both ascending pathways would be mainly ipsilaterally organized, although crossed connections are very extensive, as shown by the generalized effects of reticular or sensory arousal. This point will be discussed again below.

3. The Isolated Hemicerebrum

In SPERRY's split brain cat (see SPERRY, 1961, for ref.) the two cerebral hemispheres are completely separated by a sagittal, midline section cutting all cross connections down into the upper midbrain. BERLUCCHI (1966) combined SPERRY's preparation with a transverse hemisection of the rostral mesencephalon. One cerebral hemisphere was therefore completely isolated by these combined surgical procedures. The EEG patterns of the isolated hemicerebrum could be compared with those of the opposite hemisphere, which was still connected to the brain stem and to the spinal cord, and the results could be related to the animal's behavior. An advantage of this preparation is that the animal may be kept alive indefinitely, under perfect general conditions.

Although this major operation was made by BERLUCCHI (l.c.) in one stage only, the depression of neighboring structures was remarkably short, possibly because i) electrolytic procedures were avoided and ii) the surgical sections were carefully made with the aid of a binocular dissection microscope. On the fourth day both EEG arousal and the sleep-wakefulness cycle were intact on the normal side. The EEG of the isolated hemicerebrum remained fully synchronized when low voltage fast activity appeared in the connected hemisphere following arousal or during a paradoxical episode of sleep, thus confirming previous results suggesting the minor importance of humoral mechanisms in sleep. After 10 days EEG desynchronization occurred spontaneously in the isolated hemicerebrum even when the normal hemisphere was synchronized. The EEG desynchronization should be interpreted as evidence of arousal, since the same result was produced, probably by olfactory stimulation, when food was offered to the animals. This time the EEG of the normal hemisphere, and also the animal's behavior, indicated the presence of the arousal phenomenon. Thus spontaneous alternation between sleep and wakefulness may occur not only in the chronic *cerveau isolé* but also in the isolated hemicerebrum. If these cycles of cerebral activation and deactivation are due to hypothalamic structures, as appears likely from experiments to be reported later (p. 74), the conclusion may be drawn that cross connections between the two sides of the hypothalamus are not necessary for the onset of both sleep and wakefulness.

4. The Isolated Cerebral Hemisphere

This preparation, introduced by RINALDI and HIMWICH (1955), is obtained by separating the cerebral cortex and parts of the basal ganglia from the rest of the brain, without impairing the vascular supply. The study made by VILLABLANCA (1967) with this preparation is quoted here because the EEG activity of the isolated cerebral hemisphere was studied for several weeks, and the results were compared with those obtained simultaneously from the intact hemisphere. The synchronous slow waves that occurred at random in the isolated cerebral cortex, have in fact little in common with the normal EEG sleep patterns. They were likely to be due to, or at least severely contaminated by, pathological conditions, a conclusion supported by the observation that as chronicity progresses the background activity tended to flatten. Without these limitations, VILLABLANCA's conclusion (l.c., p. 290) that the cat's isolated hemisphere does not exhibit spontaneous electrocortical synchronization and desynchronization as seen in the *cerveau isolé* would be of the greatest importance, since its obvious implication is that the sleep-waking rhythm arises in the diencephalon. Interestingly enough, other lines of evidence, to be discussed later (see pp. 74–84), lead exactly to the same conclusion.

5. Conclusions

Our attention will be concentrated on the main conclusions which may be drawn from the experiments of midbrain hemisection. Their importance lies not so much in the demonstration of the ipsilateral arrangement of the ascending activating pathways, a generalization which will be qualified later, but in the strong support provided by certain observations on hemisected cats in favor of the conclusions we reached in our discussion on the acute and chronic *cerveau isolé*.

The striking EEG asymmetry, with marked tendency toward synchronized activity ipsilaterally to the lesion, is undoubtedly due to the withdrawal of an ascending activating influence. In our opinion the hypothesis that the midbrain transection might act mainly through non-specific mechanisms related to the surgical trauma—a statement which we have already discussed and dismissed in a previous section (p. 23)—can be regarded as disproved simply on the basis of the experiments of unilateral inactivation. Obviously, uncontaminated experiments can hardly be expected, but it appears extremely unlikely that any non-specific alteration produced by the surgery should remain so strictly localized to the ipsilateral hemisphere. This conclusion alone would justify acute experimentation.

The EEG asymmetry produced by midbrain hemisection in the midpontine pretrigeminal preparation and its remarkable fast recovery have also important theoretical implications. We have concluded from these findings i) that the activating influence, or at least the part of it which appears to be critically important for the maintenance of a waking state, arises above the pretrigeminal transection; ii) that the abolition of the sleep-waking cycle is due to an imbalance of antagonistically oriented structures of the brain; and iii) that the recovery of the cycle in the chronic preparation is due mainly to compensation of this imbalance. Thus once more we come to the conclusion that the recovery processes are basically similar to those underlying the compensation from spinal shock. The waking centers of the brain gradually recover a tonic activity. which had probably been depressed immediately after the hemisection.

The strictly unilateral synchronization might appear as a surprising phenomenon if one considers that the EEG manifestations of wakefulness and arousal are always bilateral. The EEG activation of the entire neocortex is in fact the usual result of any unilateral stimulation of the reticular formation (Moruzzi and Magoun, 1949). Only occasionally a strictly ipsilateral disappearence of slow cortical rhythms was obtained by stimulating the medullary reticular formation on one side; but the cat was then under chloralose anesthesia and liminal repetition rates were used (see l.c. Fig. 1).

A distinction should be made between *tonic* and *phasic* activation, underlying respectively the *state of wakefulness* and the *phenomenon of arousal*. What the experiments on EEG asymmetries show is that a *tonic* flow of activating reticular impulses reaches each hemicerebrum through *predominantly ipsilateral* channels. At the level of the upper midbrain, in other words, the ascending reticular system is separated in two major channels. Cross connections are likely to be present, and functionally very important, at least in the unanesthetized animal, both at the level of the brain stem and of the hypothalamus. These connections amply explain why the entire brain is usually awake or is normally aroused as a whole. After precollicular hemisection the effects of brain stem cross connections are of course no longer detectable, while those related with pathways connecting the two halves of the cerebrum are still available. The remarkably fast recovery of the sleep-waking cycle which occurs when a precollicular hemisection is made after complete prepontine transection, shows that the cross connection of the brain are in fact present and active. If the precollicular hemisection is made in an otherwise normal cat, the influence of these cross connections becomes detectable (as shown by the compensation of the EEG asymmetry) when the responsiveness of the activating neurons lying above the hemisection reaches a critical level. This is the main expression of the process we call "recovery".

C. Localized Brain Stem and Cerebellar Lesions

This approach has led to results which are much wider in scope than the theme of our review. However discussion will be limited to the experiments strictly related to sleep physiology. The reader is referred to review articles by SPRAGUE (1967) and by BERLUCCHI et al. (1970) for the results of mesencephalic lesions which are related to other neurophysiological fields.

1. Reticular Formation

When it became known that the EEG arousal could be reproduced by electrical stimulation of an ascending reticular system (MORUZZI and MAGOUN, 1949), the hypothesis was made "that the presence of a steady background of less intense activity within this cephalically directed brain stem system, contributed to either by liminal inflows from peripheral receptors or preserved intrinsically, may be an important factor contributing to the maintenance of the waking state, and that absence of such activity in it may predispose to sleep" (l.c., p. 470). As a corollary of this hypothesis a new interpretation of BREMER's experiment on the *cerveau isolé* became possible, namely that the sleep syndrome produced by the acute mesencephalic transection was not due to deafferentation in the strict sense of the word, but rather to the elimination of the waking influence of an *ascending reticular system* (l.c., p. 470). The hypothesis was immediately tested by LINDSLEY, BOWDEN, and MAGOUN (1949) with acute experiments on the *encéphale isolé* cat.

We have already recalled (p. 10) that a sleep-waking cycle is present immediately after cervical transection of the spinal cord, in BREMER's *encéphale isolé* cat (see BREMER, 1937, 1938 for ref.). This statement does not imply that the tonic flow of impulses ascending from the spinal cord is without any influence on the sleep-wakefulness rhythm (see Ho et al., 1960; HODES, 1962, 1964). There is little doubt, however, that the structures which are of *critical* importance for the maintenance of wakefulness are localized within the encephalon. The striking differences in EEG and behavioral patterns between *encéphale isolé* and *cerveau isolé* led BREMER to the hypothesis of sensory deafferentation. According to this interpretation the olfactory and visual channels, the only afferent pathways still available in an isolated cerebrum, would be inadequate for the maintenance of wakefulness. The flow of impulses reaching the *encéphale isolé* through the other cranial nerves would be of critical importance for the ability to remain awake, and therefore to maintain a sleep-waking rhythm. This was the situation when LINDSLEY et al. (1949) started their experiments.

LINDSLEY et al. (1949) showed that *medial* mesencephalic lesions, interrupting most of the midbrain reticular formation but sparing the classical ascending pathways, resulted in a synchronized EEG; while large *lateral* lesions, sparing in the reticular formation but interrupting the classical specific pathways, did not abolish the sleep-waking cycle. They reached therefore the important conclusion that the syndrome of the acute *cerveau isolé* was due to the interruption of a tonic flow of impulses arising from, and coursing along, the reticular formation. The hypothesis (MORUZZI and MAGOUN, 1949) of the tonic activating influence of the ascending reticular formation was thus confirmed.

It should be stated here that such a hypothesis by no means conflicts with the observation of Roger et al. (1956), that the EEG of the *encéphale isolé* becomes synchronized following bilateral gasserectomy. These authors interpreted their result as a proof that the tonic activity of the ascending reticular system is critically maintained, in the *encéphale isolé*, by a steady flow of trigeminal impulses. In fact this flow is not necessary for the maintenance of waking EEG and ocular patterns, as shown by the experiments on the pretrigeminal cat. The trigeminal impulses are simply necessary to counteract, in the acute *encéphale isolé*, the EEG synchronizing influence of the lower brain stem (see p. 31).

The study of the effects of selective mesencephalic sections was extended to chronic animals by Lindsley et al. (1950), who worked on cats, and by French and Magoun (1952), who utilized rhesus monkeys for their experiments. The cat experiments showed that lesion of the lateral midbrain, sparing most of the mesencephalic reticular formation, did not produce the sleep syndrome of the *cerveau isolé*. The animals could remain awake and were in fact able to stand and walk. Electrolytic lesions involving the reticular formation, and also the diencephalon, but sparing the lemniscal system, produced the classical deactivated EEG. However the synchronous EEG patterns were disrupted, and a *short lasting* arousal occurred, following strong sensory stimulations. In monkeys the symptoms produced by reticular lesions were more severe. The animals were in fact completely comatose, and no clear-cut sensory arousal could be obtained when the entire midbrain tegmentum had been destroyed.

In summary, there is no doubt that neither sleep nor coma are produced when the classical ascending pathways are interrupted. The effects of reticular lesions require some comment. The destruction was very extensive in the antero-posterior direction and both the diencephalon and the brain stem were involved. The situation was therefore entirely different from that of the *cerveau isolé*, which is characterized by the interruption of the ascending flow of both lemniscal and reticular impulses at a given midbrain level, with full integrity of the ascending reticular neurons lying above the section and, obviously, of the entire diencephalon. The interpretation of the results would have been easier had the midbrain tegmentum been simply interrupted. This is the approach of the experiments we are now going to report. Obviously when the specific sensory systems are spared, a transection of the midbrain tegmentum will not prevent the activation of the overlying structures by collaterals of the classical pathways.

Whitlock et al. (1953) interrupted the midbrain tegmentum, leaving intact both the classical sensory pathway and the pyramidal tract. Their "pyramidal" cats were mainly utilized for microelectrode studies of the pyramidal discharges during different types of arousal. The investigations was limited, however, to the acute period, during which the cat clearly behaved as an acute *cerveau isolé*.

The same approach was utilized in 1956 in chronic experiments made on cats by GENOVESI et al. (1956). A thin curtain of electrolytically destroyed tissue interrupted, at midbrain level, either the medially placed ascending reticular system or the specific sensory pathways coursing in the lateral part of the mesencephalon. When the third nerve nuclei had not been damaged by a thin intercollicular lesion, ocular behavior provided, with the EEG, adequate information on the state of the cerebrum. Chronic interruption of the lemnisci did not prevent wakefulness. Subsequent (after 4 to 15 days), partial interruptions of either the medial or lateral part of the midbrain reticular formation were also without any clear effect, a result that may be due to the diffuse organization of the ascending reticular system. Only when the outcome of the multistage chronic lesion had been the complete section of both reticular and specific systems were the EEG and behavioral patterns of coma observed, as in the acute *cerveau isolé*. Since the *pes pedunculi* was bilaterally spared, the animal might be regarded as a "pyramidal" cat, a preparation entirely equivalent to the *cerveau isolé* with respect to the complete interruption of ascending flow of specific and reticular impulses. From 4 to 10 days after the last lesion, EEG and behavioral signs of wakefulness were observed. Thus these experiments provided the first evidence that a recovery of the ability to maintain wakefulness may occur in a preparation which is functionally equivalent to the chronic *cerveau isolé* (see p. 16). The interruption of the specific sensory paths bridges the gap between the experiments of chronic reticular lesions and the chronic *cerveau isolé*. In the experimental conditions choosen by GENOVESI et al. (1956), the recovery could not be attributed to sensory impulses reaching the activating structures lying rostrally to the intercollicular lesion.

The results obtained with incomplete brain stem lesions are less interesting, because it is difficult to say how much of the recovery process was related to the midbrain reticular gates that had remained open. This group of results has been reviewed at length by SPRAGUE (1967), who rightly pointed out that electrolytic lesions may produce vascular effects on neighboring tissues, thus increasing the severity of the syndrome. In some of the cats studied by ADAMETZ (1959) most of the midbrain tegmentum caudal to the meso-diencephalic junction had been destroyed by a multistage procedure, but the raphe and the periaqueductal gray had been spared. One animal survived two months after the initial lesions. The recovery of responsiveness and of the sleep-waking cycle was good. When electrolytic lesions had been placed at one sitting, survival was shorter (12 days) and there was no recovery from coma.

The extensive studies of SPRAGUE and his colleagues led to important developments in the midbrain physiology, particularly with regard to the superior colliculi (see SPRAGUE, 1967; SPRAGUE et al., 1971). We shall be concerned here only with the effects of midbrain tegmental lesions. On the whole, extensive medial lesions of the rostral midbrain, sparing the long ascending and descending pathways, led to the usual sleep syndrome, which was interrupted by short lasting arousal following intense sensory stimulation. Behavioral and EEG signs of sleep diminished over the next two weeks, although the cats were frequently drowsy when not stimulated. Catatonic behavior was frequently observed during a survival period of two years.

Chronic midbrain lesions were frequently combined with detailed behavioral investigations and conditioned reflex studies.

These works have been fully reviewed by SPRAGUE (1967). It would be difficult to say to what extent the striking alterations of instinctive behavior observed after these lesions were due to interruption of an ascending influence acting upon the cerebrum, and to what extent they were due to lesions of descending, extrapyramidal pathways mediating the hypothalamic discharges which are responsible for instinctive behavior.

MALLIANI et al. (1963) clearly saw the main issue when they pointed out that the rage outbursts of their fully decorticated cats were abolished by the same midbrain tegmental lesion which yielded the comatose syndrome in the otherwise normal animal, and by that lesion only, although outbursts of rage could still be obtained by electrical stimulation of the hypothalamus. At least in these experimental conditions, therefore, the syndrome produced by the midbrain tegmental lesion was due to the interruption of an ascending flow of impulses reaching the hypothalamus, not to its deefferentation following lesion of descending pathways. These investigations are important because for the first time the activity of a hypothalamic center concerned with a typical instinctive behavior, the defense-aggression behavior, was shown to require an ascending flow of reticular impulses. ZANCHETTI (1966, 1967) has devoted two reviews to the problem of emotional behavior.

2. Raphe Nuclei

After the first group of experiments on the midpontine pretrigeminal cat (BATINI et al., 1958, 1959a), four lines of investigation were started in an attempt to localize the EEG synchronizing structures of the lower brain stem and to define their physiological properties.

Two of these approaches, closely related to the original line of work, have already been reviewed. The experiments were concerned with i) the proof that a genuine state of wakefulness is present in the midpontine pretrigeminal cat (pp. 25–31) and ii) the EEG asymmetries produced by pontine hemi-sections (p. 54). There are reviews mainly or exclusively devoted to the mid-pontine preparation (MORUZZI, 1960, 1963, 1964; ZERNICKI, 1964, 1968); while two articles deal exclusively with sleep induced by sensory stimulation (POM-PEIANO, 1965) and with its possible relationship to the EEG synchronizing structures of the lower brain stem (MORUZZI, 1960).

Two independent lines of investigations were concerned i) with the medullary structures exerting a damping influence on reticular arousal, a field of study started by BONVALLET and her colleagues (see BONVALLET, 1966), shortly recalled in other section of this review (see p. 36) and ii) with the sleep inducing function of the raphe nuclei, a field of physiological investigation closely related with neurochemical studies, one that will be extensively reviewed by JOUVET in these reviews. Our discussion will be concerned exclusively with the physiological aspects of the results obtained by raphe lesions and their possible relation with the works of BONVALLET et al.

JOUVET and RENAULT (1966) reported that subtotal destruction of the raphe system, from the upper medulla to the ponto-mesencephalic junction, was followed by 3–4 days of complete sleeplessness. Sleep reappeared later in these cats, but was strikingly reduced in duration, by as much as 80 %. Moreover, sleep with desynchronized EEG did not appear until the phase characterized by EEG synchronization attained 15 % of the nychthemeron. The authors pointed out (l.c.) that the most rostral part of the raphe nuclei which had been destroyed is located in the upper pons, i.e. above the level of a midpontine pretrigeminal section. This fact does not imply, in our opinion, that the insomnias of the pretrigeminal and of the raphectomized cat are unrelated phenomena. The increased tendency to wakefulness of the pretrigeminal cat might also be due to the withdrawal of an "hypnogenic" or "EEG synchronizing" influence arising from pontine raphe nuclei. First of all patterns of wakefulness are present also after a prepontine section, provided the midbrain is not damaged (see ZERNICKI, 1968). Second, the imbalance in the sleep-waking mechanisms of the isolated cerebrum produced by a pontine transection might be expected to produce hyposomnia and a striking tendency toward activation provided two conditions are fulfilled, i) maximal deafferentation of the cerebrum with respect to the EEG synchronizing structures and ii) minimal deafferentation with respect to the activating reticular system. Clearly the level of the brain stem transection producing the highest degree of activation of the isolated cerebrum, probably the midpontine pretrigeminal section, cannot be expected to coincide with the upper border of the EEG synchronizing system, as determined by the method of localized lesions.

In the split brain cat of MICHEL and ROFFWARG (1967) the vertical section included the midbrain, as well as the pons and the upper medulla. The nychthemeral percentage of EEG and behavioral alertness was quite remarkable in these animals: 80–92 % during the first two weeks, and 75–80 % later. Sleep with EEG desynchronization was again more strongly affected than the slow phase, and was in fact entirely absent during the first week. MICHEL'S (1967) *encéphale dédoublé*, a combination of SPERRY's split brain with the split brain stem preparation, also led to striking, chronic hyposomnia.

Mancia, Desiraju, and Chhina (1968) followed the same approach and confirmed, in the monkey, the results on experimental hyposomnia. The pons was splitted in their experiments, and the midline section was limited to its rostral portion, from the upper border to midpontine level. Mancia (1969) confirmed and extended the results of the monkey experiments by making longitudinal sections in the cat. He obtained chronic hyposomnia following pontine but not after midline bulbar split. Midbrain splitting did not clearly change the amount of wakefulness and of slow sleep, although the paradoxical episodes were significantly reduced. Physiological data and the study of retrograde changes after pontine splitting led to the conclusion (l.c.) that hyposomnia was due to interruption of ascending axons arising in the medulla or the pons and crossing in the rostral pons.

The alternative explanation that the lesion of raphe nuclei is directly responsible for this type of hyposomnia is supported by several facts: i) the long ascending fibers of the reticular formation are crossed as well as uncrossed, the latter component being the largest (Brodal and Rossi, 1955), and there is no evidence of a prefential crossing at the level of the pons (see Brodal, 1969); ii) quantitatively the most important contingent of efferent fibers from the raphe nuclei is the ascending one and many axons of raphe nerve cells pass to or beyond the mesencephalon (Brodal et al., 1960); iii) neurochemical evidence, to be reported by Jouvet in these reviews, suggests a control of the sleep-waking cycle by the raphe nuclei. According to this explanation, which fits Jouvet's hypothesis, the insomnia produced by raphe lesions is due to destruction of nerve cells localized in this area.

The stimulation experiments by Magnes et al. (1961 a, b) could not unveil the existence of a tonic deactivating influence, but important lesion experiments by Bonvallet and her associates (Bonvallet and Bloch, 1961; Bonvallet and Allen, 1963) would seem to support the hypothesis that besides the raphe neurons other structures of the lower brain stem have an ascending tonic influence, which is antagonistic to the action of the reticular activating system.

Bonvallet and her colleagues (l.c.) reported that the *phasic* arousal produced by reticular stimulation—as shown by the EEG activation and by the inhibition of the Edinger-Westphal parasympathetic neurons—was greatly enhanced and prolonged following very limited lesions of the medulla. They suggested that phasic responses of the reticular activating system may be inhibited by impulses arising in the region of the nucleus of the solitary tract; a hypothesis that would also account for the striking depressive effect exerted by a strong baroceptive barrage. The visceral afferents of the 9th and 10th pair of the cranial nerves reach in fact this bulbar region (see for ref. Moruzzi, 1963 b, p. 259; Bonvallet and Allen, 1963, p. 984). What should be known is whether this system, so clearly concerned with a negative feedback control of phasic arousal, also depresses, or in some way counteracts, the tonic activity of the ascending reticular system, thus explaining the striking tendency to activation in the midpontine pretrigeminal cat. Bonvallet and Bloch (1961) utilized prebulbar transections, but BONVALLET and ALLEN (1963) reproduced the release phenomena with small transversal sections of the medulla, placed 2–3 mm laterally from the midline, while small medial transverse sections were without effect. The latter observation as well as novocaine controls (BONVALLET and ALLEN, 1963) demonstrate that these effects were not due to irritation.

The result of ROSSI et al. (1963) might be quoted against the hypothesis that a tonic EEG synchronizing influence arises in the area of the nucleus of the solitary tract. They were unable to produce asymmetries of sleep synchronization with hemisections made at caudo-pontine levels. However, their observations started at least 24 hours after the *hemisection*, and the hypothesis may be made either i) that the bulbar asymmetrical influence was slight, and therefore easily and soon compensated for, or ii) that the prevalent ipsilateral organization of the ascending pathways starts only at pontine level. It should be recalled here that the chronic experiments by BONVALLET and ALLEN (1963) were based on *bilateral* bulbar coagulations. No one has yet made an attempt to repeat the experiments of BONVALLET and ALLEN (l.c.) using a cat which has partially recovered from the effects of the midline section of the pons

3. Brain Stem Regions Concerned with Paradoxical Sleep

These lines of experimental endeavor will be recalled only briefly, since details and discussions may be found in other reviews to be quoted below.

a) JOUVET (1962) abolished the EEG and somatic signs of paradoxical sleep in the cat with lesions destroying the *nucleus reticularis pontis oralis* and, partially, the *nucleus reticularis pontis caudalis*. CARLI and ZANCHETTI (1965) by studying the effect of a number of localized lesions came to the conclusion that the pacemaker was localized in the mediolateral portion of the middle, and perhaps the posterior third of *nucleus reticularis pontis oralis*. CANDIA et al. (1967) are of the opinion that localization is less precise, since all the structures located in the rostral half of the pons appear to contribute to the origin of the paradoxical episodes. This group of studies has been reviewed by ZANCHETTI (1967b).

b) JOUVET (1962) has suggested that the ascending influence responsible for the neocortical and hippocampal manifestations of paradoxical sleep is mediated by midbrain-limbic pathways. CARLI and ZANCHETTI (1965) and CARLI et al. (1965) showed, however, that lesions interrupting the ascending limb of the midbrain-limbic circuit do not prevent these manifestations, a conclusion confirmed also by HOBSON (1965). Thus the diffuseness of the ascending pathways appears to be shared both by the activating reticular system and by the pacemakers responsible for those episodes of sleep characterized by EEG synchronization. ZANCHETTI (1967b) has reviewed this field of investigation.

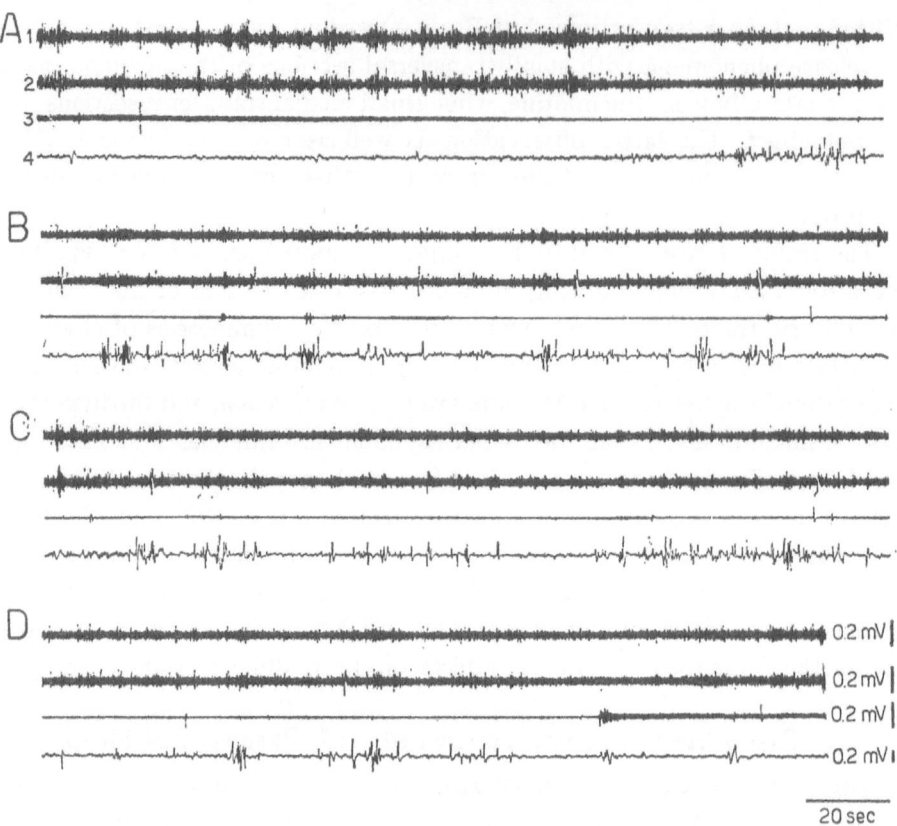

Fig. 21 A–D. *Typical patterns of REM and of cervical atonia during paradoxical sleep in the intact animal.*
Unrestrained, unanesthetized cat. Experiment made 4 days following chronic implantation of the electrodes.
In this and the following figure the bipolar records are the following: *1* left parieto-occipital (EEG);
2 right parieto-occipital (EEG); *3* posterior cervical muscles (EMG); *4* ocular movements (electro-oculo-
gram: EOG). A–D Episode of paradoxical sleep. Note the occurrence of large bursts of rapid eye movements (REM) when the EMG of the posterior cervical muscles becomes silent. See also the appearance
of myoclonic twitches in the posterior cervical muscles, frequently but not constantly related with the
bursts of REM. (From Pompeiano and Morrison, 1965)

c) We have seen (p. 6, 137) that the episodes of paradoxical sleep are characterized by a striking eruption of phasic events, all related in time with the
outbursts of rapid eye movements. Pompeiano (1966, 1967a, b) has reviewed
the works he made with several collaborators (Bizzi et al., 1964; Pompeiano
and Morrison, 1965, 1966; Morrison and Pompeiano, 1966a, b), which led
to the conclusion that these phasic events are selectively abolished when the
medial descending vestibular nuclei of the cat are destroyed in their entire
rostrocaudal extent (Figs. 21, 22). These effects are not reproduced by bilateral
section of the 8th nerve, i.e. after deafferentation of the labyrinthine nuclei,
nor by complete cerebellectomy, the last observation showing that the effects
of the vestibular lesion are not due to interruption of cerebellofugal fibers
passing through these nuclei. Moreover, bilateral and symmetrical lesions of
the superior and lateral vestibular (Deiters') nuclei are ineffective. The tonic

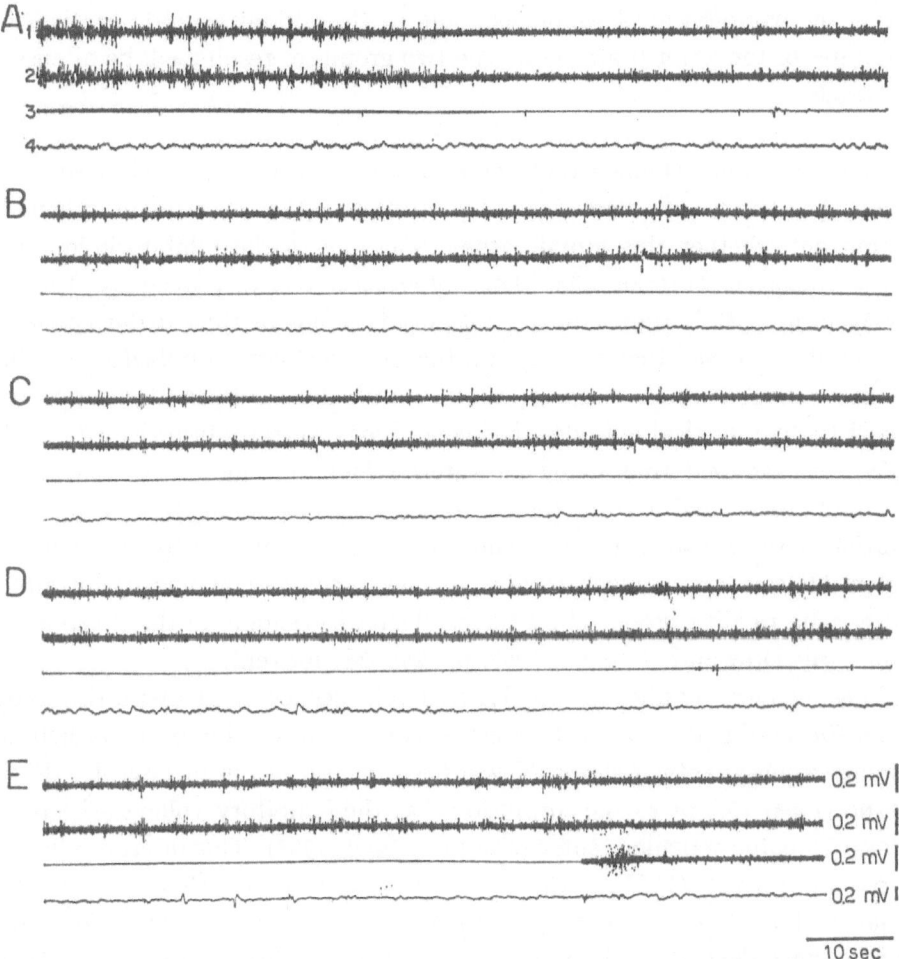

Fig. 22 A–E. *Abolition of the bursts of REM and persistence of cervical atonia during paradoxical sleep following complete bilateral destruction of the vestibular nuclei.* Same animal as in Fig. 21. Experiment made 9 days following chronic implantation of the electrode and 3 days after complete destruction of the four vestibular nuclei of both sides. A–E Episode of desynchronized sleep, characterized by desynchronized EEG patterns and cervical atonia. Note complete absence of bursts of REM. Only slow ocular movements and isolated ocular jerks are observed. Myoclonic twitches not related in time with ocular movements may affect the posterior cervical muscles. (From POMPEIANO and MORRISON, 1965)

manifestations of the paradoxical episodes, i.e. the EMG silence of the antigravity muscles and the EEG desynchronization, remain when the phasic outbursts are abolished by the vestibular lesion. These facts and electrophysiological studies on the lateral geniculate body (MORRISON and POMPEIANO, 1966b) show that the medial and descending vestibular nuclei are triggered by a pacemaker, which is probably localized in the pons.

4. Cerebellum

The experiments to be reviewed in this section, all made on cats, were concerned with the cerebellar control of i) the postural phenomena occurring during the paradoxical phase of sleep and ii) the activating and deactivating structures of the lower brain stem. The two groups of results will be discussed separately.

When it became known that the *extensor tonus* was strikingly depressed during the episodes of paradoxical sleep (Jouvet, 1962), an attempt was made to find out whether this hypnic atonia was due to the inhibitory function of the cerebellar anterior lobe. This hypothesis was disproved by Jouvet (1962), in his EMG studies on the antigravity tonus of the cat during sleep. Jouvet (l.c.) showed that even within the first week after cerebellectomy the extensor rigidity, which characterizes Luciani's dynamic period, disappeared during the paradoxical episodes. His conclusions were confirmed by Hobson (1965), who showed that following anterior lobe ablation neither the sleep atonia of the normal cat nor the cataplexic episodes of the decerebrate preparation were abolished. Hobson concluded (l.c., p. 56) that the mechanisms responsible for sleep atonia are able to overcome the combination of alpha and gamma rigidity produced by the cerebellar topectomy in the decerebrate animal (see Dow and Moruzzi, 1958, pp. 255–283 for ref.).

These converging lines of experimental evidence, combined with the study of the effects of partial spinal transections on the hypnic inhibition of heteronymous monosynaptic reflexes (Giaquinto et al., 1964a, b), have led Pompeiano (1965, 1967b) to the conclusion that the inhibitory volleys arise from Magoun's bulbo-reticular center (Magoun, 1944, 1963). This interpretation is shared by Jouvet (1967a), who has rightly pointed out that the reticulospinal inhibitory neurons must in turn be driven by descending pontine discharges, since the cataplexic episodes are absent after low pontine transection.

Although sleep hypotonia may still occur during the paradoxical episodes in the cerebellectomized cat, it should be pointed out that a cerebellar contribution to the hypnic regulation of postural tonus is not disproved by these experiments. Marchesi and Strata (1970) have actually come to the conclusion that the cerebellum is involved in the postural manifestations of the paradoxical phase. Microelectrode recording from the anterior lobe, in the free moving unanesthetized cat, has shown that the rate of firing of the climbing fiber complex (CF) increases at the onset of the episodes of paradoxical sleep characterized by atonia, when the phasic outbursts are absent. Since the resulting Purkinje cell activation is inhibitory in nature, the authors (l.c.) have suggested that the increased CF discharge may contribute to the onset of sleep atonia. The study of the antigravity tonus of the cervical muscles in the totally cerebellectomized cat has led Guglielmino and Strata (1971) to the same conclusions. The EMG silence is observed only in coincidence with strong bursts of rapid eye movements, while in the lulls between them the cervical antigravity tonus is indeed strikingly reduced, but not abolished as in the normal animal. The cats have been studied so far for 3–4 weeks, therefore after the end of the release syndrome that characterizes LUCIANI's dynamic period.

Before discussing the *EEG effects of cerebellar lesions*, attention should be called to an important cause of error revealed by MASSION et al. (1965). These authors have shown that EEG changes can be produced, in a purely aspecific way, by the loss of the cerebro-spinal fluid. Old behavioral (see Dow and MORUZZI, 1958, p. 349) and EEG investigations (see l.c., p. 330) were probably contaminated by factors of this kind, and probably even by anatomical damage to the brain stem.

In conditions under which this cause of error was adequately controlled, the problem of the cerebellar contribution to the maintenance of the electro-cortical "tone" was studied by FADIGA et al. (1967). In their experiments, EEG synchronization was seen to follow the bilateral electrocoagulation of the fastigial nuclei or the severance of the inferior cerebellar peduncles, but was absent after electrocoagulation of the vestibular nuclei or interruption of the superior cerebellar peduncles. These acute experiments, performed on curarized, artificially respirated cat, were probably biassed by a marked tendency to EEG synchronization. Chronic experiments on free moving cats have led to less dramatic but more complex effects (GIANNAZZO et al., 1969). Only during the first three days following bilateral fastigial destruction did the cats present an increase of sleep with EEG synchronization, with no abolition of the sleep-waking cycle or of the alternation between slow and paradoxical episodes. However the transient effects were reversed in sign as chronicity progressed, and an increase of wakefulness was present during the rest of the survival period (up to 27 days after the lesions). The last EEG changes were not an aspecific consequence of surgical trauma since they were not duplicated by extrafastigial or dentate lesions. Further experiments by the same group have led to other findings which also deserve to be considered by the sleep physiologist.

MANZONI et al. (1968) showed that inactivation of the fastigial nuclei, by cooling, in the free moving unanesthetized cat, produced the classical release of the antigravity mechanisms, and also EEG and behavioral manifestations of arousal, provided of course that the experiment was performed on a background of drowsiness (Fig. 23). The same cooling technique, when applied to the decerebrate cat, was followed by the typical, striking enhancement of the extensor rigidity, in confirmation of old findings (CAMIS, 1923). Thus the arousing effect was not due to irritation, but to functional inactivation of the cooled structures. The proprioceptive barrage elicited by the enhancement of antigravity tonus was not responsible for EEG arousal since the electrocortical phenomena were the first to appear (Fig. 23). All these

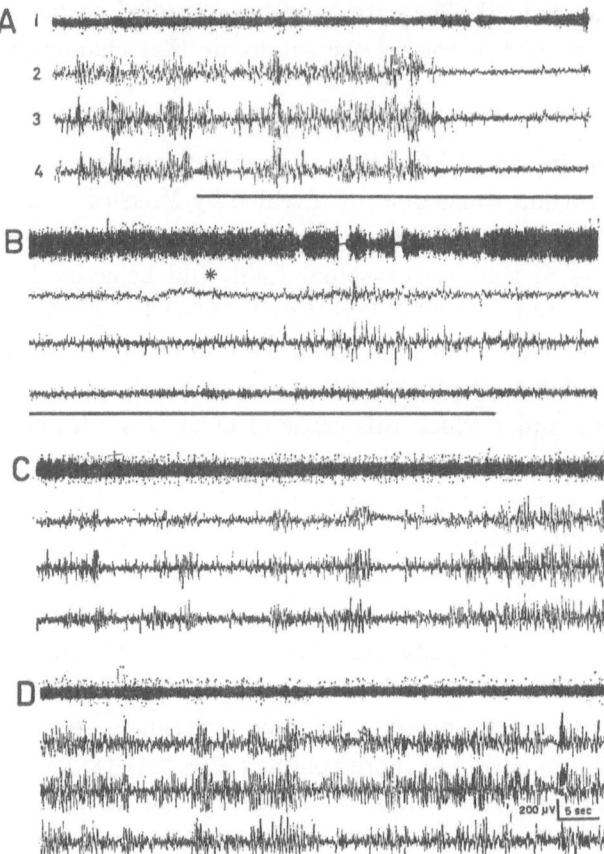

Fig. 23 A–D. *EEG desynchronization produced by cooling of the fastigial region in an unanesthetized, un-restrained cat*. The experimental session took place 8 days after preparation surgery. *1* EMG of neck muscles (bipolar recordings); *2* right occipital; *3* left occipital; *4* right-left fronto-frontal. The heavy line under-neath the records A and B marks the time of fastigial cooling (3.5 min: one minute elapsed between A and B, during which cooling was maintained). From B to C and from C to D, 5- and 7-minute intervals, respectively. The asterisk in B signals the beginning of the extensor hypertonus. Note that the "ascending" effects *precede* the appearance of the "descending" effects on muscle tone. (Fom Manzoni et al., 1968)

data suggested to the authors that the fastigial system exerts "a tonic, syn-chronizing influence, not very different from that already envisaged for other rhombencephalic structures" (l.c., p. 70).

It remained for Fadiga et al. (1968) to provide a convincing explanation of these results, which might appear difficult to reconcile with those obtained by surgical lesions. They showed that an EEG activating influence, clearly tonic in nature, arises in the fastigial nuclei, at least in the *encéphale isolé cat*. This influence, however, is probably overwhelmed—in the intact, free moving animal—by an opposite EEG synchronizing effect, also arising from the fastigial nuclei. Only the activating influence had been previously revealed by stimulation experiments (Moruzzi and Magoun, 1949, see p. 458). Fadiga et al. (l.c.) showed that a localized lesion of the rostro-lateral portion of the fastigial nuclei was in fact followed by EEG synchronization, while a rostro-medial lesion produced clear-cut EEG desynchronization. Unfortunately the short duration of the experiments in these curarized, artificially ventilated cats prevented a study of the sleep-waking cycle. However, several well conceived controls (l.c.) suggest that two antagonistic influences arise from the rostro-medial and rostro-lateral part of the fastigial nucleus. These structures are also known to exert opposite effects on postural tonus as well (MORUZZI and POMPEIANO, 1957; BATINI and POMPEIANO, 1958; see DOW and MORUZZI, 1958, pp. 85–86, 131–134).

The cerebellar approach to sleep physiology deserves the attention of the experimenter, who may be misled by the dominant view which regards the cerebellum as a structure mainly concerned with the control of movements and posture. Such a control is exerted also by the brain stem reticular formation and yet an ascending reticular influence on the cerebrum is well documented. In fact this association of functions is not surprising since postural adjustments and movements are largely involved in any behavioral manifestation of sleep and wakefulness. Cerebellar stimulations have striking effects on the discharge of reticular neurons (MOLLICA et al., 1953; VON BAUMGARTEN et al., 1954; GAUTHIER et al., 1956; see MORUZZI, 1954), which do not necessarily belong, in all instances, to the descending reticular system. In fact the anatomical separation between descending and ascending system is far from being complete (SCHEIBEL and SCHEIBEL, 1958; MAGNI and WILLIS, 1963, 1964; see BRODAL, 1969, pp. 314–315).

5. Conclusions

Our discussion will be centered around the effects of lesions of the activating and deactivating structures of the brain stem.

Since in the chronic *cerveau isolé* there is a remarkable recovery of the ability to maintain wakefulness and of the sleep-waking cycle, it is not surprising that similar and actually faster recoveries may be observed when only the ascending flow of reticular impulses is interrupted at midbrain levels. The problem of the activation of reticular neurons by impulses impinging upon them through collaterals given off by the long sensory paths has been widely discussed (see ROSSI and ZANCHETTI, 1957, p. 322; BRODAL, 1969, pp. 315–316).

In summary, the results of chronic reticular lesions fit the old assumption that the coma of the acute *cerveau isolé* results mainly from the withdrawal of the ascending reticular barrage.

The second point which deserves all our attention is the demonstration (MALLIANI et al., 1963) that the ascending flow of reticular impulses is critically important also for the appearance of the sham rage outburst in the thalamic cat. We shall see later (p. 74) that the "centers" necessary for wakefulness are located in the posterior hypothalamus, just as are the structures responsible for the sham-rage outbursts of the thalamic cat (BARD, 1928) and probably also for the "mood" of defense-aggression in the normal animal (HESS,1968b). The experiments of ZANCHETTI and his colleagues (see MALLIANI et al., 1963) have shown, on the other hand, that both the tonically active center of wakefulness, and the phasic eruptions of activity occurring in the center for the instinctive behavior of defense-aggression, are conditioned by a flow of impulses ascending in the activating reticular system. A working hypothesis which deserves consideration is that the same hypothalamic structures may be responsible for the state of wakefulness, by means of their mild tonic activity, and for fits of rage, during periods when their discharges are phasically enhanced: this hypothesis will be discussed again at the end of this review. For the anatomical background of these influences SCHEIBEL and SCHEIBEL (1967, see p. 83) and BRODAL (1969, pp. 333–334) should be consulted.

An impressive amount of experimental evidence has been collected during the last ten years on the EEG synchronizing structures of the lower brain stem. Although their existence seems now to be certain, several problems remain, namely: i) the relation between the results of experiments on the raphe nuclei and those on the region of the nucleus of the solitary tract and ii) the mechanism of these EEG synchronizing effects. Important lines of evidence, especially the recent electrophysiological experiments by BREMER (1970a, b) ,suggest an inhibitory influence on the activating reticular system.

D. Localized Brain Lesions

1. Decortication

The classical studies made by GOLTZ (1892) on chronically decorticate dogs brought the first behavioral evidence that the alternation between sleep and wakefulness may be observed even after complete ablation of the cerebral cortex. KLEITMAN (1963) has reviewed the physiological literature, while clinical findings have been reported and discussed in articles by OSWALD (1962), BARRET et al. (1967) and ORTHNER (1969). In summary, there is unanimous agreement of opinion that the sleep-waking cycle may be present when the cerebral cortex is either absent or inactive.

Among the physiological papers only the study of KLEITMAN and CAMILLE (1932) will be quoted, because of its theoretical importance. These authors confirmed the existence of an alternation between sleep and waking behavior in four dogs which survived up to one year after a two-stage decortication. With a simple mechanical device, the authors recorded the walking movements that occurred during wakefulness: inactivity was used to define periods of rest.

Their showed that five or six periods of activity alternated each 24 hours with periods of rest, during which the animals appeared to be behaviorally asleep. While the normal dog, a monophasic animal, can be awake throughout the day and asleep through the entire night, these decorticated animals had about six periods of sleep during the nychthemeron. Thus adults dogs had been transformed by decortication into polyphasic animals with respect to sleep. They behaved as young puppies. This result is of great theoretical significance, since it may now be regarded as proof that the regulation of the sleep-waking cycle is influenced by the cerebral cortex.

JOUVET (1962) made an attempt, which turned out to be successfull in six cats, to limit the ablation to the neocortex. By combining electrophysiological records with behavioral observations, he identified both the slow and the paradoxical phases of sleep, but pointed out that during the periods corresponding to slow sleep of the normal animal spindles could no longer be recorded from subcortical leads. He concluded that subcortical synchronization is driven by corticofugal volleys.

It should be recalled that in the chronically decorticated animals the extent of damaged or malfunctioning tissue is always greater than the extent of the original ablation. The problem of retrograde thalamic degeneration following decortication has been recently reviewed by MACCHI (1969).

2. Thalamectomy

Old theories (see KLEITMAN, 1963, p. 360), HESS' stimulation experiments to be reviewed later (see p. 90) and, finally, modern electrophysiological investigations on thalamic driving of electrocortical EEG spindles (PURPURA et al., 1966; see PURPURA and YAHR, 1966, for review articles) have centered the attention of sleep physiologists on the thalamus. The old literature on the effects of thalamic lesions was reviewed by KNOTT et al. (1955). Several among the old experiments were contaminated by lesions of the hypothalamus and are therefore of little interest for the physiology of sleep.

Important results may be found in a recent paper by NAQUET et al. (1965, 1966), who were able to follow six completely thalamectomized cats for up to 25 days. They showed that in their animals the sleep-waking cycle was still present, although the EEG spindles were absent during behavioral sleep. Clearly sleep may occur even in the absence of the thalamus, an important conclusion which will be discussed when dealing with the stimulation experiments (see p. 112). The abolition of the EEG spindles fits the classic notions on thalamic driving of cortical neurons.

The experiment of partial thalamic lesion (see ANGELERI et al., 1969, for ref.) will not be reviewed. Suffice it to say that even after extensive chronic lesion of the midline non-specific nuclei of the cat's thalamus, no statistically significant changes were observed in the distribution between wakefulness and sleep, both with and without EEG synchronization (l.c.).

3. Hypothalamic Lesions

The anatomo-clinical observations made by von Economo (see 1926, 1929, for ref.) on cases of sleep-sickness were the first to call the attention of sleep physiologists to the central gray and the hypothalamus. However only in the thirties, when Ranson reintroduced the Horsley-Clarke technique in neurophysiology, did an experimental approach become possible.

The old clinical and experimental literature has been extensively reviewed (Harrison, 1939). The recent clinical literature is reported in Orthner's article (1969). We shall be concerned here only with the experimental literature, from Ranson's works to the present time.

Ingram et al. (1936) reported that lesions of a region between the mammillary bodies and the third nerves, thus involving the caudal hypothalamus and the upper part of the mesencephalic tegmentum, were followed by a state resembling catalepsy. During the first days after the operation they observed i) profound somnolence, ii) loss of motor initiative and iii) remarkable plastic hypertonus of the muscles. The plasticity of the limbs, as shown by good acceptance of passively induced poses and by marked rigidity, were attributed to damage of extrapyramidal structures. These cardinal symptoms of the cataleptic syndrome generally disappeared within one week. Our attention will be concentrated on those symptoms which suggest an ascending influence of the structures that had been destroyed.

Somnolence is obviously the most important manifestation for the theme of our review. It might rather be defined as *lethargy*, since the cats of Ingram et al. could be aroused by sensory stimulations, although the threshold was high and the duration of the arousal was rather short. By reading the protocols one is impressed with the striking similarity between experimental lethargy and the clinical symptoms of the sleep sickness, as decribed by von Economo (see 1926). Perhaps what was defined as "a general attitude of stupidity" (Ingram et al., 1936, p. 1184: cat 5) was mainly the result of a far reaching disorganization of cat's instinctive behavior. Other quotations support this statement: "When placed before a group of barking dogs the cat pricked up it ears and looked mildly interested, but showed no other signs of emotion" (l.c., p. 1184: cat 5, 33 days after operation); "no interest was shown when a rat crawled over the cat's back. It made no effort to keep itself clean" (l.c., p. 1185: cat 7, 17 days after operation). All these symptoms persisted longer that the motor and postural ones, although habits of cleanliness and sexual behavior occassionally reappeared. However the animal remained "calm and unemotional and showed little interest in their surroundings" (l.c., p. 1187) for the entire survival period.

A few years later RANSON (1939) described, in a fundamental paper, the effects of hypothalamic lesions made on rhesus monkeys (Macaca mulatta). His observations are particularly important because the midbrain and often the thalamus had not been encroached upon. Bilateral lesions in the lateral hypothalamic areas, extending to the caudal border of the mammillary body, were most effective in producing a lethargy syndrome. During the first days the monkey woke only when handled, and then quickly fell asleep again. Clearly this state was not one of coma. Later, the animal became progressively less sleeply, and finally only a state of drowsiness prevailed. However the monkeys had become tame and unafraid, while their faces remained immobile and masklike. Lack of motor initiative and striking symptoms of catalepsy were also reported. When the lesion was placed more dorsally, at the junction of the thalamus and the hypothalamus, so that the lateral hypothalamic area was spared, the sleep syndrome was absent, and the animals were not much tamer than before the operation.

In his discussion RANSON rightly emphasized that

"properly placed lesions in the hypothalamus produce results which are the opposite of those caused by hypothalamic stimulation. Emotional reactions are decreased or abolished, and in place of excitement there is drowsiness or somnolence. It is reasonable to assume that elimination of the excitation caused by hypothalamic activity is responsible for the somnolence and that under normal conditions the hypothalamic drive plays a large part in maintaining the waking state (RANSON, 1939, p. 18)."

Here again we are confronted with a remarkable coincidence between the localization of the hypothalamic structures responsible for an important instinctive behavior, the defense-aggression behavior, and the region whose destruction produces the lethargic syndrome. RANSON suggested (1939, p. 18) that "the active hypothalamus discharges not only downward through the brain stem, spinal cord and peripheral nervous system into the body, but also upward into the thalamus and the cortex". Later on RANSON and MAGOUN (1939), in their review article for *Ergebnisse der Physiologie*, suggested (l.c., p. 119) that "this upward discharge may well be associated with emotion as a conscious experience". In their opinion, however, "the excitation of the cerebral cortex by the hypothalamus is not essential for maintaining the waking state" (l.c., p. 119). They stated "that the excitation, spreading from hypothalamus and preventing sleep, probably acts chiefly on lower centers in the subthalamus, brain stem and spinal cord" (l.c., p. 119).

The experiments reported by NAUTA (1946) were concerned with the chronic effect of transverse sections of the hypothalamus in the albino rat. The sleep-waking cycle was not influenced, at least from a behavioral standpoint, when the hypothalamus was only unilaterally encroached upon by an incision which extended to the base of the brain. When the transverse section was situated in the immediate vicinity of the mammillary bodies the lethargic syndrome was observed throughout the survival period of four to

eight days. It was a real state of lethargy, not coma, as shown by the following description.

"They lay curled up on one side, breathing regularly, and could promptly be awakened
by sufficiently strong stimuli, *i.e.* by pinching the tail or handling. When they were
left alone afterwards, they would *yawn* and *stretch* and settle down in a comfortable
position to go to sleep again" (l.c., p. 292).

The striking alteration in the sleep-waking cycle was attributed to the
disconnection of *a waking center from the underlying brain stem*. Nauta recalled
that electrical stimulation of the posterior hypothalamus yields widespread
sympathetic discharges, which give an appearence of intense emotional activity.
He suggested "a partial identity between the waking center and the ortho-
sympathetic centre in the hypothalamus" (l.c., p. 297). After the experimental
findings of Hess (see 1968a, b) we know that these hypothalamic structures
correspond to an important center of instinctive behavior, responsible for
defense and aggression.

When the transverse section was situated in the rostral half of the hypothalamus the alteration of the sleep-waking cycle was exactly opposite in sign.
The normal rat is a polyphasic animal, sleep being distributed over 10 periods
for a total of 14 hours for the nychthemeron (Szymanski, 1918, 1920). The
operated rats were awake for the entire survival time, averaging three days,
"after which the exhausted animal fell into a state of coma, which soon
ended in death" (p. 303). Nauta noted that unilateral sections at the same
level were without effect, and rightly pointed out that this fact alone disproved
the irritative hypothesis. He thus concluded that the rostral half of the hypothalamus, roughly corresponding to the suprachiasmatic and preoptic area,
is the site of a nervous structure which is of special importance for the capacity
of sleeping. In fact more rostral transverse lesions were never followed by
insomnia.

In summary, "exclusion of the waking center can be obtained by transection of the mammillary part of the hypothalamus, whereas the sleep center
seems to be put out of action by a similar lesion in the suprachiasmatic region"
(l.c., p. 303). Nauta suggested an inhibitory action of the sleep center on the
waking center, which would activate the cortex through the lateral hypothalamic area.

The studies on the chronic effects of hypothalamic electrolytic lesions made
by Ranström (1947) on cats confirmed Ranson's work on monkeys. Bilateral
destruction *within* the posterior part of the hypothalamus was followed by
the usual lethargic syndrome, while complete unilateral lesion was without
effect, in agreement with Nauta's results on the rat. However, when the
bilateral lesions were made under nitrous oxide anesthesia, the gas supply
ceasing immediately after the last electrolysis, the cats were awake for a
few hours and the lethargic syndrome started only later. This earlier phase

was obviously missed when the operation was performed under barbital anesthesia.

This puzzling observation was confirmed on the rat by McGINTY (1969). Both investigators (RANSTRÖM, 1947, p. 43–44; McGINTY, 1969, p. 77) suggested that the primary lesion was incomplete and that only in the next few hours did the hypothalamus become totally inactivated. This is by itself not unlikely (see RANSON, 1939, p. 13–14, and SPRAGUE, 1967, p. 167), although it is hard to believe that the caudalmost part of the hypothalamus had never been encroached upon during the original surgery in all cats and rats that had been operated by these investigators. One cause of error which may be of some importance in the interpretation of these acute observations has not been considered so far. The injury of the posterior hypothalamus that unavoidably occurs at the moment of the acute lesion is likely to produce an irritative barrage of impulses which will reach the preganglionic neurons of the thoracic segments through the caudally projecting pathways controlling the sympathetic system, thus producing a hormonal discharge from the adrenal medulla. The problem is to know the effect of adrenaline and noradrenaline on EEG and behavior after hypothalamic lesion. We know that the reticular system is strongly activated by adrenaline (BONVALLET et al., 1954), although the mechanism of the adrenaline or noradrenaline arousal (MANTEGAZZINI et al., 1959) is still controversial (see BAUST et al., 1963).

Several experiments on the effects of hypothalamic lesions were prompted by the investigations on the ascending reticular system. Unfortunately the lesions frequently involved both the brain stem reticular formation and diencephalic structures, often referred to as the cephalic portion of the ascending reticular system. This is an unfortunate terminology, and one that possibly led to a disregard of the remarkable difference between the *lethargic syndrome* caused by hypothalamic lesion and the *coma* following acute midbrain transection.

All these works are related with the problem of the neural mechanism of the *phasic arousal*, to be reviewed by LINDSLEY in *Ergebnisse der Physiologie*. Anatomical (SCHEIBEL and SCHEIBEL, 1967) and physiological (SCHLAG and CHAILLET, 1963; WEINBERGER et al., 1965) observations have suggested that the EEG desynchronization produced by reticular stimulation is mediated by the ventral leaf of the ascending reticular projection, which runs through the subthalamus and the hypothalamus. However after nearly complete lesion of the posterior hypothalamic region, a midbrain reticular stimulation still produces EEG activation, although behavioral arousal is lacking (FELDMAN and WALLER, 1962). Hence EEG activation may also be mediated through the dorsal leaf of the ascending reticular projection, which influences the intralaminar nuclei of the thalamus (see SCHEIBEL and SCHEIBEL, 1957); a conclusion supported by McGINTY (1969) with experiments to be reported below.

In the recent years combined behavioral and EEG studies were carried out by several investigators. Only the effects of chronic lesions, mainly localized to the hypothalamus, will be selected for review.

NAQUET et al. (1965, 1966) reproduced RANSON's lethargy by destroying a region at the border between the subthalamus and the posterior hypothalamus. During the first 8–10 days their cats were almost constantly asleep, but could be aroused by sensory stimulation, although only for a very short time. The paradoxical episodes of sleep were always present, as shown by the EMG and the EEG records. Hobson (1965) produced "extensive damage" to the posterior and lateral part of the hypothalamus, and also to neighboring structures. The cat survived 21 days and presented the typical lethargy syndrome, just as the animals operated by Naquet et al. (l.c.). Important additional findings were i) the observation that the animal "lay motionless in a variety of normal appearing sleep postures"; ii) the presence of short-lasting EEG and behavioral arousals following strong sensory stimulation; and iii) the occurrence of paradoxical sleep episodes, as shown by the usual behavioral (atonia) and electrophysiological signs (EEG desynchronization, and erratic waves in the pons and lateral geniculate body). Both in the experiments of Naquet et al. (1965, 1966) and in those of Hobson (1965), the hypothalamic lesions were less extensive than in the rat experiments of McGinty (1969). This fact may explain the discrepancy to be reported below. Later investigations by Swett and Hobson (1968) showed that neither medial nor lateral lesions of the hypothalamus reproduced the somnolence syndrome resulting from destruction of the whole posterior hypothalamus. Ranson's cataleptic syndrome was reproduced by a lesion limited to the lateral hypothalamus, but not by medial lesions alone.

In the recent years important results have been obtained with lesions of the *preoptic-basal forebrain area*. McGinty and Sterman (1968) confirmed the conclusions reached by Nauta (1946) in the rat, namely that the cat's sleep-waking cycle (see Sterman et al., 1965) is severely altered after lesion of the anterior hypothalamus. Complete sleeplessness was found only in two cases. In most of the instances, however, a severe imbalance of the cycle was observed during the first four weeks, with striking reduction of the sleep phase characterized by EEG synchronization ("quiet sleep"). Complete disappearence of the episode with EEG desynchronization ("active sleep") was observed when the amount of synchronized sleep fell below $15 \pm 5\%$ of the recording time. A significant recovery occurred by 4 to 8 weeks. Fig. 24 summarizes these results. Aphagia and adipsia as well as thermoregulatory disorders were reported, but even after these symptoms disappeared, insomnia was still present. In an attempt to explain the total, temporary disappearence of active sleep, the authors suggested (see also Sterman and Clemente, 1970) that the lesion releases a process which inhibits the pontine structures that trigger paradoxical sleep. However, exactly the same results were reported by Jouvet and Renault (1966) during the insomnia produced with raphe lesions. With the possible exception of the clinical cases of narcolepsy (see Dement and Rechtschaffen, 1908; Hishikawa, 1968; Passouant, 1968), the paradoxical sleep episodes are always preceded by a stage characterized

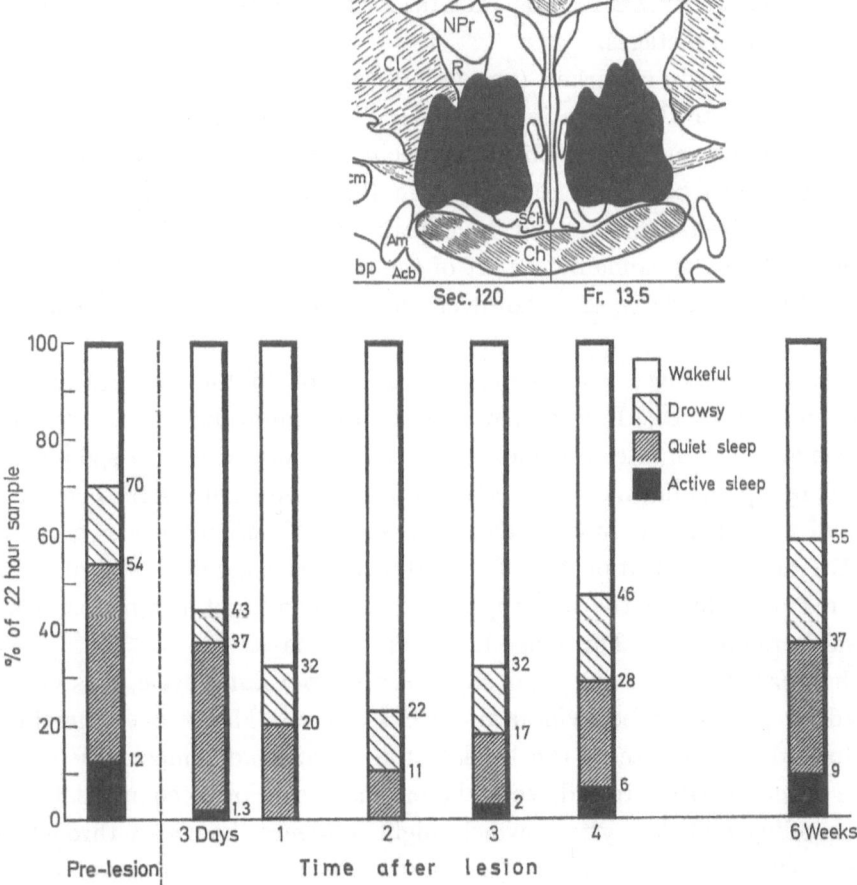

Fig. 24. *Experimental hyposomnia produced by preoptic-basal forebrain lesions in the cat.* The extent of the lesion is given in the upper scheme. The percentages of wakefulness, and of "quiet" and "active" sleep (number besides in bars indicate cumulative percentage) at different intervals from the operation show that hyposomnia developed gradually, reached a peak after 2 weeks, and showed gradually recovery thereafter. (From McGINTY and STERMAN, 1968)

by EEG synchronization. It would not be surprising, therefore, that a critical amount of slow sleep be required for the onset of the paradoxical episodes. We have already pointed out that the pontine structures triggering paradoxical sleep are probably inhibited during wakefulness by the activating reticular system. Thus the release of structures which inhibit the brain stem mechanisms of paradoxical sleep (see STERMAN and CLEMENTE, 1970) would be in fact a release of the activating reticular system. This interpretation is supported by recent findings of BREMER (1970a, b), to be reported later (p. 106).

The work of McGINTY (1969), on the chronic effects of bilateral electrolytic lesions of the dorsal posterior hypothalamus in the rat, deserves a careful discussion, also with respect to the conclusions reached in 1946 by NAUTA on the same animal. The behavioral investigation was complemented by the

study of EEG and EMG records. McGinty classified the effects of his lesions
into four temporal stages.

We have already remarked (pp. 76–77) that the *first stage*, lasting only
a few hours after the operation and characterized by EEG activation and
behavioral hyperactivity, is obviously related to the period of early wakefulness described by Ranström (1947) in the cat, after similar lesions. An
attempt has been made (p. 77) to explain this period as the consequence
of the fact that the caudalmost part of the hypothalamus was missed by the
original lesion. We have seen, however, that this is not the only explanation
(p. 77).

McGinty's *second stage* obviously corresponds to the classical lethargy
syndrome of Ranson. It was characterized by somnolence, which was first
continuous and then later became *interrupted* by short periods of spontaneous
arousal, occupying not more than 5 % of the total time. During the continuous
phase of somnolence, paradoxical sleep, as shown by simultaneous occurrence
of EEG desynchronization and EMG atonia, was altogether absent; it reappeared only during the interrupted phase, when short lasting periods of
arousal were observed. This stage lasted up to 4 days.

The *third stage*, showing clear signs of a sleep-waking cycle, was characterized by periods of behavioral and EEG arousal. The recovery might be
due to i) disappearance of the local factors which had temporarily thrown
out of function structures adjacent the original lesion or ii) compensation by
the ascending reticular system, which might influence the cortex through the
midline nuclei of the thalamus (see p. 97). Recovery was always complete
within 8 days.

In order to understand the significance of the *fourth stage*, which appeared
after complete recovery from somnolence and which was characterized by
hyposomnia, one should recall that the pathways arising from Nauta's sleep
center pass through the posterior hypothalamus. They had obviously been
interrupted by the lesion. Thus the hypothesis might be made that the hyposomnia is a release effect, a phenomenon that should be expected to emerge
only after full recovery of the waking ability. While dealing with the preceding
stage, we have tentatively ascribed this recovery to a compensatory influence
by ascending reticular neurons. It is an interesting coincidence that the full
recovery from a deficit in the ability to maintain wakefulness is reached,
in the rat, 8 days after caudal diencephalic lesions; and that about the same
interval of time, one week, is required, again in the rat, for the truly remarkable recovery of eating, drinking, and grooming behavior after precollicular
decerebration (Wood, 1964). The elaborate patterns of behavior observed in
the rat after chronic decerebration have never been observed in the decerebrate cat. The use of rats probably explains why hypersomnia symptoms
were so striking in McGinty's experiments.

The experimental insomnia produced by lesions of the preoptic region (and also of the basal portion of the diagonal band of BROCA) were confirmed by MADOZ JÁUREGUI (1969). A review article of the experimental work prompted by the experiments on the sleep-inducing structures of the preoptic region, also with regard to the possible neural pathways, is now available (STERMAN and CLEMENTE, 1968).

4. Neuroendocrine Disorders as Related to Brain and Brain Stem Lesions

The main difficulty in the interpretation of the chronic effects of hypo-thalamic lesions is related to the fact that the damage to the neurosecretory systems and to the hypophysis has been usually disregarded. It would have been possible, of course, to control for this objection by performing chronic hypothalamic lesions on previously hypophysectomized, possibly hormonally treated, animals; but it turns out that these experiments have never been made.

BARD and MACHT (1958), WOODS and BARD (1959), HALÁSZ and PUPP (1965; see also SZENTÁGOTHAI et al., 1968) have shown that the hypothalamico-hypophyseal system may be functionally normal after complete isolation from the rest of the brain. A behavioral study of the cat after complete neural isolation of the hypothalamus has been made recently by ELLISON and FLYNN (1968). They state that the syndrome may be described as one of extreme inactivity rather than of "somnolence". In fact, well-integrated aggressive, drinking, and feeding behavior may be obtained with appropriate stimuli, in sharp contrast with the complete lack of any spontaneous behavior. Thus the severe syndrome produced by hypothalamic lesions may be contaminated by neuroendocrine alterations or by other kinds of damage which are less strongly produced, or not produced at all, by surgical isolation. Behavioral and EEG studies on the sleep-waking cycle were unfortunately not reported. While several kinds of appetitive and consummatory behaviors may be related also with structures lying outside the hypothalamus, it is unlikely that future neuroendocrine investigations will substantially modify the conclusions drawn from chronic hypothalamic lesions. NAQUET et al. (1965) reported that moderate cooling of a region at the lateral border between the posterior hypothalamus and the subthalamus was followed by unilateral EEG synchronization without any behavioral signs of sleep; and that bilateral EEG synchronization and a somnolent behavior occurred when the thermode was placed more medially (Fig. 25). These results could be reproduced on a cat several times in a single session, and in a fully reversible way. The following quotation which emphasizes the *fast* temporal course of the phenomena permits us to dismiss the hypo-thesis that the syndrome was the consequence of secondary neuroendocrine disorders.

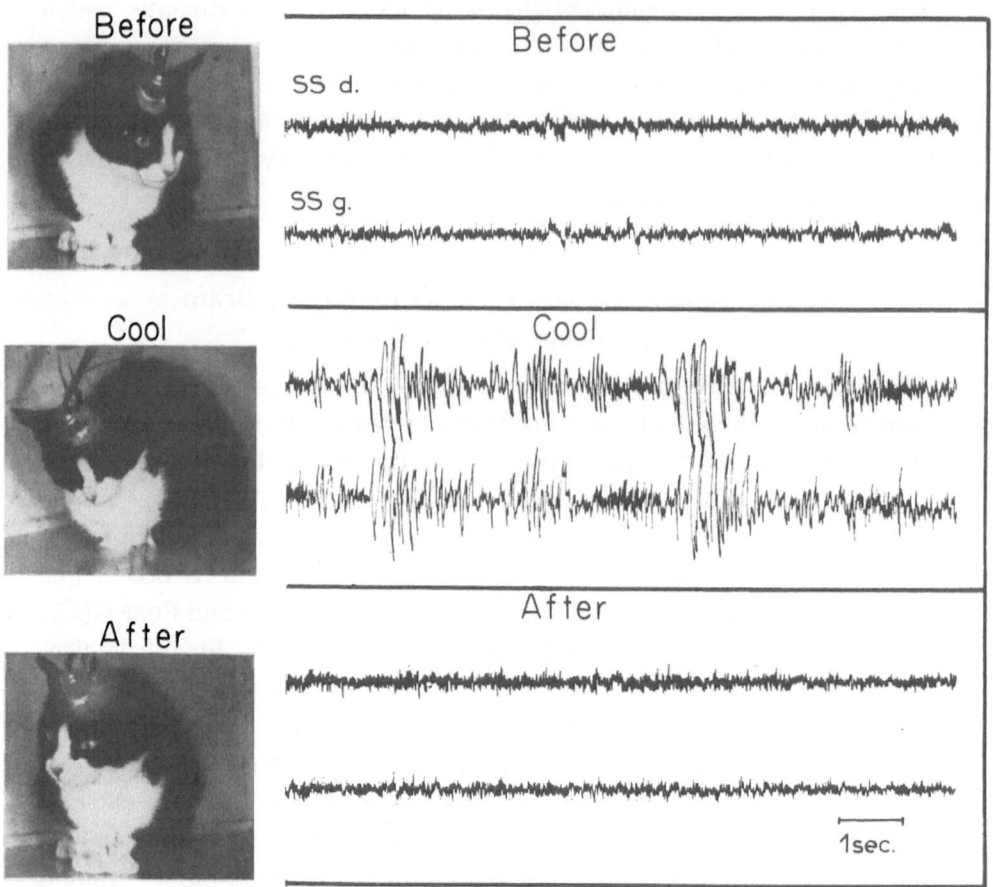

Fig. 25. *Effect of hypothalamic cooling on behavior and EEG in the free moving cat.* Cooling to 10–20° C of the posterior hypothalamus, at the border with the subthalamus, produces behavioral and EEG signs of drowsiness. (From Naquet et al., 1965)

«Dans les trois premières minutes qui suivent l'installation du refroidissement localisé, l'animal, qui était éveillé, interrompt l'activité en cours et cherche une position de repos; les paupières closes et les membranes nictitantes relevées, il s'endort parfois. L'animal tout en restant éveillable par des stimuli sensoriels est devenu toutefois beaucoup moins sensible aux bruits ambiants. Trente secondes après l'arrêt du refroidissement, la tracé cortical et le comportement de l'animal redeviennent, en tout, semblables à ce qu'ils étaient auparavant. Si les refroidissements sont répétés fréquemment (toutes les 5 à 10 minutes), le sommeil comportemental induit devient chaque fois plus franc mais garde toujours son aspect physiologique. Après quelques heures de repos, l'animal étant à nouveau très éveillé, un nouveau refroidissement provoque des effets analogues à ceux obtenus au début de l'experience. La réversibilité est donc parfaite» (Naquet et al., 1965, p. 115).

Concerning the experimental hyposomnia obtained by McGinty (1969) with lesions of the anterior hypothalamus, a purely neural interpretation of the phenomenon is supported by stimulation experiments to be reviewed later (p. 101).

These conclusions, however, by no means imply that the classical hypothalamic syndrome was not contaminated by hypophyseal disorders, particularly in the acute stage. In fact this possibility deserves careful consideration. The reader is referred to a recent review of the literature (MARTINI and
GANONG, 1966–67; see particularly vol. I, pp. 301–306) and to reports of the
original experiments on the effects of midbrain transection (ANDERSON et al.,
1957). However, the EEG asymmetries after unilateral midbrain section (see
p. 51) and the effects of mesencephalic cooling (NAQUET et al., 1966, p. 160)
show that the main deficiency syndrome is basically neural in character.

5. Conclusions

The main conclusion of the experiments of decortication and of thalamectomy is the convincing demonstration that the sleep-waking cycle does not
arise within the cortex and the thalamus. This should not be taken as implying
that sleep is unaffected by these lesions. In fact, i) the pacemaker of cortical
synchronization is localized in the thalamus (ANDERSEN et al., 1967a, b; see
PURPURA and YAHR, 1966) and ii) the corticofugal volleys impinging upon
subcortical structures and the brain stem reticular formation, during the EEG
"spindles" (ADRIAN and MORUZZI, 1939; WHITLOCK et al., 1953), certainly
have an important significance for the regulation of slow wave sleep. On the
other side, paradoxical sleep is likely to be affected by any alteration occurring
in the slow phase, since i) in normal conditions it always appears on a background of EEG synchronization and ii) in cases of experimental insomnia
the paradoxical episodes disappear altogether whenever the sleep with EEG
synchronization falls below a critical level (see p. 78). Finally, the discussion
of the nychthemeral distribution of cataplexic episodes in the chronic decerebrate cat and of paradoxical episodes of sleep in the normal animal has
led to the conclusion that the pontine pacemaker is probably inhibited by
the awake brain (see pp. 43–44).

The experiments concerning the effects of hypothalamic lesions have also
led to other important conclusions. At the end of the chapter devoted to
the effect of complete brain stem transections (pp. 45–51), we stated that
all the available evidence shows that the sleep-waking rhythm is likely to
arise within the cerebrum and that it is simply influenced or modulated by
the brain stem. We know now that rhythmicity arises within the hypothalamus.
Sleep-inducing neurons are localized in its anterior part, while the waking
center is localized in its posterior part. The inherent ability to alternate
between sleep and wakefulness is related i) to the reciprocal connections
existing between sleep and waking centers and ii) to a proper equilibrium
between activating and deactivating influences arising from the brain stem.
The effects of an imbalance in these ascending influences and of the recovery

from this imbalance have been analyzed in the chapter devoted to the chronic effects of brain stem transections.

IV. Stimulation Experiments

A. Peripheral Stimulations

1. Pavlovian Sleep

Pavlov (1927) showed that a "positive conditioned stimulus may be transformed into a negative or inhibitory one by the simple method of repeating it several times in succession without reinforcement" (l.c., p. 68). He called this phenomenon internal inhibition. Since sleep was frequently the final result of this experimental situation, Pavlov suggested that internal inhibition and sleep might be two aspects of the same inhibitory mechanism. This process would simply be extremely localized when only internal inhibition is present, thus permitting wakefulness; while sleep would be the final result of the irradiation of inhibition to both cerebral hemispheres, and possibly to the midbrain (l.c., p. 50).

A source of confusion is represented by the fact that when Pavlov developed his theories, very little was known about the general physiology of central inhibition. We now realize that a distinction must be made between inhibited and inhibitory structures; that inhibition is always produced, actively, by inhibitory neurons; that its outcome must be the decrease or the abolition of the "spontaneous" or evoked firing of the inhibited nerve cells. Evarts' microelectrode studies on single cortical neurons of the free moving, unanesthetized animal have clearly shown that the concept of a generalized inhibition of the cerebral cortex in sleep is untenable (Evarts, 1962, 1964, 1965; see 1967). Before the electrophysiological era, however, the term "inhibition" was used to refer to any reversible loss of a specific function. This loss might also be the outcome of a *disorganization*, or a different organization, of the interrelations between nerve cells, a situation leading to a disorder of integrated activity. Thus Pavlov's theories (Pavlov 1953–1956) may still retain their validity if his concept of inhibition is intended in this loose way. While true inhibition is always actively produced by inhibitory neurons, several mechanisms may lead to a disorganization or to a different organization of the integrated activity of a population of nerve cells. The usual statement that Pavlov was the forerunner of the active theories of sleep is therefore probably not correct. In fact Pavlov accepted, towards the end of his life, the possibility of both active and passive mechanisms of sleep (see Moruzzi, 1960, p. 244, for ref.).

The attempt at an interpretation of the Pavlovian studies led to doctrines which are of general interest for sleep physiology. Konorsky (1948) followed the passive deafferentation theory. He stated (l.c., p. 50) that "monotonous stimuli are less of an obstacle to our falling asleep than diversified stimuli, this being because of the adaptation of the organs receiving the given stimulus, and also because of the extinction of the orienting reaction to it". When reinforcement was abolished, he stated, the sensory volleys "*do not hinder sleep, or hinder it only insignificantly*" (l.c., p. 50). Hence sleep would occur spontaneously. In Pavlov's experimental situation the onset of sleep would simply not be prevented by the sensory volleys.

Moruzzi (1960) took a different view and suggested an active mechanism in an attempt at explaining the reflexly induced sleep. He started (l.c., p. 243) from the assumption, which is supported by several lines of experimental evidence, that two antagonistically oriented systems are present in the brain stem, and suggested that the activating reticular system might be easily driven by any new sensory stimulation, while the EEG synchronizing structures would respond with a progressive increase of their sleep inducing activity only when repeatedly impinged upon by a regular sequence of stereotyped sensory volleys. Thus during monotonous sensory stimulation the activating reticular system would be gradually overwhelmed by the antagonistically oriented EEG synchronizing structures. This hypothesis might explain how monotonous sensory stimuli produce sleep quite independently of any conditioning procedure. However Moruzzi (l.c., pp. 247–250) recalled several experimental data of Pavlovian physiology which could not be easily explained purely by mechanisms restricted to the brain stem. When sensory stimuli are reinforced during the conditioning procedure, corticofugal discharges impinging upon the activating reticular system might counteract the inhibitory effect of the EEG synchronizing structures, which would be otherwise driven by the repetitive unconditioned stimuli. In other cases sleep is elicited extremely quickly, and specifically, by the conditioning procedure, an observation that was explained with the assumption that corticofugal volleys may drive the EEG synchronizing structures of the lower brain stem.

Konorski (1967b, p. 300) has recently developed a new interpretation, which implies (in our opinion) an active origin of sleep occurring during conditioning procedures. He starts from a clear distinction between the unconditioned and conditioned mechanisms leading to both somnolence and actual sleep. "Such external stimuli as lying or half-lying position, relaxation of the antigravitational muscles, monotony of the environment, and softeness of the bed we are lying upon" (l.c., p. 300) would represent the unconditioned stimuli leading to, and possibly also maintaining, sleep. The conditioned stimuli, on the other hand, "generally established to time, according to our daily stereotype of going to sleep" (l.c., p. 300) may also lead to drowsiness and sleep (sleep as conditioned reflex, C.R.). In order to explain conditioning of a state apparently opposite to arousal, such as sleep, Konorski has made some considerations which deserve out attention. To quote:

> We do not agree that somnolence is less active state than any other drive. On the contrary, a sleeply animal looks actively for some place to fall asleep, just as the hungry animal looks for food, and certainly he is very alert to all those stimuli which suit this aim. Therefore, *somnolence is as good for activating the association involved in sleep C.R., as hunger for activating the association involved in the food CR*[5] (l.c., p. 301).

In our opinion Konorski's hypothesis implies the existence of structures which *actively promote sleep*. The basal forebrain and the lower brain stem,

---

5 Italics are ours.

probably two closely associated structures, are the most likely candidates. Their unconditioned activity would be involved in the onset of sleep; just as the activity of an hypothalamic center is responsible for the behavioral manifestations of hunger. However, appropriate activity in the centers for both sleep and hunger can be conditioned, in suitable experimental conditions. In fact experiments by Clemente, Sterman, and Wyrwicka (1963), to be reviewed later (see p. 104), have shown that EEG and behavioral signs of sleep can be obtained by pairing auditory stimuli with hypnogenic electrical stimulation of the basal forebrain area.

The sharp distinction between unconditioned and conditioned responses is of great didactic value, but in fact a separation between innate and learned behavior is made with great difficulty in *normal* conditions. This distinction is easier to make when conditioned reflexes are *intentionally* created in the laboratory with Pavlovian procedures, which standardize, control, and emphasize the eastblishment of plastic changes in the central nervous system.

Spontaneous conditioning may explain several habits which are regarded as necessary or convenient procedures in order to fall asleep. Their significance is mainly to facilitate sleep, which would occur anyway through unconditioned mechanisms, may be a little later. Even the awakening at a given time in the morning that occurs in several individuals with such an amazing precision, may be explained by early, unnoticed processes of conditioning of the arousal mechanism.

The evolutionary theory of sleep and wakefulness has been reviewed at length in Kleitman's book (1963, pp. 363–370) and in Berlucchi's article (1970, pp. 185–191). Suffice it to recall here that the ontogenetic evolution from *wakefulness of necessity* to *advanced wakefulness* permits "the consolidation of several short periods of sleep into a long night sleep" (Kleitman, 1963, p. 369), thus causing the transition from the primitive polycyclic rhythm of the newborn infant to the monocyclic rhythm of the adult man (l.c., p. 367). This transition requires the maturation of the cerebral cortex (l.c., p. 369), but there is little doubt that learning factors have great importance in the gradual shift into the adult circadian sleep-waking rhythm, under the influence of the alternation of day and night and of social influences from the family (see Berlucchi, 1970, p. 187). In summary, the adult sleep-waking cycle of each man results from both inborn factors—basically the development of the cerebral cortex and of its interrelations with the brain stem—and learning processes.

The importance of conditioning experiments in sleep studies lies mainly in the fact that the physiologist can thereby drive *selectively*, by *natural stimuli*, the activity of sleep inducing structures which could otherwise be excited only in a rather crude way by electrical pulses.

2. Natural Unconditioned Stimuli

It is well known that sleep may be produced in man by monotonous sensory stimulation (see Gastaut and Bert, 1961), quite independently of any intentional conditioning procedure. Reflexly induced EEG synchronization and sleep-like behavior has often been reported in animal experiments.

The data to be discussed below will show, however, that the appearance of EEG synchronization does not necessarily coincide with the onset of genuine sleep.

a) Strong afferent volleys from arterial baroceptors may produce sleep-like behavior (KOCH, 1932) and EEG synchronization (BONVALLET et al., 1954; MAZZELLA et al., 1956; GARCIA AUSTT, 1965), an effect observed without contaminations in the *encéphale isolé*, since the reflex fall of arterial pressure is avoided (BONVALLET et al., 1954; see DELL et al., 1961). The deactivating influence of visceral afferent volleys may be related with the EEG synchronizing effects elicited by electrical stimulation of vagal afferent fibers (BONVALLET and SIGG, 1958) and of the region of the nucleus of the solitary tract (MAGNES et al., 1961). As it will be pointed out later, it is not certain, however, that short-lasting EEG synchronization necessarily implies the onset of a true episode of slow wave sleep (see MORUZZI, 1969, pp. 196–201).

b) Photic stimulation may produce EEG synchronization even on the background of enhanced alertness that characterizes the midpontine pretrigeminal cat (ARDUINI and HIRAO, 1960; MANCIA et al., 1959). Repetitive flashes produce EEG and behavioral signs of sleep in man (GASTAUT and BERT, 1961).

c) Several types of sensory stimulations may lead to EEG and behavioral manifestations of deactivation in the unanesthetized rabbit (TAKAGI, 1957; TAKAGI et al., 1959; VAN REETH and CAPON, 1962; KUMAZAWA, 1963). TAKAGI's skin pressure reflex is a typical example of this phenomenon (l.c.). However, the final outcome of these manipulations if often not sleep but animal hypnosis (see STEINIGER, 1936; GILMAN and MARCUSE, 1949; CARLI, 1969, for the literature), a behavioral state related to, but different from, the slow and paradoxical hypnic phases (GEREBTZOFF, 1941; KLEMM, 1966; CARLI, 1969).

3. Electrical Stimulations of Somato-Sensory Nerves

Following a line of experimentation developed by PAVLOV's school, ROITBACK (1960) found that unanesthetized cats and dogs which were submitted to low rate electrical stimulation of the skin gradullay became drowsy, and eventually fell asleep, while typical slow waves appeared on the EEG. It was obviously impossible to determine, with this approach, what cutaneous afferent fibers were responsible for the hypnogenic effects. However the results of ROITBACK are important, since they provide convincing proof that combined EEG and behavioral signs of sleep may be produced by *peripheral* electrical stimulation.

POMPEIANO and SWETT (1962a, b) studied the effect of electrical stimulation of cutaneous and muscular afferent fibers in the free moving, unanesthetized cat. Chronic electrodes attached to the skull, muscles and nerves permitted them to record both the EEG and the cervical EMG during painless electrical stimulation. In confirmation of ROITBACK's findings, they showed that stimulation of cutaneous fibers, and of these fibers only, led to behavioral and EEG patterns of sleep. They demonstrated, moreover, that this deactivating effect was due to volleys coursing along Group II cutaneous fibers.

A *single* train of 0.1–1 msec rectangular pulses, 3 to 8/sec pulse rate, lasting from 2 to 4 seconds—for stimulus strengths exciting the lowest threshold

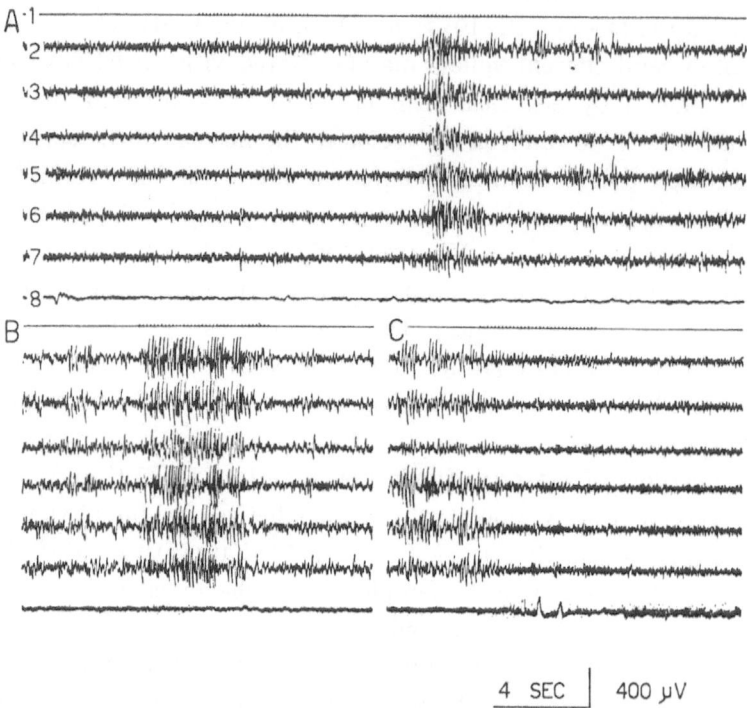

4 SEC   400 µV

Fig. 26 A–C. *EEG synchronization and arousal elicited by electrical stimulation of a cutaneous nerve at different voltages.* Intact, unanesthetized cat. Experiment made 5 days after the implantation of the electrodes. Stimulation of the right superficial radial nerve (6/sec; 0.5 msec pulse duration). Bipolar records. *1* Stimulus marker; *2* medial and lateral part of the left posterior sigmoid gyrus; *3* left parieto-temporal; *4* left temporo-occipital; *5* medial and lateral part of the right posterior sigmoid gyrus; *6* right parieto-temporal; *7* right temporo-occipital; *8* right neck EMG. A Stimulation produced no effect with 0.11 V (left). A slight increase of voltage to 0.12 V reached the threshold for EEG synchronization (right). B Stimulation with 0.13 V induced marked EEG synchronization. C Further increase in voltage (0.17 V) caused desynchronization of the EEG and an increase in EMG activity. (From POMPEIANO and SWETT, 1962)

(Group II) cutaneous fibers—elicited a typical arousal reaction. The tonic cervical EMG showed an increase in activity, while the cat's behavior was that of an animal during the orienting reflex. However, when *several* such pulse trains were applied at intervals of 30 to 60 seconds, an habituation of the arousal response was soon observed; while slow, high voltage bursts appeared on the EEG, and were time-locked with the stimulus train (Fig. 26B). This *induced EEG synchronization* appeared only on a background of relaxed wakefulness or of drowsiness. This electrophysiological response did not imply, however, the onset of a true episode of slow sleep. Sleep appeared in fact only after several trains, and its onset was heralded by the persistence of response to an individual train outlasting the period of stimulation. As was the case with induced EEG synchronization, induced sleep never occurs on a background of strong arousal or of paradoxical sleep. Induced EEG synchronization, followed later by the onset of sleep, was obtained for stimulus intensities ranging between 0.4–0.9 times the threshold for arousal, as determined by applying the same repetitive trains of electrical pulses on a background of slow sleep.

In summary, the experiments of POMPEIANO and SWETT provide convincing evidence, one coming also from other lines of endeavor (see MORUZZI, 1969, pp. 196–201), that a distinction should be made between induced EEG synchronization and induced sleep. Both require a background of relaxed wakefulness or of light drowsiness, but only the repetition of trains of afferent volleys shift EEG and behavioral background towards the low levels of activation which are required for the onset of the slow phase of sleep. The agreement with HESS' classical experiments of central stimulation is truly remarkable.

Other investigations (POMPEIANO and SWETT, 1962a, b) showed that the induced synchronization occurred when many group II cutaneous fibers, conducting at rates from 80 to 42 m/sec, were activated. An arousing effect, obviously overwhelming the synchronized response, appeared when the intensity of stimulation was increased to levels which also activated a substantial number of group III fibers, conducting at rates of 36–20 m/sec. The synchronized response was unaffected by section of the dorsal column and/or total cerebellectomy.

These and other experiments made along similar lines, were reviewed by POMPEIANO (1965). Only a few conclusions will be reported here. The group II cutaneous fibers, whose low rate stimulation is responsible for induced synchronization, are probably related with touch, pressure and hair receptors. Group III and IV fibers are mainly nociceptive in nature, although at the threshold for arousal no sign of pain is observed. Group I muscle afferents do not affect the EEG, and an arousing effect may be produced only by low rate stimulation of group III muscle afferents.

Unit recording from the brain stem of the decerebrate cat (POMPEIANO and SWETT, 1963), showed that most of the neurons driven by Group II cutaneous afferents—for electrical parameters of stimulation inducing EEG synchronization in the normal animal—were localized in the medulla and in the caudal part of the pons, the regions which have been shown by lesion experiments to be mainly antagonistic with respect to the activating reticular system. Additional activation of Group III cutaneous afferents influenced the majority of the units localized in the pontine and mesencephalic reticular formation.

B. Central Stimulations

For reasons to be elaborated in the discussion which will close this chapter (see pp. 107–111), we are more interested in the general patterns of the hypnic responses produced by central stimulations than in an attempt to localize sleep inducing structures by applying trains of electrical pulses to different districts of the encephalon. The latter approach is unlikely to be rewarding, as will be shown later (see l.c.). Therefore only three lines of investigations which may lead to results of general interest will be selected for discussion.

The review will start with a detailed analysis of HESS' work, both because of its historical importance and in view of the great physiological significance of some observations. Only a short summary of the results obtained by stimulating the brain stem will follow, since most of this area of investigation will be reviewed by LINDSLEY in an article for *Ergebnisse der Physiologie* devoted to the arousal phenomenon. The third part of our report will be concerned with a detailed analysis of the works on the basal forebrain area, which recent experiments have shown to be an extremely promising region for sleep studies.

The reader is referred to recent anatomical (AKERT, 1965) and physiological (AKERT, 1961; PARMEGGIANI, 1962, 1964, 1968) reviews for problems of localization. The experiments of chemical stimulation will be discussed by JOUVET in these reviews.

1. Hess' Work

In 1927 HESS presented at the Frankfurt Meeting of the German Physiological Society his preliminary report and a film on sleep produced by brain stimulation in the free moving, unanesthetized cat (HESS, 1927). The complete list of the papers published during the following forty years may be found in a recent monograph (HESS, 1968b), which contains also a penetrating analysis of HESS' life work. The present review is based on the full paper of 1944 (HESS, 1944b), on reports and discussions published in the proceedings of two Symposia (HESS, 1950, 1954); and, finally, on three monographs (HESS, 1949, 1968a, b). References to a few other papers will be found in the following pages.

HESS has shown that low rate stimulation of a region lying just lateral to the massa intermedia elicits a progressive decrease of activity, with clear behavioral signs of drowsiness, followed by true physiological sleep. To quote from his report at the St. Margherite Symposium (ADRIAN et al., 1954).

"The animal sits or lies down and curls up in its natural sleeping position. Eyelids and nictitating membranes are closed. But physiological stimuli, such as loud noises or the smell of meat, will rouse the animal. Nevertheless some tendency to go to sleep again persists as a rule" (HESS, 1954, p. 120).

Thus the cat's behavior was entirely normal during the stage preceding sleep and during sleep itself (Fig. 27), as shown also by the possibility of arousing the animal by natural sensory stimulation (Fig. 28). In summary, electrical stimulation of the midline nuclei of the thalamus reproduced a genuine, well-integrated physiological pattern characteristic of sleep.

The significance of HESS' discovery cannot be fully understood if the results are viewed only from the perspective of sleep physiology, as is usually done. In fact other

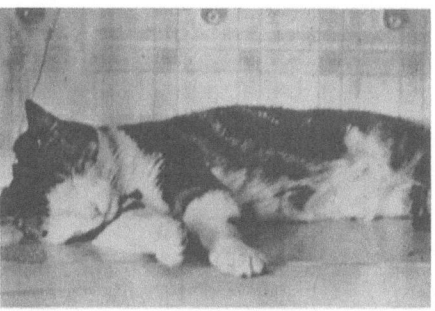

A   B

Fig. 27 A and B. *Lying postures in Hess' induced sleep.* Two typical sleep postures obtained by electrical stimulation of the massa intermedia. (From HESS, 1944b)

A   B

Fig. 28 A and B. *Olfactory arousal from Hess' induced sleep.* Before (A) and after (B) placing meat before the nose. (From HESS, 1944b)

complex, fully-integrated types of physiological behavior may be obtained by stimulating other parts of the brain, as shown by HESS himself (see 1949, 1968a). He found that typical patterns of feline instinctive behavior could be reproduced when the same stimulation techniques were applied to well defined districts of the hypothalamus. AKERT (1961), JUNG (1967), HASSLER (1967), and PLOOG (1964) have reviewed this field of investigation, which is on the border line between ethology and physiology, and the relation of instinctive behavior to sleep has been the theme of a recent article (MORUZZI, 1969).

The phenomenon of induced sleep should be regarded as an example of what may be obtained by triggering with electrical pulses the system concerned with the recovery processes, HESS' *trophotropic system.* The instinctive behavior of defense-aggression is, on the other hand, a striking example of the activation of an antagonistically oriented *ergotropic system,* which is responsible for the display of any fully integrated response to an animal's environment. One need not discuss the two phenomena separately just because sleep is considered a behavior of rest, while rage is characterized by a violent outburst of activity. The similarities between stimulation experiments which produce various types of integrated behavior in the free moving, unanesthetized cat are by far more important than the observed differences of localization or behavioral pattern.

Hess published his theoretical concepts in 1925, before the discovery of electrically induced sleep, and developed them in 1931 and 1933 (Hess, 1925, 1931, 1933). Unfortunately, the tendency among physiologists has generally been to consider Hess' sleep studies as a line of work entirely separate from his experiments on the ergotropic responses. It should be noted that the difference between trophotropic and ergotropic systems is based upon a physiological concept which does not necessarily imply a strict anatomical division. Perhaps the physiologists were misled by Hess' emphasis on the im-

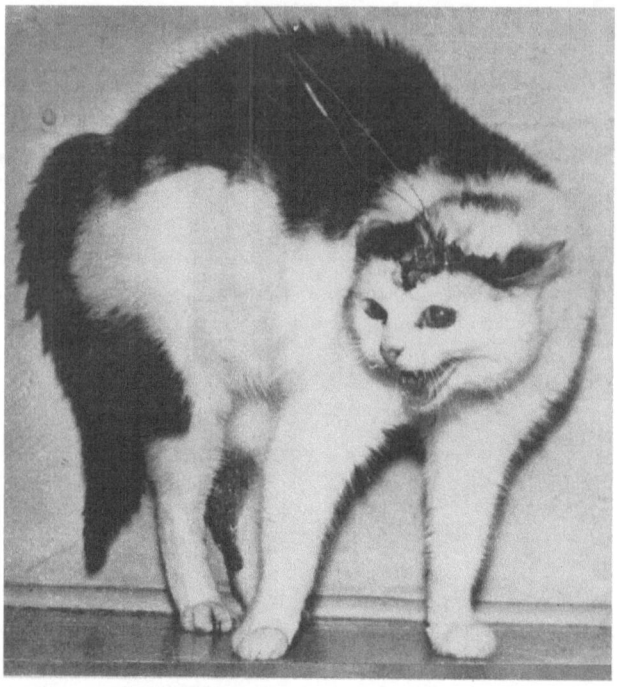

Fig. 29. *Threat response elicited by electrical stimulation of the hypothalamus.* The response is characterized by hissing, lowering of the head, flattening of the ears, hunching of the back, dilatation of the pupils, and piloerection. The dramatic increase of blood pressure and of heart rate are the visceral aspects of this response, which Karplus and Kreidl (1927) detected in their anesthetized animals.
(From Hunsperger et al., 1964)

portance of the parasympathetic and sympathetic systems which, in fact, merely represent the final common paths utilized by the trophotropic and ergotropic discharges acting upon the autonomic sphere.

When Hess started his work, most of the experiments of central stimulation had been made on anesthetized animals, with the exception of a few pioneer works with chronically implanted electrodes (Simonoff, 1866; Ewald, 1898; Talbert, 1900; Baer, 1905), and therefore only those fragments of complex behaviors which survive general anesthesia had been elicited. Thus the striking increase of blood pressure, which Karplus and Kreidl (1927) had produced by stimulating the hypothalamus, is now regarded as merely one of the autonomic components (Hess, 1968a, see p. 13) of the threat and flight behavior that may be easily obtained by appropriate stimulation of the same region (Fig. 29) in the free moving, unanesthetized cat (Hess, 1949, 1968a, b; Hunsperger et al., 1964).

Although it was surprising in a way that such complex and beautifully integrated responses could be obtained by artificial stimulations, Hess' results on the ergotropic effects elicited by electrical stimulation of the hypothalamus were immediately accepted. The main reason was probably that these observations fitted the results of the stimulation experiments of classical neurophysiology, since the responses appeared at the onset of the stimulus and rapidly subsided at the end of it. In contrast, many physiologists were somewhat reluctant to accept the sleep experiments since the time parameters for this response were so strikingly different from those associated with the usual motor and autonomic responses.

HESS' results on sleep induction were unexpected for experimenters accustomed to observing a response immediately after stimulation. HESS used trains of low rate (about 8/sec) pulses, lasting as long as 30 seconds, and he reported that it was exceptional to obtain sleep after just one period of stimulation. More frequently the trains had to be repeated, at intervals of 30 seconds; occasionally several times. Thus, "latencies" of the order of several seconds and even of a few minutes were observed quite commonly. The after-effects of the stimulation were also unusually long by classical neurophysiological standards, thus revealing an extraordinary inertia of the response. In some cases the cat remained asleep even for hours, although it could be aroused by the usual sensory stimuli. Awakening could also be obtained by increasing the strength of the electrical pulses, a procedure that probably led to co-activation of neighboring arousal structures.

BREMER stressed these difficulties as early as 1931, and again in 1953 (BREMER, 1931, 1954). It is obvious, moreover, that RANSON and MAGOUN (1939) were led to suggest their interpretation of HESS' experiments mainly by the unusual temporal course of induced sleep. However their objection (l.c., p. 107) that small electrolytic lesions produced by the stimulating currents might have been the cause of sleep has now only historical interest. The sleep obtained by HARRISON (1940) with such lesions was shown by HESS (1949) to be an entirely different phenomenon, which he called hypothalamic adynamia. The full reversibility of thalamically induced sleep is by itself the best evidence that we are confronted with a true physiological phenomenon, actively produced by excitation of nerve cells or axons; one quite different in nature from the lethargy following hypothalamic lesions. HESS' results were repeated and confirmed several times (AKERT et al., 1952; see PARMEGGIANI, 1968), and the behavioral studies were later complemented by EEG recordings (HESS JR. et al., 1950, 1953; CASPERS and WINKEL, 1954; AKIMOTO et al., 1956; HÖSLI and MONNIER, 1962, 1963; YAMAGUCHI et al., 1964; ROITBACK and ERISTAVI, 1966). The problem, therefore, is not the existence of HESS' phenomenon but its interpretation.

Any attempt to interpret HESS's results must be concentrated on two phenomena which represent the main difference between trophotropic and ergotropic responses: the long "latencies" of induced sleep and its striking tendency to outlast the stimulation. The instinctive behaviors produced by stimulation of the hypothalamus, all belonging to the ergotropic sphere, will also be recalled in an attempt to find out the differences and the similarities between trophotropic and ergotropic responses.

Hess called attention to the tremendous inertia of the sleep response, both with regard to its onset and to its persistence after the end of the stimulation. It is not easy to understand, he pointed out, why the induced sleep might last several hours after the stimulation, provided that the animal was left undisturbed; nor why after an arousing stimulation the cat eventually fell asleep again, as if some tendency to sleep had remained. He postulated that perhaps some humoral factor contributed to these long lasting responses. When sleep was not interrupted by intentional sensory stimulation, it had a tendency to perpetuate itself also through lack of sensory reverberations. Sleep generates sleep, Hess stated, and he considered this to be the substance of the deafferentation theory. To quote:

> In Tat und Wahrheit ist dem Schlafeffekt eine große Trägheit eigentümlich, nicht nur im *Entstehen*, sondern auch im *Verharren*. Mit Rücksicht auf die zeitlichen Verhältnisse beim physiologischen Einschlafen ist ersteres verständlich. *Nicht ohne weiteres zu begreifen ist hingegen, daß der künstlich angeregte Schlafzustand die Reizung erheblich, u. U. mehrere Stunden überdauert, sofern das Tier nicht gestört wird*[6]. Ja bei stärkerem Schlafeffekt ist es sogar die Regel, daß sich das Tier nach dem Aufwachen durch äußere Weckreize auch ohne neue Reizungen bald wieder legt, um weiter zu schlafen. Der Weckreiz wird wohl adäquat beantwortet; *er läßt aber eine gewisse Schlafneigung bestehen*[6], wenn er nicht kräftig genug, u. U. mehrfach appliziert wird. Vielleicht kommt in dieser Verhaltungsweise die *Mitwirkung* eines humoralen Faktors zum Ausdruck. Was das Festhalten am einmal eingeleiteten Schlafzustand betrifft, so ist auch in Betracht zu ziehen, daß die Reizschwellen stark erhöht sind. Infolgedessen ist das Tier von einem tiefen Schweigen umgeben und weder aus der akustischen, optischen noch taktilen Sphäre werden Erregungen wirksam, welche zu einer ergotropen Umstimmung Anlaß geben könnten. Schlaf zeugt Schlaf! Dies ist der wahre Inhalt der sog. Reizmangeltheorie (Hess, 1944b, p. 336).

It must be conceded that once an imbalance is produced, by any kind of stimulation, in the neural mechanisms controlling the sleep-waking cycle, the predominance of the sleep mechanisms is likely to increase, or at least to consolidate itself, through feedback produced by both central and peripheral loops. The last lines of our quotation from Hess (1944) as well as Bremer's concept (1954) of the deactivation *en avalanche* utilize this possibility. It is hard to understand, however, how a well established imbalance may survive even a short lasting arousal, so that "some tendency to sleep remains" (Hess, l.c.). That the sleep inducing structures might act by inhibiting the activating reticular system had been postulated by Hess (1954) and has been recently shown by Bremer (1970a, b). However when thinking along classical neurophysiological lines it may appear difficult to understand how some inhibitory effects remain after an interval of arousal. Only the hypothesis of humoral mechanisms or, better, of the local accumulation of chemical mediators may account for the persistence of a tendency to sleep. These aspects of sleep

---

6 Italics are ours.

physiology are reviewed at length by JOUVET in his article for *Ergebnisse der Physiologie*.

We shall now discuss separately the problems of the enormous "latencies" and the long aftereffects observed in the experiments on HESS' induced sleep.

HESS reported that his cats looked around, searched for a corner in which to lie, and finally curled up in their natural sleep position. Thus he reproduced with electrical stimulus what, from a purely ethological standpoint (TIN-BERGEN, 1955), may be regarded as genuine appetitive behavior. The extremely long and inconstant "latencies" observed by HESS included, of course, the time spent during the appetitive phase (MORUZZI, 1969, see p. 184). The response to the electrical stimulation was already building up during this phase, probably by mechanisms of temporal summation combined with those of central facilitation *(Bahnung)*.

The phenomenon of the primary (central) facilitation (see GRAHAM BROWN, 1927) has been usually studied in stimulation experiments on the motor cortex (see GRAHAM BROWN, 1915a, b; 1927), but is also present in the related subcortical structures (GRAHAM BROWN, 1915), following repetitive trains of electrical shocks. Its time course is of the order of several seconds (l.c.). HESS' long "latencies" may be the result of processes of central facilitation occurring not only within the midline nuclei of the thalamus, which are directly stimulated (primary facilitation), but also in other structures, probably localized in the basal forebrain region (see p. 101), which are impinged upon by the induced thalamic volleys (secondary facilitation).

It is likely, moreover, that the appetitive behavior of sleep, which HESS observed as the earliest phase of the process leading to induced sleep, was related to a genuine feeling of drowsiness. It would not be easy, of course, to give a direct proof of this statement, since sleep induced by low rate midline thalamic stimulation is a rare event in man (HASSLER and RIECHERT, 1961, p. 105; HASSLER, 1961). However in the few cases where sleep was produced by such stimulation during surgical operations, it was always preceded by a feeling of tiredness with slowing down of mental processes, loss of spontaneous speaking and movements, and repeated yawning (HASSLER, 1970). Evidence that behavioral responses produced by electrical stimulations of other parts of brain are accompanied by subjective feelings is provided, also, by other experiments which involve stimulation of the hypothalamic centers controlling defense-aggression and the regulation of water content of the body.

HESS and several of his colleagues (1968b, pp. 94–100) have recently reviewed an impressive amount of data (see particularly HUNSPERGER et al., 1964) suggesting that the threat and flight behavior produced by hypothalamic stimulation in the free moving cat is related to a real defence-attack mood. HESS has pointed out (HESS, 1968b, p. 37), moreover, that true sensation of thirst is likely to occur when the hypothalamic center for osmotic regulation is electrically stimulated, as shown by the experiments of ANDERS-SON and WYRWICKA (1957), who in fact reproduced in their goats the clinical man-ifestations of centrogenic polydipsia. As KONORSKI (1967a) has rightly pointed out, "the phenomena denoted in psychology as *emotions* or *drives*, such as fear, anger, hunger or thirst are controlled by definite, anatomically delineated regions in the central nervous system" (l.c., p. 267). The discovery of their functions becomes possible when the electrical stimulation "is administered not in a physiological vacuum, but in a meaningful biological environment, that is, in the presence of specially selected stimulus-objects or in specially trained animals" (l.c., p. 267). Several other lines of evidence, reviewed by Zanchetti (1966, 1967a) and Doty (1969), clearly show that motivational effects are produced by hypothalamic, but not by neocortical, stimulation. Indeed, changes in mood (Hassler, 1961; Hassler and Riechert, 1961) and psychical responses of high order (Penfield and Perot, 1963) may be produced by electrical brain stimulations in man.

It is of course extremely unlikely that synchronized firing of any single group of brain neurons may be *directly* responsible for changes of "mood" or for complicated behavioral responses. Actually stimulation may be acting "as a functional lesion, by scrambling the otherwise orderly traffic of impulses" (Goddard, 1964, see p. 28), a conclusion supported by the observation (see Penfield, 1966, pp. 220–221) that aphasia is produced by electrical stimulation of the speech cortex. Doty (1969) has stated this point very clearly:

". . . current neurophysiological knowledge makes it essentially axiomatic that the neural organization responsible for such complex outcome lies remote to the neurons stimulated. This principle or axiom follows from the fact that any coordinated movement involving spatiotemporal sequencing of action in motoneuronal pools requires equally precise phasing of excitatory and inhibitory control, as does most sensory experience. It is impossible to achieve this by electrical stimulation. The current cannot impose upon neurons the spatiotemporally coded and integrated output they normally achieve; it can only drive them in bizarre and nonsensical synchrony. Thus any subtle, highly integrated neural effects resulting from the stimulation must ensue only because the neural systems downstream (or, possibly, to a slight degree upstream) receiving the nonsense signal are able to transform it into an effective neural code. Were the stimulation inserted into the midst of neurons required to organize a complex action, it could only interfere with, not produce, this action" (l.c., p. 292).

Thus it is unlikely that Penfield's "anatomical record of the stream of consciousness" is localized in the areas of the brain where electrical stimulation elicits recollection of past experiences. It is probable that "the stimulating electrode sets up a sequence of neuron activity at a distance by means of axonal conduction from the point of stimulation" (Penfield, 1966, p. 221). Hess has come to a similar conclusion in interpreting the elementary affective reactions related to instinctive behavior (1968b, p. 98).

We have dealt at length with these concepts because we believe that they may be utilized for explaining Hess' sleep experiments. One usually says that sleep is *produced* by electrical stimulation of the thalamus. This statement tacitly implies that we are confronted with a simple phenomenon, analogous to inhibition of the antigravity tonus yielded by electrical stimulation of the cerebellar anterior lobe in the decerebrate cat. Such assumptions are misleading, as shown by the observation that sleep occurs spontaneously after thalamectomy (Naquet et al., 1965). It would perhaps be better to state that sleep is *triggered* by electrical stimulation of the thalamus. In other words electrically induced volleys arising in the midline thalamic nuclei would be said to activate inborn neural patterns responsible for sleep behavior.

Sleep can no longer be attributed simply to a cessation of activity, although is does require, and is in fact causally related to, a process of reticular de-activation (MORUZZI and MAGOUN, 1949; BREMER, 1954; HESS, 1954; DELL et al., 1961; MORUZZI, 1969); which in turn may be due to inhibitory influences arising from EEG synchronizing structures (BREMER, 1970a, b). Sleep is a complex sequence of events, a process that is unlikely to be *driven* by the synchronized discharge of a group of midline thalamic neurons, comprising the so-called "sleep center". It may simply be *triggered* by electrical stimu-lation of the thalamus, just as instinctive behaviors may be triggered by hypothalamic stimulations.

With regard to the effects of thalamic stimulation a distinction should be made between:

i) the recruiting potentials *driven* by a *single* train of low rate electrical pulses of the midline nuclei of the thalamus (MORISON and DEMPSEY, 1942; see JASPER, 1949, 1961), which reach a maximum of amplitude within less than one second; and

ii) the induced sleep, which may be *triggered* by stimulating with repetitive pulse trains the same structures (HESS, 1927, 1944), a phenomenon fully developed only after "latencies" on the order of one minute or more.

The electrophysiological and behavioral effects of midline thalamic stimulation may thus be dissociated.

Low rate stimulation of the thalamic intralaminar nuclei gives rise to recruitment potentials, which may be led from the cerebral cortex (see JASPER, 1949, 1961) and also from the brain stem reticular formation (SCHLAG and FAIDHERBE, 1961). However, the cortical synchronization produced by *low rate* stimulation of midline thalamic nuclei remains after precollicular transection (SCHLAG and CHAILLET, 1963), while the arousing responses elicited by high rate stimulation (see JASPER, 1949; HÖSLI and MONNIER, 1962, 1963) are abolished (SCHLAG and CHAILLET, 1963). The cortical synchronization is due to, or is at least related to, thalamic internuclear inhibition (see PURPURA et al., 1966; see PURPURA and YAHR, 1966).

Cortical recruitment potentials, and related intrathalamic events, may coexist with a state of wakefulness in both cat (EVARTS and MAGOUN, 1957) and man (HOUSEPIAN and PURPURA, 1963). It is also true, however, that repetitive trains of recruiting stimuli may eventually lead to sleep, as in HESS' original experiments (see also HUNTER and JASPER, 1949; YAMAGUCHI et al., 1964; ROITBAK and ERISTAVI, 1966), by producing a progressive decrease in the background of activation. However, recruiting stimulations may also lead, in this case suddenly, to an entirely different behavioral response, the arrest reaction, which is characterized by the *sudden* cessation of all spontaneous activity (l.c.). Clearly it is not necessary to synchronize the entire neocortex in order to produce sleep behavior; which, incidentally, is not markedly changed, at least in low mammals, by decortication (see p. 72).

In another review (MORUZZI, 1969, p. 201) we have emphasized the im-portance of making a sharp conceptual distinction between EEG synchroniza-tion and synchronized sleep. Reticular deactivation releases the thalamo-cortical circuits, and thus leads necessarily to EEG synchronization. A critical change in the level of reticular activation, with an appropriate time course, would be required, however, in order that the neuronal patterns responsible for sleep may be triggered. A short lasting fall of the reticular tone may lead to the appearance of brief bursts of synchronous waves without a clear effect on behavior; while a sudden, dramatic fall of reticular activation may lead to coma. Both of these phenomena have very little, if anything, in common with genuine physiological sleep (l.c., pp. 201–202).

In summary, several types of inborn behavior may be *triggered* by electrical stimulation of the diencephalon: sleep behavior, aggression-defense reactions, instinctive behaviors related with food or water intake. These triggered response patterns have many features in common. In any event, they are entirely different from the *fragments* of motor behavior which are *driven* by the stimulation of the motor cortex in the free moving, unanesthetized animal; or from the inhibition of postural tonus produced by electrical stimulation of the anterior lobe of the cerebellum. The term "sleep center" is misleading mainly because the distinction between triggered and driven responses is generally not made clear enough in the discussion of the stimulation experiments.

Once the typical behavioral and EEG patterns of sleep have been triggered by thalamic stimulation, it is not surprising that these patterns outlast the stimulus, and that a tendency to sleep remains even when the animal has been temporarily aroused. This is, in fact, the typical behavior of the domestic cat, an animal which certainly oversleeps with respect to its needs. This tendency to oversleep might be prevented only by an homeostatic regulation of sleep recovery, but these neural mechanisms are unlikely to be present, or at least to be critically important, particularly in the cat (see MORUZZI, 1969, pp. 185–188).

Any typical instinctive behavior, such as the threat-flight response, can be elicited only in the presence of appropriate *external* stimuli, such as a stuffed (HUNSPERGER, 1964) or living cat (ADAMS et al., 1969), which may become the object of the induced attack. In the same way, induced sleep behavior develops fully only when appropriate *internal* conditions are present. These internal conditions are probably represented by chemical changes within the triggering neural structures and may be produced or improved by sleep deprivation. Fortunately for the experimenter, the domestic cat has a tendency to oversleep, and therefore induced sleep may be produced even when the chemical situation is not optimally appropriate. Given suitable internal conditions, however, the induced sleep may outlast the stimulation for an even longer period of time.

2. Brain Stem Stimulations

Electrical stimulation of the *activating reticular system* produces EEG (MORUZZI and MAGOUN, 1949) and behavioral (SEGUNDO et al., 1955a, b; FAVALE et al., 1960) arousal, has marked effects on evoked cortical potentials (see BREMER, 1960) and on perception (FUSTER, 1958; see LINDSLEY, 1961), abolishes thalamic recruiting potentials (MORUZZI and MAGOUN, 1949) and influences in a striking manner the activity of cortical (KLEE, 1960) and thalamic (PURPURA et al., 1966) neurons. In brief, reticular stimulation reproduces the arousal elicited by any sensory stimulation on a background of relaxed wakefulness, drowsiness or sleep. These *phasic activating responses* elicited by reticular stimulation are outside the scope of our review, and are dealt with in LINDSLEY's article on the arousal phenomenon to be published in these reviews. Also, the studies on EEG arousal produced by sensory stimulation and its habituation (SHARPLESS and JASPER, 1956); or on the orienting reflex, the behavioral response produced by any new sensory stimulation (see SOKOLOV, 1969); as well as the effects of sensory stimulation on the discharge of reticular neurons (see MORUZZI, 1954) are related to the phasic phenomenon of arousal and will not be reviewed here.

FREDERICKSON and HOBSON (1970) have used prolonged electrical stimulation of the activating reticular system in order to test the passive or "deactivation" theory of sleep. According to this hypothesis a long lasting, intense arousal maintained for three hours by electrical stimulation of the brain stem reticular formation should produce fatigue of the activating neurons, thus increasing the tendency towards reticular deactivation. The authors compared the subsequent sleep period with that of cats which had been kept awake for three hours by natural means. They were unable to find a change in the latency to slow wave sleep, in the total amount of slow wave sleep, or in the total amount of waking during the following 21 hours. Only prolonged stimulation of the pontine tegmentum was followed by a reduced latency to paradoxical sleep and an increase in the total amount of paradoxical sleep. The authors concluded that their experiments do not confirm the hypothesis that reticular "fatigue" might explain the marked tendency to sleep after prolonged wakefulness. Their evidence against the passive theory of sleep fits the conclusion reached through other lines of investigation, which have been reported on several occasions in this article. The difference between midbrain and pontine stimulation is unexplained.

The opposite *EEG synchronizing responses* elicited by trains of low rate electrical pulses are also in themselves phasic phenomena. These responses, however, gradually shift the general level of activation towards drowsiness and sleep, or at least their repetition is frequently associated with the onset of *tonic deactivation*. For this reason the studies of *the EEG synchronizing* and possible *sleep inducing structures* of the brain stem will be recalled in this review article devoted to the sleep-waking cycle.

During early experiments on the brain stem reticular formation it had been shown that the EEG and behavioral responses to reticular stimulation were occasionally opposite in sign (see FAVALE et al., 1961, for ref.), since synchronization appeared on the EEG when stimulation was applied during periods when the animal was behaviorally awake.

However, systematic studies were undertaken only after the demonstration that a tonic EEG synchronizing influence arises in the lower brain stem (BATINI et al., 1958, 1959a). Only two groups of experiments will be reviewed in some detail.

7*

Favale et al. (1961) worked on unanesthetized, free moving cats. Simply by changing slightly the position of the stimulating electrodes (Fig. 30), they obtained from the brain stem reticular formation opposite effects on the EEG, namely either synchronization or arousal. Low rate electrical pulses were used, but the response was related not only to the parameters of the stimulation, but also to the specific properties of the stimulated structures. Confirming Hess' results (1944), however, behavioral sleep could be elicited only after

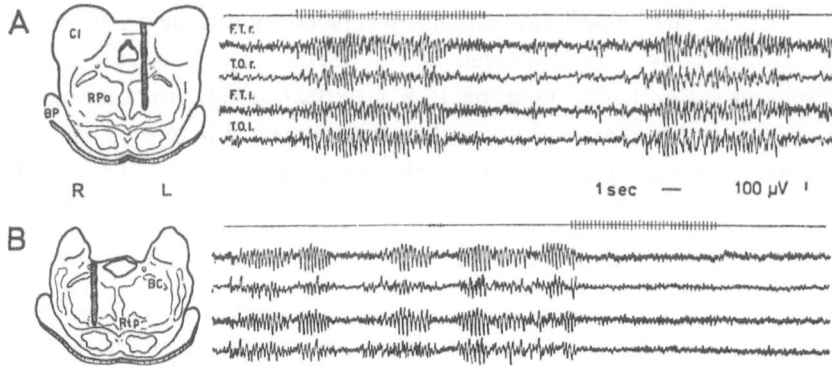

Fig. 30A and B. *Opposite EEG effects produced by low frequency stimulation of two different reticular sites.* Intact, free moving cat. Bipolar EEG records from right (*r.*) and left (*l.*) fronto-temporal (*F.T.*) and temporo-occipital (*T.O.*) areas. A 0.1 msec, 5/sec, 5 V to the left nucleus reticularis pontis oralis (*RPo*) produces EEG synchronization. B (2 min after A) 0.1 msec, 5/sec, 5 V to the most ventral part of the same reticular nucleus in the right side produces EEG activation. (From Favale et al., 1961)

several trains of low rate pulses. Short lasting, stimulus-locked, bursts of EEG synchronization were in fact never associated with behavioral changes suggesting the onset of sleep. Only when EEG synchronization began to outlast the reticular stimulation were the first behavioral signs of sleep observed; when full sleep was reached, the EEG was of course synchronized also in the intervals between the pulse trains. A review on this and related lines of investigation has been prepared by Rossi (1965).

Magnes et al. (1961a), worked on *encéphale isolé* cats, thus avoiding the contaminating effect of changes in blood pressure and respiration which are usually produced by the stimulation of the medulla. This approach involves a drawback with respect to studies using unanesthetized, free moving animals, since body posture cannot be used as a behavioral sign of sleep. It was shown that low rate electrical stimulation of the solitary tract induced EEG synchronization (Fig. 31A, C), an effect that markedly outlasted the stimulus on a background of drowsiness (Fig. 31A); one, however, that was absent on a strong background of activation (Fig. 31B). The localization of the EEG synchronizing structures was found to be critical, in that displacement of the electrode of 1 mm dorsally or ventrally from a point eliciting synchronization caused the response to disappear completely. Upon moving the electrode back

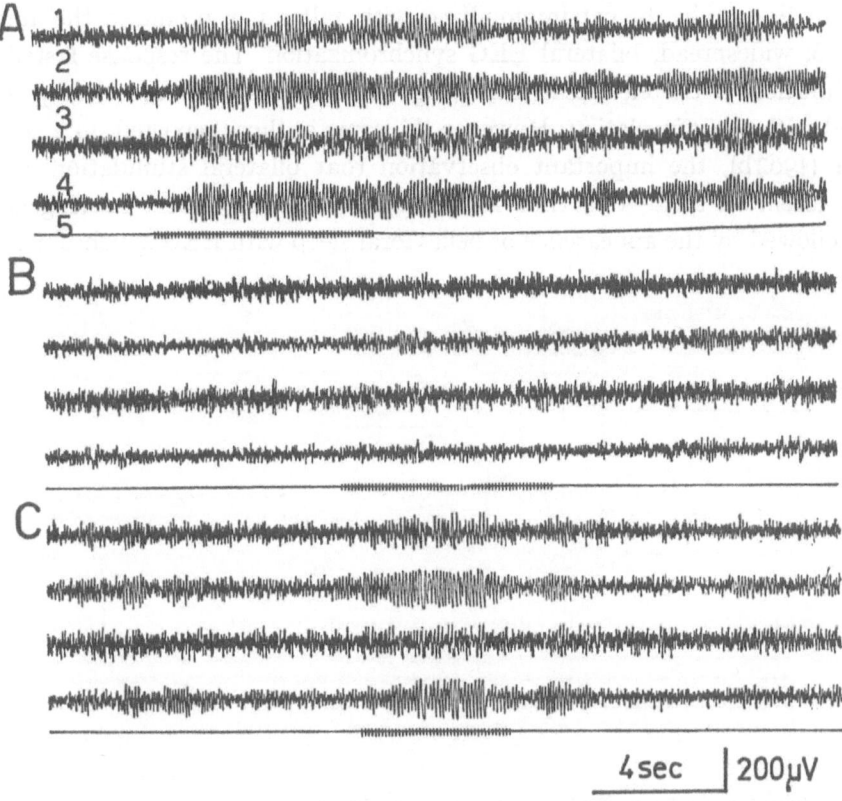

Fig. 31 A–C. *Synchronization of EEG elicited by low rate stimulation of region of nucleus of solitary tract.* "Encéphale isolé" cat. Stimulation with rectangular pulses of 3 V, 5 msec. Bipolar records: *1* right parieto-temporal; *2* right temporo-occipital; *3* left parieto-temporal; *4* left temporo-occipital; *5* stimulus marker. A Stimulation at 10/sec: EEG synchronization outlasting stimulus. B Same stimulation at 10/sec, 30 sec after acoustic stimulus: no effect. C Continuation of B; same stimulation at 10/sec; response with shorter after-effect. (From MAGNES et al., 1961)

to its original position the responses once more appeared. Since the same low rate pattern of electrical stimulation occasionally produced EEG arousal when applied to neighboring brain stem loci, it is clear that the induced synchronization was not due simply to the stimulation parameters.

Reviews of the deactivation responses elicited by stimulation of the brain stem after spinal transection appeared in the early sixties (MAGNES, 1961 b; DELL et al., 1961). Deactivation responses have been obtained also in the unanesthetized rabbit following low rate midbrain stimulation (HÖSLI and MONNIER, 1962, 1963); however no genuine sleep was produced.

3. Basal Forebrain Stimulation

STERMAN and CLEMENTE reported in a short preliminary note (1961), and in a full paper published during the following year (1962a), that unilateral low rate stimulation of the lateral preoptic region and adjacent diagonal band of BROCA, in the cat immobilized with gallamine, produced the abrupt onset of widespread, bilateral EEG synchronization. The response lasted for the duration of the electrical train (Fig. 32) and was therefore clearly time-locked with the stimulation. However the same authors reported, in a second paper (1962b), the important observation that bilateral stimulation of the same forebrain area—this time in the unanesthetized, freely moving cat—was followed by the appearence of behavioral sleep with EEG synchronization

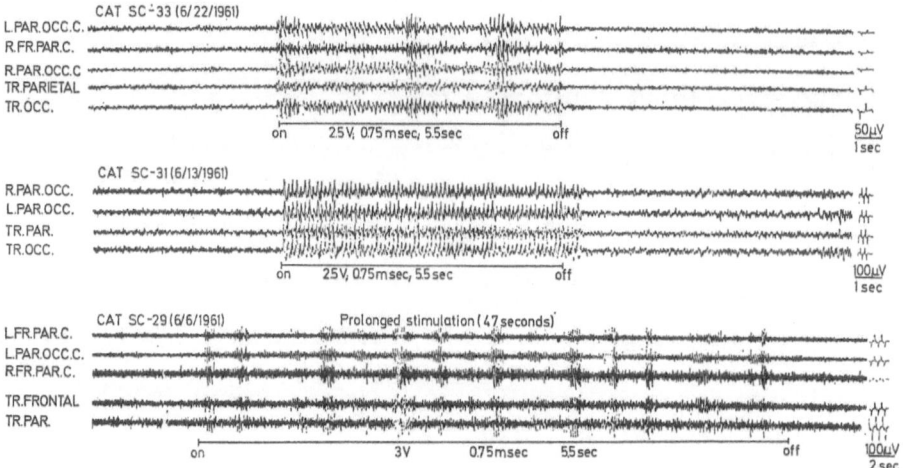

Fig. 32. *Diffuse but short lasting EEG synchronization upon low rate stimulation of left basal forebrain synchronizing area.* Cat immobilized with gallamine. Sudden onset of EEG synchronization, not outlasting the stimulus, upon low rate, unilateral stimulation of the forebrain area. Signals with stimulation parameters below each record. Left (*L*) and right (*R*) parieto-occipital (*Par.occ.*) and fronto-parietal (*Fr.par.*) leads. Transfrontal (*Tr.frontal*), transparietal (*Tr.par.*) and transoccipital (*Tr.occ.*) records were also taken.
(From STERMAN and CLEMENTE, 1962a)

(Fig. 33). Time parameters were markedly different in this case, since EEG synchronization occurred after a latency of some seconds and the behavioral signs of drowsiness and sleep appeared with even longer latencies, some time after the EEG synchronization.

The results obtained by STERMAN and CLEMENTE (1962b) on free moving cats are different from those of all previous investigators in two respects. First of all EEG synchronization and behavioral sleep could be elicited also with high rate stimulations (Fig. 34), which always produce arousal when applied to the midline nuclei of the thalamus and to the brain stem. Second, the "latency" of the sleep effect was frequently, though not constantly, lower than that observed in HESS' experiments. Only occasionally "latencies" of the order of those reported in HESS' works (up to 3 minutes) were found. The average time was only 30 seconds, and occasionally sleep appeared after delays as short as 5 seconds (l.c., p. 108). The importance of these results will become apparent in the final discussion.

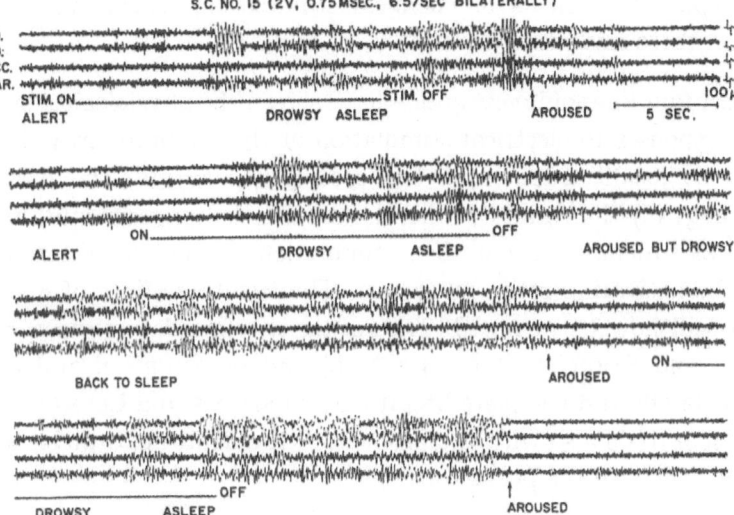

Fig. 33. *Slow onset of an EEG synchronization outlasting the low rate stimulation, with behavioral signs of sleep, upon bilateral stimulation of basal forebrain.* Unanesthetized free moving cat. Note long latency and long aftereffects of EEG and behavioral signs of sleep. Bilateral low rate stimulation (2 V; 0.75 msec, 6.5/sec). Abbreviations and signals as in Fig. 32, but behavioral responses are indicated below each record.
(From STERMAN and CLEMENTE, 1962 b)

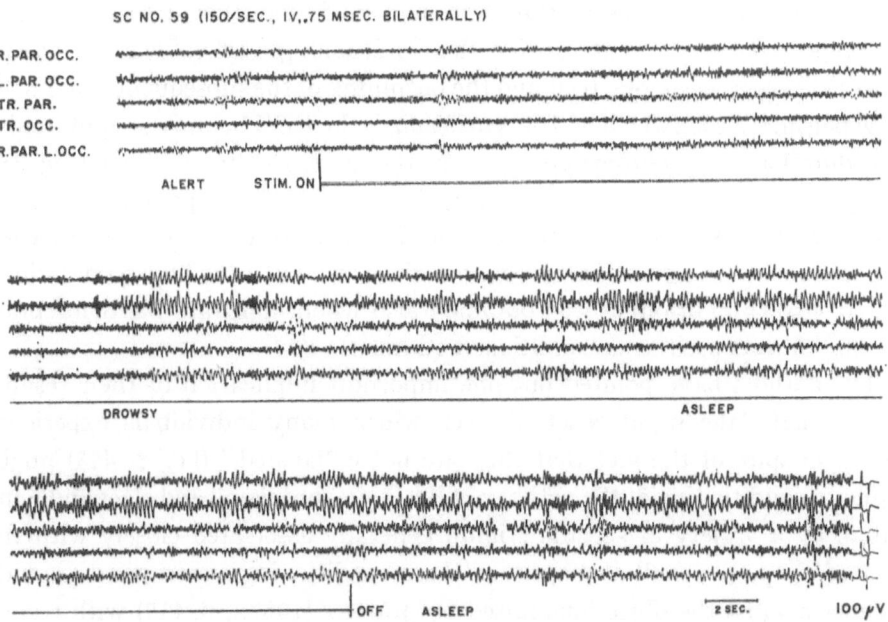

Fig. 34. *Widespread EEG synchronization produced by high rate stimulation of the basal forebrain.* The bilateral, high rate (150/sec), forebrain stimulation (pulse duration 0.75 msec; 1 volt) was followed by EEG synchronization and by behavioral signs of drowsiness and sleep. The prolonged delay preceding the onset of EEG synchronization was characteristic of this animal, but is not seen typically. Abbreviations and signals as in Figs. 32, 33. (From STERMAN and CLEMENTE, 1962 b)

The work of HERNANDEZ-PÉON (1962) confirmed the sleep inducing influence of electrical stimulation of the basal forebrain area, and led to a study of the hypnogenic effects of chemical stimulation, which are reported by Jouvet in these reviews.

At least four lines of investigation have shown the physiological significance of these responses to electrical stimulation of the forebrain area: i) insomnia produced by localized lesions in this area (see p. 78); ii) the possibility of conditioning; iii) the results of thermal stimulation experiments; and finally iv) Bremer's demonstration that stimulation of the basal forebrain area inhibits the activating reticular system. The last three lines of experimental endeavor will now be reviewed. The reader is referred to review articles by Clemente and Sterman (1963, 1967a, b) and by Sterman and Clemente (1970) for details and complete literature. Sterman's and Clemente's recent review (1968) deals also with the problem of the neural pathways, which will not be considered in the present article.

a) Clemente, Sterman, and Wyrwicka (1964) showed that both the EEG synchronization and the behavioral sleep patterns produced by basal forebrain stimulation could also be obtained by means of a conditioning procedure.

A tone of a given frequency was presented for 10 seconds before basal fore-brain stimulation and was continued throughout it, for other 20 seconds. Trials were repeated every minute during the conditioning session, with an interval of 30 seconds between one trial and the beginning of the subsequent tone. After a few pairings, a tone which had desynchronized the EEG before the conditioning procedure had a synchronizing effect on the alert, free moving cat (Fig. 35). "In addition, the alert animal may suddenly show sleep preparatory behavior to the onset of the tone. It reclines, drops its head to its paws, closes its eyes, and remains in this position throughout the period of tone and stimulation" (l.c., p. 407). Conditioning was obtained also when high rate basal forebrain stimulation was used as unconditioned stimulus.

The authors have pointed out one important implication of their results, namely that "the rapid onset of sleep which many individuals experience nightly in spite of the fact that they are not exhausted" (l.c., p. 415) might be due to conditioning of the activity of EEG synchronizing and sleep inducing centers by a variety of specific stimuli generally associated closely with the onset of sleep. We shall also see that these conditioning experiments provide evidence against the objections raised by Jouvet (1967a, p. 135) with regard to the use of stimulation techniques as an approach to sleep investigations.

b) Roberts and Robinson (1969) showed that diathermic warming of the preoptic region and of the anterior hypothalamus produced sleep-like postural changes in the free moving cat. With high diathermic currents the animals lowered their heads to the floor or onto their paws, and assumed a posture very similar to that described by Sterman and Clemente following electrical stimulation of the same region (see p. 102). Myosis, relaxation of the nictitating membrane and eye closure completed the sleep syndrome. The EEG became synchronized, although the high voltage waves were slower than those following electrical stimulation of the same area. The latencies for these effects (from 20 to 150 seconds) were of the same order as those reported by STERMAN and CLEMENTE. Thermically induced sleep was fully reversible, just as was the electrically induced sleep in HESS' experiments. Control experiments showed

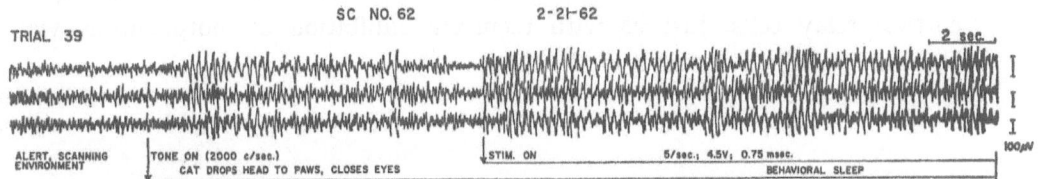

Fig. 35. *Conditioning of EEG and behavioral sleep patterns elicited by basal forebrain stimulation.* Conditioning was obtained by presenting a tone of 2000 cps 10 sec in advance of basal forebrain stimulation, and was continued throughout an applied 20 sec period of forebrain stimulation. After a few pairings, the tone evoked sleep preparatory behavior accompanied by a shift in the EEG to slow-wave synchronous activity; although an arousing effect was obtained before conditioning procedure. Note similarity between this conditioned EEG response and the unconditioned effect elicited later by forebrain stimulation. (From CLEMENTE et al., 1963)

that the sleep response was due to specific stimulation of thermoreceptors. The temperature threshold for sleep was lower than that for panting, the typical thermoregolatory effect produced by localized heating of the hypothalamus (MAGOUN et al., 1938; see THAUER, 1939, 1967).

Apparently the thermoregulating and sleep inducing structures overlap. Stimulation of these *central* receptors by changes in blood temperature is likely to be an important source of impulses driving the sleep-inducing structures of the basal forebrain. Of course reflexive activation of the central structures sensitive to heating by *peripheral* thermal receptors is also likely to be important in the facilitation of sleep produced by a warm environment (see ROBINSON and LEE, 1946; ROBERTS et al., 1969). We know that there is an interaction between central and peripheral mechanisms in the control of thermoregulation (see HENSEL, 1952; HARDY, 1961). The literature on the interrelations between sleep and thermoregulation has been recently reviewed (PARMEGGIANI and RABIN, 1970).

A forgotten observation made by chance by PAVLOV (1923) during classical conditioning procedures may be recalled at the end of this section. He stated (l.c., p. 43) that when thermal stimulation of the skin was used as a conditioned stimulus, it was particularly easy to obtain sleep. He noted, moreover, that when the interval between conditioned and unconditioned stimulus was rather long, from 30 to 60 seconds, the animal became drowsy and fell asleep every session, even during the *first* application of the thermal stimulation. The volleys starting from the thermal receptors probably acted on the basal fore-brain area.

c) Post-synaptic inhibition of thalamo-cortical relay neurons may be revealed in the ventro-basal nuclei of the anesthetized cat either i) by intra-cellular recording or ii) by observing with unipolar semi-microelectrodes the *positive potential fields* generated by the hyperpolarizing currents (Andersen et al., 1964a, b). An important inhibitory response is probably "mediated by inhibitory interneurons excited from axon collaterals of the thalamo-cortical relay cells, just as with recurrent inhibition of motoneurons via Renshaw cells" (Andersen et al., 1964a, p. 366).

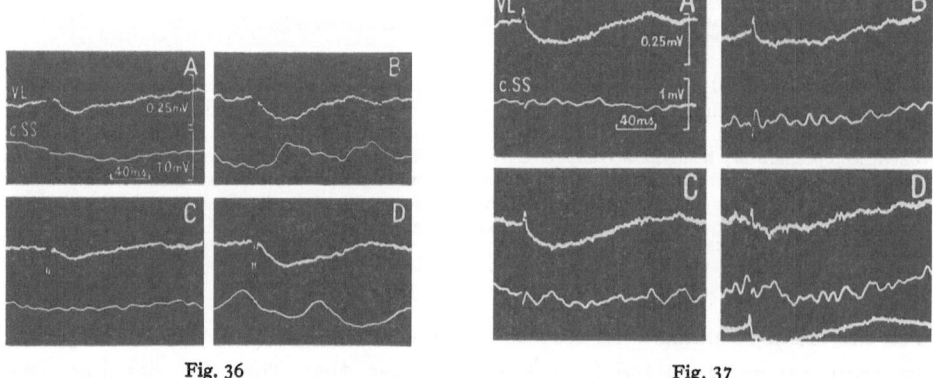

Fig. 36   Fig. 37

Fig. 36 A–D. *Enhancement of intrathalamic inhibition during sleep.* "Encéphale isolé" cat with ocular and EEG signs of alternation between relaxed wakefulness (A, C) and sleep (B, D). Enhancement during sleep (B, D) of the positive field potential of the ventro-lateral nucleus (upper record) elicited by a single electrical pulse applied to precruciate gyrus. Lower record: surface potential from middle suprasylvian gyrus. Positivity corresponds to downward deflection. (From Bremer, 1970a)

Fig. 37 A–D. *Depression of intrathalamic inhibition by reticular volleys.* The positive field potential produced and recorded as in Fig. 36, is clearly present during behavioral and EEG drowsiness (A), disappears when the electrical pulse applied to the precruciate gyrus is preceded (interval: 100 msec) by a volley of 5 electrical pulses at 100/sec applied to the midbrain reticular formation (B). The response is again present when the reticular effect has disappeared (C), to the abolished completely by a volley of 10 pulses at 300/sec applied to the same point of the reticular formation (D). (From Bremer, 1970a)

Bremer (1970a) has recently recorded these intrathalamic positive potential fields in the *encéphale isolé* cat, hence in the absence of anesthesia. He has shown that spontaneous alertness (Fig. 36) or reticular arousal (Fig. 37) result in a reduction of the amplitude of these responses, while the opposite effect appears during spontaneous sleep (Fig. 36). A marked, permanent enhancement of the positive field potential is observed when the preparation is transformed into a *cerveau isolé* by a midbrain transection. Thus brain deactivation increases intrathalamic inhibition, possibly because during wakefulness the de-polarizing influence of the tonic, ascending reticular barrage counteracts the hyperpolarizing effect of the intrathalamic inhibitory neurons. Bremer's experiment may be regarded as the first proof that active inhibitory processes occur within the cerebrum during sleep, possibly as a consequence of de-facilitation.

However, reticular deactivation may also be produced by inhibition. BREMER (1970b) has shown that positive field potentials, similar to those generated in the thalamus by summed hyperpolarizing IPSP's, are led from the midbrain reticular formation of the *encéphale isolé* cat when an electric pulse is applied to the preoptic region which CLEMENTE and STERMAN have

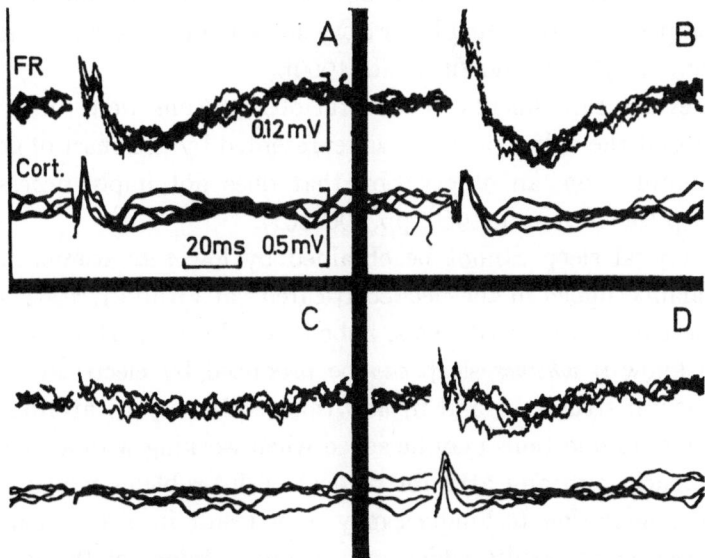

Fig. 38 A–D. *Positive field potentials elicited in the midbrain reticular formation by electrical pulses applied to the basal forebrain area.* "Encéphale isolé", showing EEG and ocular signs of drowsiness. Arousal as easily produced by brief stimulation of midbrain reticular formation. A single rectangular pulse (0.3 msec, 2 V) applied to the basal forebrain area of STERMAN and CLEMENTE is followed by a positive field potential in the midbrain reticular formation, as shown by the downward deflection (*FR*). Lower record: mono-polar lead from ectosylvian cortex (*Cort.*). The reticular field potential elicited by a single pulse (A) dis-appears when the stimulating electrode is raised by 3 mm (C), to reappear again when the electrode is placed in the old position (D). The response is larger when two pulses are applied at an interval of 3 msec (B). (From BREMER, 1970b)

shown to be an hypnogenic area in the intact cat (Fig. 38). There is a facilitatory enhancement of this reticular response when two pulses are applied at appropriate intervals, and the location of the stimulating electrode appears to be as critical as in CLEMENTE and STERMAN's experiments. According to BREMER (1970a, b), stimulation of other hypnogenic structures might act by inhibiting the activating reticular system.

C. Conclusions

1. On the Validity of the Experimental Approach

The following criticisms have been raised against stimulation experiments.

i) Sleep has been produced by stimulation of not one but several structures of the encephalon, and also by excitation of several kinds of receptors and sensory fibers (for the literature, see Jouvet, 1967a, pp. 135–136). It is unlikely that all of these central structures are hypnogenic, or that so many sensory mechanisms are routinely involved in the onset of physiological sleep.

ii) Induced sleep appears more easily on a background of relaxed wakefulness, usually after long "latencies". Hence the objection may be raised that the animal would have gone to sleep spontaneously, even without the stimulation (see Jouvet, 1962, p. 135). It may be added, moreover, that most of these experiments were made on the cat, an animal which is frequently asleep when placed in the usual laboratory situation (see Batini et al., 1959a; Sterman et al., 1965; Angeleri et al., 1969).

iii) In several experiments, behavioral observations were difficult or impossible, so that the only evidence was represented by the onset of generalized EEG synchronization, an observation that does not imply necessarily the onset of sleep (see Moruzzi, 1969, pp. 196–201).

iv) Behavioral sleep cannot be obtained by low rate stimulation of the midline thalamic nuclei in the neodecorticated cat (Jouvet, 1962, p. 169).

That these are serious criticisms, nobody would deny. However, the main point is to know i) *whether* sleep *can* be produced by electrical stimulation of central neural structures and ii) *how* this result may be attained. Only a limited number of questions may be asked when working with a single experimental approach, and any attempt to deal with problems which cannot be solved with stimulation techniques may only result in a sceptical attitude, even with respect to results which are, in our opinion, of the greatest importance for sleep physiology.

Before discussing what we regard as the permanent achievements of this experimental approach, we may try to examine the main causes of error, and delimit the areas of uncertainty.

a) The "latency" of induced sleep must in fact be of the order of several seconds, if we are really concerned with the experimental reproduction of the physiological phenomenon, with its long lasting appetitive phase (see p. 128). Moreover, the long "latent times" of sleep induction may also be explained by the fact that the repetitive volleys elicited by stimulation of receptors, sensory nerves, or of central neural structures will recruit progressively the activity of the sleep inducing neurons, through the usual processes of temporal summation and *Bahnung* (see p. 95). Of course the recruited hypnogenic neurons may be far away from those we are stimulating. This is the reason why no conclusions about localization can be made with stimulation experiments alone, while the causes of error are much less when we are concerned with the reproduction of the usual types of instinctive behavior occurring in the ergotropic sphere. The sleep physiologist cannot look at the stimulation experiments without taking into account the effect of lesions. In fact the insomnia produced by pretrigeminal transection (Batini et al., 1958, 1959a), or by lesion of raphe nuclei (JOUVET and RENAULT, 1966) or of the forebrain area (STERMAN and CLEMENTE, 1962b) shows that the hypnogenic structures we are investigating are *tonically* active. For this reason no attempt has been made to review the literature on every kind of central stimulation which produces behavioral and EEG signs of sleep. Our attention has been centered around the lower brain stem, the basal forebrain area and the midline thalamic nuclei. In fact no clear-cut insomnia is produced by thalamic lesions, and behavioral sleep occurs in the thalamectomized cat (see p. 73). HESS' results have been quoted extensively, nonetheless, not only for their obvious historical importance, but also in view of the fact that their relationship with the in-stinctive behavior produced by hypothalamic stimulations is of great physio-logical significance (see pp. 91–93). In summary, any attempt to reach conclusions on *localizations*, by utilizing *only* sleep elicited by either central or peripheral stimulations, would imply the asking of questions that cannot be answered.

b) Electrical stimulation excites both neurons and axons, a drawback that is probably not shared by the experiments of chemical stimulation. Moreover, it is almost impossible to avoid costimulation by the electrical pulses of neighboring or intermingled activating fibers or neurons. In fact low rate pulses have been preferentially used, since HESS' pioneer work (see p. 93), mainly in order to avoid overwhelming of the sleep effect by the opposite arousal response. Were we able to carry out *clean* stimulations, by exciting *selectively* the sleep inducing structures, it would probably not be so important to work on a background of relaxed wakefulness, nor even to utilize low rate electrical pulses. These precautions are made necessary by the fact that contaminations by activating effects is apparently less likely to occur on a background of relaxed wakefulness. We are probably nearer to the ideal of a "clean" stimulation experiments when we apply our electrical pulses to the forebrain aera, whence sleep can be obtained also with high rate stimulation. Here is one of the main reasons for the importance of the experimental approach developed by STERMAN and CLEMENTE.

c) It is not difficult to understand why induced sleep was missed when thalamic stimulation was made in the neodecorticate animal. JOUVET's experi-ments (1962) were made on cats, an animal presenting the classical sham rage following ablation of the cerebral cortex (CANNON and BRITTON, 1925; see ZANCHETTI, 1966, 1967a, for the literature). In this experimental situation the ergotropic influences have a great tendency to overwhelm the sleep effect. In fact, the spells of behavioral sleep recalling the slow wave phase of the normal animal, as shown by the persistence of cervical EMG activity, last only 3–4 minutes in the neocorticated cat (JOUVET, 1962, p. 142). After this operation 90% of the time spent behaviorally asleep is represented by the paradoxical phase (l.c., p. 144).

In short, most of the objections raised against the stimulation experiments are related to questions which cannot be solved by means of this approach, or at least by means of this approach alone. Hence the only objection that deserves to be discussed at length is that the animal might have gone to sleep even without the stimulation.

There is little doubt that *EEG synchronization* may be produced by electrical stimulation of cutaneous group II fibers (see p. 88 and Fig. 26), of the region of the solitary tract (see p. 100 and Fig. 31) or of the forebrain area (see p. 101 and Figs. 32–34), since the EEG response begins and ends with the stimulus train, in a thoroughly predictable manner. There is little doubt, moreover, that these effects are related to the specific action of neurons or fibers, since the opposite response, EEG arousal, may be obtained if Group III fibers are excited (see p. 89) or whenever the tip of the stimulating electrodes is moved slightly up or down (see p. 100). Thus the existence of a cause-effect relationship between stimulations and bursts of EEG synchronization is certain, and we may confidently state that the induced volleys drive EEG synchronizing structures.

The uncertainty arises when, after several trains of electrical pulses, we observe that the EEG synchronizing response outlasts the stimulus train. This is the first sign that the background level is moving towards deactivation, whose final outcome will be behavioral sleep. When the background activity moves in the opposite direction even the synchronizing effect of the stimulation may disappear, submerged by an overwhelming tendency toward arousal. What we need to discuss, therefore, is simply whether the shift in the background of activation towards a state of sleep is produced, or better triggered, by the volleys yielding EEG synchronization; or whether it arises spontaneously, for unknown reasons. The problem is to decide between a causal relationship and an occasional coincidence.

Before the investigations on the forebrain area, the problem was basically one of statistical evaluation. This study has been made by comparing sham experiments with stimulation experiments. As expected, there is high probability that also non-stimulated animal fall asleep; however, they fall asleep faster following stimulation, and the difference is statistically significant (Roelofs et al., 1963). There is little doubt, however, that errors may be made when one looks for an effect that occurs easily also in natural conditions, particularly if the response is characterized by long latency and gradual onset.

These difficulties are lessened when one stimulates the basal forebrain area because i) the average latency of the induced sleep is shorter (30 sec) and delays as brief as 5 sec may be found (Sterman and Clemente, 1962b); ii) the response is not critically bound to low rate stimulation, but may be obtained also with high frequency pulses (l.c.); iii) sleep is not elicited by unilateral stimulation (l.c.); finally iv) EEG and behavioral sleep responses can be evoked by a tone which had been paired in time with electrical stimulation of the forebrain area in the cat (CLEMENTE et al., 1963). The conditioned response to tone may appear a few seconds after the beginning of the auditory stimulation, before either the high or low frequency basal forebrain stimulation, acting as unconditioned stimulus, is started. An observation such as that reproduced in Fig. 35 leaves little doubt about the validity of the statement that sleep may be produced by volleys arising in the basal forebrain area.

This statement does not imply that the sleep inducing neurons are localized within the basal forebrain area, a point to be developed later (see p. 114). In fact, pairing occurs between unconditioned volleys arising in the basal forebrain area and conditioned auditory volleys. Both trains of impulses act on some structure lying downstream to the fibers or neurons which are driven by the electrical stimulus. The reader is referred to DOTY's reviews (1961, 1969) for the conditioned reflexes formed and evoked by brain stimulation. What matters for the formation of a conditioned reflex is the temporal association between conditioned and unconditioned volleys. If the sleep occurring after forebrain stimulation were a chance effect, not causally related with volleys arising in the stimulated area, conditioning after a few trials and the occurrence of the response after a few seconds would hardly be possible.

The conclusions drawn from stimulation experiments are strengthened by the fact that they fit the results of lesion experiments and of electrophysiological recordings. In fact these conclusions are almost unescapable when we consider the striking insomnia produced by a lesion of the basal forebrain area (see p. 78)—an observation implying the tonic activity of the sleep inducing structures—and the results of BREMER's experiments (1970), which suggest that the activating reticular system is inhibited by preoptic volleys.

It is difficult to believe that this statement should not also be valid for other stimulation experiments, particularly in those cases where the production of experimental insomnia by lesions shows that a *tonic* sleep inducing influence is normally present in districts which, when electrically stimulated, lead to sleep. The striking similarity between the results of HESS and those of STERMAN and CLEMENTE have already been stressed (see MORUZZI, 1969, pp. 198–200). But in fact the patterns of the responses are always similar whenever sleep is the final outcome of the stimulation: a synchronized response first lasts as long as the stimulus train, then outlasts the electrical stimulation, while behavioral signs of drowsiness and sleep gradually appear. It is extremely likely, therefore, that our conclusions have a general validity.

2. An Attempt at an Interpretation

The main results of lesion and stimulation experiments will now be discussed, in an attempt to reach a satisfactory interpretation of the neural mechanisms underlying the alternation between sleep and wakefulness.

There are three systems which may be regarded with some certainty as responsible for, or at least contributing to, the onset of sleep. They are localized respectively in the basal forebrain (pp. 78, 101), in the lower brain stem (pp. 31, 100), and in the midline nuclei of the thalamus (p. 90).

According to Koella (1967) Hess' thalamic sleep center would be the *head ganglion of sleep*, to use Sherrington's terminology. Koella brings several interesting considerations in support of his hypothesis (l.c., p. 123–124). Probably the most important among them is the striking similarity between natural sleep and Hess' induced sleep. Koella regards the hypnogenic regions of the basal forebrain, including the anterior hypothalamus, and of the lower brain stem, as subordinate sleep controlling structures. He recalls that Hess (1944b) obtained a syndrome of generalized muscular relaxation, without typical sleep behavior, from an area closely related to the basal forebrain.

Koella's interpretation would imply, in our opinion, that whenever central or peripheral stimulation leads to genuine physiological sleep, the hypnogenic neurons of the midline thalamic nuclei should be involved in one way or another. They would represent, to borrow again the Sherrington terminology, the *final common path* of all the neural mechanisms actively leading to sleep. The main difficulty with this interpretation lies in the fact that natural sleep occurs after thalamectomy (see p. 73), while the tonic activity of the hypnogenic structures of the basal forebrain and of the lower brain stem is shown by the striking insomnias produced by their lesion (p. 78), or by disconnecting the pons from the cerebrum (p. 31). Although quantitative studies would be desirable, there is not yet evidence of a marked hyposomnia produced by lesion of the midline thalamic nuclei.

In our opinion the criterion of the existence of *tonic activity* is one of major importance in any attempt to localize the key structures involved in the active production of sleep. We must therefore consider the opposite hypothesis that sleep occurs following stimulation of the massa intermedia whenever neurons related with the basal forebrain are recruited by the thalamic volleys, through temporal summation or central facilitation *(Bahnung)*. Also other hypnogenic stimulations of the cerebrum (see Parmeggiani, 1968) would act through the same mechanism.

The structures endowed with a *tonic* sleep inducing activity are localized within, or are in some way functionally related to, the basal forebrain and the lower brain stem. They are allied, and probably functionally interrelated, structures. Both of them are antagonistically oriented with respect to the activating reticular system.

We have now convincing evidence, from Bremer's recent experiments (1970a, b), that the ascending reticular system, the waking system, may be inhibited by volleys arising in the sleep inducing structures of the basal forebrain (p. 107). We have no direct evidence that the opposite interrelation exists, but it seems an attractive hypothesis, indeed even a likely one, that sleep inducing structures may be inhibited by the activating reticular system.

A reciprocal inhibition between the two antagonistic systems would account for the alternation between sleep and wakefulness. Several observations might be easily explained if we make the hypothesis that the situation is, on the general lines, similar to that characterizing the reciprocal inhibition of the flexor and extensor half-centers of the spinal cord. As soon as the tonic activity of the sleep inducing structures increases, it would progressively overwhelm the activating reticular system. Reticular deactivation would increase "en avalanche" by the mutual play of inhibition and release from inhibition. The same process would proceed in the opposite direction at the moment of arousal. The time course of these processes leading to the reversal of the previous imbalance would determine the speed of the onset of sleep in the night and the velocity and the efficiency in the development of full arousal during the next morning. The time course of these opposite processes is typical for each man and presents strong individual variations.

Both sleep and waking structures may be driven by sensory volleys and also by impulses coming from other parts of the central nervous system. We have reviewed at length the experimental evidence showing that the sleep response to stimulation of the basal forebrain may be conditioned (p. 104). It is likely that these structures are driven during any conditioning process whose final outcome is sleep. The experiments of thermal stimulation suggest, moreover, that the same group of neurons may be driven by central or peripheral thermoreceptors, thus contributing to drowsiness and the onset of sleep in a warm environment.

It is likely that the sleep inducing structures of the lower brain stem are driven by several kinds of unconditioned impulses, arising for example from carotid sinus baroceptors or from tactile receptors related with Group II cutaneous fibers (see pp. 86–89). Sleep produced by monotonous stimulation may also be mediated by the structures of the lower brain stem. The lower brain stem might exert its hypnogenic functions by acting upon the basal forebrain system. However the alternation of activity recalling the sleep-waking rhythm that occurs after chronic decerebration shows that some kind of reciprocal organization between hypnogenic and activating structures may occur also within the brain stem, independently from the cerebrum. On the other hand the reappearence of sleep-waking rhythm in the chronic *cerveau isolé* suggests that reciprocal organization between cerebral activating and hypnogenic mechanisms may occur also in the absence of the brain stem. Of course chronic experimentation may reveal only potentialities of central organization. The interrelations between brain stem and cerebrum are likely to be extremely close, and of great physiological significance, in the normal animal.

The hypothesis of reciprocally organized activating and deactivating structures would not be disproved if it were shown that in some cases the lesions were not concerned with neurons, but rather with bundles of fibers arising elsewhere. This problem has been discussed at length for the insomnia produced by lesion of the raphe nuclei (see p. 64). BREMER (1970b) has suggested that the basal forebrain area might contain a bundle of inhibitory reticulopetal fibers, rather than the somata of sleep inducing neurons, which might be localized elsewhere. In fact the results obtained with either lesions or stimulations are the same whether we are dealing with somata or fibers. The issue axon vs. soma might be resolved by intracellular recording, but no attempt has been made so far to utilize this approach in the study of the free moving animals, for obvious technical reasons.

These considerations and the inadequacy of our information on the behavior of single units concerned with sleep and wakefulness lead us to emphasize another major gap in our knowledge, one that has been recently pointed out by BERLUCCHI (1970). There is in fact no clear-cut evidence that single unit firing, in the brain stem or the diencephalon, is *consistently* and *predictably* characterized by the patterns of discharge one would expect from structures concerned with the maintenance of either sleep or wakefulness. Units behaving occasionally in this way, with respect to the waking state or the slow wave phase of sleep, have in fact been described in the reticular formation (MOLLICA et al., 1953; CASPERS, 1961; HUTTENLOCHER, 1961). However the behavior of other neighboring neurons may be entirely different. Moreover, the identification of the nature of the recorded units is difficult. Finally, the paradoxical phase of sleep is characterized by entirely different patterns of discharge. More satisfactory results are obtained by multiunit recording and averaging (SCHLAG and BALVIN, 1963), possibly because the background activity includes the spike discharges of units which are less easily recorded. Small neurons are likely to be missed by single unit recording.

To summarize the results of both lesion and stimulation experiments we may draw the following conclusions:

i) The ascending reticular system and a group of neurons lying in the posterior hypothalamus are endowed with a tonic activating influence. They are probably concerned with the maintenance of wakefulness.

ii) The lower brain stem and the basal forebrain area contain structures with an opposing function, which exert a tonic deactivating influence and lead ultimately to sleep.

V. Discussion

1. The Concept of a Level of Brain Activity

The term *vigilance* was introduced by HEAD (1923) to indicate the degree of physiological efficiency of an animal or experimental preparation. He pointed out that a "high state of physiological efficiency differs from a pure condition of raised excitability; for although the threshold value of the stimulus is not of necessity lowered it is associated not only with an increased reaction but with highly adapted responses" (l.c., p. 133).

For about thirty years the concept of vigilance had a negligible influence on neurophysiological thought, possibly because the term had been used by HEAD in such a broad sense as to appear almost meaningless. He had quoted, in fact, as examples of different levels of vigilance i) the spinal preparation in its acute and chronic conditions, ii) the changes of extensor rigidity occurring in the decerebrate animal and, finally, iii) the levels of consciousness in the intact animal. To quote:

Consciousness stands in the same relation to the vigilance of the higher centres as adapted and purposive reflexes to that of those of lower rank in the neural hierarchy. When vigilance is high, mind and body are poised in readiness to respond to any event external or internal (l.c., p. 143).

The notion of "vigilance" was bound to be revived when terms such as "ascending reticular system" and "diffuse projection system" became familiar in the field of neurological sciences. It was realized that the central nervous system was concerned not only with the performance of specific tasks, but also with maintaining, or with changing, *a level of activity*, and that the two functions were attributable to different neural structures. DELL (1958, p. 199) pointed out that by accepting the notion of vigilance, two fields of research, neurophysiology and ethology were linked, although unfortunately not very closely. "In the case of bodily needs" he stated (l.c., p. 198) "the vigilance creates first the conditions for the organism to be actively interested in the outside world, and afterwards the more and more specific readiness to receive only some kinds of stimuli to the exclusion of others, and to perform the motor patterns of one definite instinct".

HINDE (1966) has rightly pointed out in his recent book (l.c., p. 157) that the concept of level of activation "is concerned primarily with showing that there are a number of consequences *common* to many and diverse motivational variables. This is not the same as saying that they reflect a general state *causal* to all possible activities"[7].

In our opinion the term "vigilance" should be restricted to the activity of the cerebrum and more especially to conscious behavior. Undoubtedly there are similarities between the effects of facilitation and defacilitation related to changes in the intensity of the ascending and descending flow of impulses arising from the brain stem. We have pointed out (p. 48) that the recovery from the coma of the acute *cerveau isolé* may be compared to the recovery from spinal shock. However, the changes of posture or the movements that occur in the chronically decerebrate cats, interesting as they may appear to be as a sign that a sleep-waking alternation of activity still occurs, are merely the postural and motor components of complex waking and sleeping behaviors.

2. Activating and Deactivating Structures

Experiments reported in the previous chapters of this review have led to the conclusion that the level or critical pattern of cerebral activity is controlled by the brain stem. A tonic ascending flow of reticular impulses would be responsible for the state of wakefulness, its complete or almost complete

---

7 Italics are ours.

interruption would lead to a state of coma; while, finally, a temporary slackening of the tonic reticular discharge might be responsible for the onset and the maintenance of natural sleep (Moruzzi and Magoun, 1949). The reticular hypothesis appears, therefore, as a new version of the old theories of passive deafferentation (Purkinje, 1846; Kleitman, 1929; see Kleitman, 1963, and Moruzzi, 1963a, 1964b). These doctrines postulated, in fact, that sleep is basically an inability to remain awake. When the tonic reticular barrage falls below a critical level the animal would sleep simply because it is unable to stay awake. Thus the ascending reticular system would be "das Organ des Wachens" which Purkinje had postulated since 1846 (Purkinje, 1846, p. 474) and had wrongly localized in the cerebrum (see Moruzzi, 1964b).

After more than twenty years since first being proposed (Moruzzi and Magoun, 1949), the concept of a tonic reticular control of wakefulness and sleep still maintains its validity. Any level of cerebral activity is probably related to, and actually maintained by, a given level of reticular activation.

This statement does not necessarily imply (Moruzzi, 1958) the acceptance of the hypothesis that the ascending reticular system exerts an unspecific facilitatory control upon the cerebrum, an effect that would be related to the purely quantitative aspect of an ill-defined "energizing" influence. We are in fact likely to be confronted with phenomena which are by far more complex than those involved, say, in the descending facilitation of the pre-ganglionic sympathetic neurons by the bulbar vasoconstrictor center. Changes in single unit (Huttenlocher, 1961) and in the integrated discharge (Schlag and Balvin, 1963; Podvoll and Goodman, 1967) of the reticular system undoubtedly occur during the sleep-waking cycle. There is little doubt, moreover, that the overall ascending reticular barrage has been enhanced, phasically, whenever an arousal phenomenon interrupts a state of slow wave sleep, or when a fit of rage occurs on a background of relaxed wakefulness. It is unlikely, however, that sleep, drowsiness, relaxed wakefulness and active wakefulness may be described only in a purely quantitative way. It is improbable, in other words, that these states are explainable exclusively in terms of different levels of the overall reticular barrage. The spatio-temporal patterns of discharge of the individual reticular neurons (see Huttenlocher, 1961) and their functional interrelations should also be considered. For this reason the term "level of reticular activation" should be understood in a rather loose way. A level of reticular activity will be defined as high when it leads to an organization of encephalic activities which mediates an alert behavior. The level of reticular activity is assumed to be higher during wakefulness than during sleep simply because the level of activation of the waking brain is higher. A major problem of sleep physiology is therefore to analyse the neural mechanisms underlying reticular deactivation (see Bremer, 1954; Dell et al., 1961; Moruzzi, 1969).

Several stimulation experiments suggest that deactivation is an active process, probably related to inhibition of the ascending reticular barrage. Structures functionally antagonistic to the activating reticular system have been discovered in the lower brain stem (see p. 31, 100) and in the cerebrum itself (see p. 107). At least two of these regions are tonically active, as shown by the fact their inactivation produces striking hyposomnia (l.c.): they are localized, respectively, in the lower brain stem (l.c.) and in the forebrain area (l.c.). Finally, in one instance electrophysiological evidence that the hypnogenic structures inhibit the reticular system has been provided by BREMER (1970a, b).

The interrelations between activating and deactivating structures, and the functional significance of the sleep inducing mechanisms of the lower brain stem and of the cerebrum have been discussed at length in the chapter devoted to lesion experiments. The fact that the ability to maintain a state of wake-fulness may reappear in the chronic *cerveau isolé* (see p. 21) shows that struc-tures endowed with an activating influence are present in the cerebrum. They are mainly, though probably not exclusively, localized in the posterior hypothalamus. It is likely that these activating structures coincide with, or are at least strongly related to, the hypothalamic center which is responsible for the outbursts of sham rage after decortication and for the defense-aggression behavior of the normal animal. Both the mild, tonic activation required for the maintenance of wakefulness and the strong, phasic discharges responsible for rage outbursts require, in fact, the support of the ascending reticular system (see p. 62). The sudden withdrawal of the activating reticular influence would determine a striking imbalance, which might be responsible for the coma of the *acute cerveau isolé* (see p. 24).

3. The Alternation between Sleep and Wakefulness

The alternation between sleep and wakefulness implies a reciprocal organiza-tion of the activating and deactivating structures. The sleep-waking cycle may be abolished (coma) or greatly disturbed (lethargy, hyposomnia) following any serious imbalance in the reciprocal interrelations between the two antago-nistic mechanisms.

Cross connections must be present within the brain stem between the activating reticular system and the deactivating structures localized in the medulla (see p. 31) and in the raphe nuclei (see p. 62), as shown by the alter-nation of postural activities occurring in the chronically decerebrate animal. The postural and motor changes observed in this preparation represent in fact the somatic component of a much more complex and fully integrated behavior occurring during sleep and wakefulness in the normal animal. Since anatomical (SCHEIBEL and SCHEIBEL, 1958; see BRODAL, 1969, p. 314) and physiological (MAGNI and WILLIS, 1963) evidence shows that several reticular neurons project both rostrally and caudally, it is not surprising that a cycle may still be detected with regard to the descending influences of the brain stem, when the ascending flow of impulses is interrupted by the decerebrating section.

Cross connections must also be present between the activating and de-activating structures of the cerebrum, as shown by the occurrence of a sleep-waking cycle in the chronic *cerveau isolé*, after precollicular section. The localization of these antagonistic structures has been discussed at length in the sections of this review devoted to lesion and stimulation experiments.

The existence of a cycle is likely to be due to the slow accumulation and dissipation of chemical products within well defined groups of neurons, an important subject of investigation to be discussed by Jouvet in these reviews. Rhythmicity is anyway *potentially* present both in the cerebrum and in the brain stem, as shown by the experiments on the chronic *cerveau isolé* and of chronic decerebration. It is more difficult to state where rhythmicity arises in the normal animal, although several considerations suggest the hypothalamus as a likely candidate.

4. The Alternation between Slow Wave and Paradoxical Stages of Sleep

The alternation between the two stages of sleep is an example of a cycle occurring within a single phase of another cycle. With the only exception of the pathological cases of narcolepsy (Dement and Rechtschaffen, 1968; Hishikawa, 1968; Passouant, 1968), behavioral sleep always starts with the synchronized or slow wave period. The paradoxical episodes occur on a background of slow wave sleep, at least in physiological conditions.

The paradoxical phase starts and ends suddenly. An accumulation followed by dissipation of chemical substances within a pontine "center" is likely to be the cause of the onset and of the end of each paradoxical episode. However, this biochemical mechanism cannot account alone for the paradoxical phase, since the triggering pontine discharges do not occur during the waking state; they must be introduced by a period of EEG synchronization. The physiological importance of the paradoxical episodes is shown by experiments of selective sleep deprivation (see Jouvet, 1967). They provide convincing evidence that there is a specific type of recovery which may occur only during the paradoxical episodes. Their selective suppression, in fact, is followed by a specific sleep debt, which can be paid exclusively during later episodes of fast wave sleep.

Ephron and Carrington (1966) have discussed the functional significance of the paradoxical phase of sleep, which they regard as a process of "endogenous afferentation". According to their homeostatic theory, the neocortex would call upon centers in the pons to supply it with necessary stimulation during sleep. There is, however, a major objection against the theory of the neocortical homeostasis. After a prepontine section of the brain stem one observes manifestations, called "cataplexic episodes", which clearly recall the paradoxical periods of the normal animal. They are characterized by a collapse of extensor rigidity (see p. 38). The average duration of each periods is about the same (6 minutes), while the circadian percentage (10%) is only slightly lower than in the normal animal, where values of the order of 15 % are found (see JOUVET, 1967a, p. 147). The main difference between a normal and a decerebrate animal is that, *after decerebration, the outbursts of pontine activity occur throughout the nychthemeron* (JOUVET, 1965b, p. 414).

Any conclusion must start from the hypothesis (JOUVET, 1962, 1967a) that the cataplexic episodes of the chronic prepontine cat are due to the same neurochemical mechanisms which are responsible for the paradoxical episodes of the normal cat. This assumption, which is supported by several convergent lines of experimental evidence, leads to the conclusion that the alternation within the pontine triggering center of periods of silence and of the outbursts of activity responsible for the paradoxical episodes of the normal animal is due to accumulation followed by dissipation of an endogenous product. This does not mean, obviously, that the functional significance of paradoxical sleep is to dissipate the endogenous products of the pontine center, any more than the physiological significance of drinking water is to reduce the stimulation of the hypothalamic osmoreceptors. The pontine neurochemical cycle still occurs after decerebration, therefore without any control from the brain. Independence of any feedback control also characterizes several biological clocks (see RICHTER, 1967) and the activity of centers for non-homeostatic instinctive behavior (see MORUZZI, 1969, p. 190–194). This point is of some theoretical interest and will be discussed again later.

The pontine discharges are probably inhibited by the waking brain, and this would explain why the paradoxical episodes do not occur on a background of wakefulness, at least in normal conditions. Hence, during the waking state there is a gradual accumulation of endogenous products, but their dissipation is prevented by the waking brain. This hypothesis may explain while the paradoxical episodes do not occur throughout the nychthemeron, as in the decerebrate animal, but are concentrated in the period of behavioral sleep.

The implications of this hypothesis may be stated as follows:

i) The episodes of paradoxical sleep are not controlled *directly* by the structures which benefit from sleep recovery.

ii) These episodes may be controlled only *indirectly* by the cortex via the cortico-reticular barrage and its influence upon the activating reticular system. However this control should be mainly concerned with the nychthemeral distribution of the paradoxical episodes, rather than with their overall duration.

iii) The activating reticular system inhibits the pontine center whose discharge triggers the paradoxical episodes.

The evidence concerning slow wave sleep is less clear. A direct control by the structures which benefit from the slow wave recovery may occur through the roundabout way of the cortico-reticular discharges and of the ascending reticular system.

5. The Functional Significance of Sleep

Almost all functions of the brain and of the body are influenced by the alternation of sleep and wakefulness. Von Economo's old distinction (1929) between sleep of the brain and sleep of the body, although justified from a phenomenological point of view, requires qualification since several bodily manifestations of sleep are really only epiphenomena.

It is easy to show (Moruzzi, 1965, 1966) that the aim of sleep is not to give a period of rest to skeletal muscles or to visceral organs, nor to permit the recovery of the spinal cord and of the autonomic nervous system. There is a general agreement that sleep reinstates or restores a condition of the brain that was present at the beginning of the previous period of wakefulness. Sleep has been defined by Sherrington (1946, p. 252) as a period of restorative activity "in what has suffered wear and tear". The main problem is to know what functions of the brain require such a prolonged period of recovery.

The symptoms of sleep deprivation (see Kleitman, 1963, pp. 215–229, and Luce, 1965, pp. 18–27) are mainly in the psychical sphere. "They suggest a fatigue of the higher levels of the cerebral cortex—the levels that are responsible for the critical analysis of incoming impulses and the elaboration of adequate responses in the light of one's previous experience" (Kleitman, 1963, p. 229).

Although the higher functions of the brain are those which apparently benefit most from the restorative function, firing goes on during sleep in all the cortical nerve cells which have been sampled so far (see Evarts, 1967). However, their patterns of discharge are clearly altered with respect to the waking state (l.c.). It would seem, therefore, that recovery of the cerebral cortex through sleep does not require an arrest of specific neuronal activities. The only possible alternative to this conclusion is that the neurons which benefit from sleep recovery have so far escaped microelectrode recording.

The nature of the recovery processes that occur during sleep has been the subject of much speculation. The hard core of the problem may be approached only if we start by making a conceptually clear distinction between the neurons whose activity is modified as a *consequence of the state of sleep* and those whose recovery is the major *task of sleep*. Sleep strongly influences, e.g., respiration, heart rate, postural tonus, but we all know that neither the recovery of the respiratory center, nor that of the vagal cardioinhibitory neurons or of the nerve cells of the cerebellum constitutes the main task of sleep. Comparative physiology shows that most of the hypnic symptoms are in fact epiphenomena (see Moruzzi, 1965, 1966): their importance is magnified by our unconscious anthropocentric attitude. Muscular atonia and eye closure are absent e.g. in oxen, and righting reflexes are clearly present in sleeping birds (l.c.).

A distinction has been made (MORUZZI, 1966) between fast and slow processes of neural recovery.

The fast recovery processes are related with conduction and synaptic transmission of nerve impulses; that is, essentially, with changes of membrane
permeability. The recovery is completed in a few milliseconds, i.e. in the interval
between two volleys of impulses or even in the time elapsing between two
action potentials. It is likely that these fast recovery processes are perfectly
adequate for the nerve cells and the synapses of the respiratory center or
of the cerebellum. A prolonged disorganization of these structures would be
either uncompatible with life or would lead to phenomena, such as extensor
rigidity or opisthotonus, which in fact are never observed during sleep.

The *slow recovery processes* take a much larger share of the life of several
mammals, almost one third of the entire human life. Several functions are
of course influenced by sleep, but only those neural activities which are in
some way related to conscious behavior appear to be deeply disorganized
during sleep.

By definition sleep is characterized in all animals by an impairment of
consciousness and/or by reduced responsiveness to the environment, while all
the other outward manifestations vary depending upon the species. The hypothesis has been made (l.c.) that *sleep concerns primarily not the whole cerebrum,
nor even the entire neocortex, but only those neurons or synapses, and possibly
glia cells, which during wakefulness are responsible for, or related to, the brain
functions concerned with conscious behavior.* Plastic changes are likely to occur,
during wakefulness, in nerve cells or synapses which are actively concerned
with learning processes, while conduction and synaptic transmission of impulses are the main occupations of neurons concerned with routine neurophysiological tasks.

Thus sleep recovery would be responsible for the stability of the physico-
chemical properties of those brain structures which are affected by, or contribute to, the plastic changes occurring during the waking state. The functional significance of sleep would be, therefore, different from that of the
neural processes concerned with classical homeostasis, i.e., with the fixity
of the *milieu intérieur*. Sleep would permit a kind of local homeostasis of the
brain structures related to higher nervous functions.

It has been claimed that the internal environment may be modified either i) by
sleep deprivation (LEGENDRE and PIÉRON, 1913; PAPPENHEIMER et ai., 1967) or ii) by
sleep obtained with electrical stimulation of the thalamus (KORNMÜLLER et al., 1961;
MONNIER and FALLERT, 1967a, b). It is puzzling that sleep-promoting substances may
appear in the blood or in the cerebral spinal fluid both as a consequence of prolonged
wakefulness and of induced sleep. Moreover, the physiological significance of these
experiments appears still uncertain, since i) no constant relationship has been found
between the sleep-waking cycles of Siamese twins (ALEKSEYEVA, 1958) and ii) the signs
of sleep and wakefulness occur independently in the cat below and above a complete
transection of the brain stem (HOBSON, 1965; VILLABLANCA, 1966a).

Whatever the final outcome of this line of experimental endeavour, it should be stated that a physiologically meaningful result would be obtained only if it might be shown that a sleep inducing factor accumulates in the blood or in the cerebral spinal fluid during a normal period of wakefulness and *disappears* during natural sleep. Were such a result obtained it would have important theoretical implications. Sleep might then be regarded as an example of classical homeostatic regulation controlled by a humoral feed-back. We have so far no proof that messages arising in brain structures which are in need of sleep reach the activating or deactivating structures of the brain stem via humoral channels, at least in normal conditions. However when the animal is prevented from sleeping for a long period of time endogenous products normally endowed with pure local action might well escape into the blood or the cerebro-spinal fluid.

The hypothesis that sleep is related to the cognitive processes occurring during the waking state appears to be supported by several lines of investigation, which have been recently reviewed by Feinberg and Evarts (1969). These authors have rightly pointed out that the idea of a positive function of sleep "implies the notion of complementarity, i.e., that the amount and intensity of sleep should be related to the intensity of those waking processes which it serves to complement" (l.c., p. 338). Among the positive functions of sleep, they quote not only the consolidation of the engrams but also the decay of certain categories of memories. They recall that the last view had been proposed by Hughlings Jackson and appears to be in agreement with the well known importance of selective forgetting of non meaningful material. Clearly sleep involves processes which are by far more complex than those involved in reinstating or restoring a condition of the brain that was present at the beginning of the previous period of wakefulness. In fact such a condition could not be restored without abolishing the effects of learning. A true steady state of the brain is impossible.

The term *local homeostasis* may still be used, nevertheless, both because to prevent the accumulation of the effects of the "wear and tear" remains probably an important aspect of the functional significance of sleep; and also in view of the fact that it would be difficult to find another terminology. It should be pointed out, however, that terms like recovery or steady state have not the same meaning in the physiology of the cerebrum and of the skeletal muscles.

6. Sleep and Homeostasis

The coordinated physiological processes which maintain a steady state in the organism have been designated by Cannon (1932) as *homeostatic mechanisms*, and *homeostasis* is the stability or the fixity obtained in this way. In so far as sleep prevents the accumulation of the effects of the "wear and tear" (Sherrington, 1946, p. 252) occurring during wakefulness, it certainly may contribute to the maintenance of a steady state in the neurons or in their complex interrelations. This process has been defined as *local homeostasis*, since we have assumed that the fixity of the internal environment is not the aim of sleep or at least not its major task.

If it is conceded that the sleep-waking cycle is not concerned, at least in physiological conditions, with the fixity of the internal environment, the conclusion may be drawn that the regulation of local processes of recovery should be basically different from that involved in classical homeostasis. Sleep is not likely to be regulated, therefore, by humoral messages coming from the structures of the brain which are in need of recovery.

A good feed-back regulation might nonetheless be obtained through purely neural channels. This kind of homeostatic regulation has been frequently postulated by reticular physiologists, as shown by DELL's (1963) and BON-VALLET's (1966) hypotheses on cortical and lower brain stem control of reticular homeostasis. The general principle is of course the same as that involved in the neural regulation of motoneural discharge by the RENSHAW's circuit (see GRANIT, 1963). A feed-back control of the activating reticular system actually appears as a likely hypothesis, since "masses of neurons sensitive to humoral factors and subjected to bombardment by many afferent systems" (DELL, 1963, p. 97) might react in a disorganized manner. DELL's and BONVALLET's hypotheses give a satisfactory explanation of the *fast acting* control of the *phasic* aspects of the ascending reticular discharge.

This statement, however, does not imply necessarily that the structures which are in need of slow processes of recovery exert a direct neural control over the *tonic* mechanisms involved in the onset and in the maintenance of sleep. Cortico-reticulo-cortical loops may in fact regulate the tonic activity of the ascending reticular system; thus controlling, indirectly, the onset and the duration of sleep (see p. 44). Mechanisms of this kind may contribute to explain the striking changes in the amount and types of sleep which appear to be in some way related to the level of cognitive processes (see FEINBERG and EVARTS, 1969).

However, a precise control through neural feed-backs is obviously impossible. Regulation of sleep is a centralized process, while neurochemical needs are probably different in different parts of the cerebrum. The need of slow recovery is in fact likely to depend both on the intensity and on the type of activity during the previous periods of wakefulness. But recovery cannot occur at random, only when needed; it must be concentrated in a given period of time, called sleep. This, incidentally, is likely to be the functional meaning of a centralized control of sleep by the brain stem and the diencephalon. Its implication is that any refined feed-back control of sleep is virtually impossible. In the majority of cases the animal inevitably oversleeps with respect to the neurochemical needs of at least some districts of its brain. The alternative solution would be worse, since without a centralized control of sleep and wakefulness we should be never really awake or completely asleep.

Osmotic pressure, blood pressure, O_2 and CO_2 tensions are regulated by homeostatic mechanisms provided with extremely precise feedback regulators. This is a strict physiological requirement, since osmotic pressure, blood pressure, O_2 and CO_2 tensions oscillate only within very narrow ranges. There is no adverse effect, however, in oversleeping and mild sleep deprivation may be well tolerated.

In summary, the steady state of the cerebrum is the outcome of processes of slow recovery occurring only during sleep. This *local homeostasis*, however, is due to mechanisms which are comparatively free of feedback control; and therefore are by far less precise than those involved in the classical homeostasis.

7. The Biological Clocks

RICHTER (1967) has demonstrated the existence of biological clocks entirely free of all feedback. They are timing devices which regulate important functions, frequently instinctive in nature, such as feeding and drinking activities. In some animals, such as the rat, sleep also appears to be entirely under their influence.

RICHTER has made an important study on the 24 hour clock of the rat. He has shown that the clock is inborn, that it is not dependent on exposure of an animal to alternating 12 hour periods of light and darkness, and that it manifests itself by alternating 12 hour periods of activity and inactivity. During the periods of activity the animal performs the instinctive behaviors related to food or water intake, hence the clock must regulate directly or indirectly the feeding and drinking centers. Starvation and high food intake, which have such a strong influence on the feeding centers—as well as severe dehydratation, which markedly affects the drinking center—have no effect on the clock. Apparently there is no humoral or neural feedback acting upon it. As RICHTER has pointed out (l.c., p. 23) there is a strict one-way relationship between the clock and these centers of instinctive behavior. Thus typical nutritional instincts are regulated, at least indirectly, by the clock, but the clock is not controlled by humoral or neural feed-backs related with food or water intake.

RICHTER (l.c., p. 24) has also shown that in the rat a one-way relationship exists between the clock and the sleep regulating center. He has pointed out that "rats spend much of the inactive period in sleep, not continuous sleep, but sleep interrupted at irregular intervals every 2 to 5 hours by a few minutes spent in eating, drinking, and grooming, and that rats wake up 10 to 30 minutes or more before the start of the active period, and spend most of this time in grooming, some in eating and drinking and resting". The clock is not affected by prolonged sleep deprivation, which influences instead, so strikingly, the centers of sleep and wakefulness. Thus the clock is not controlled, via humoral or neural feed-backs, by the structures which are in need of sleep recovery.

The rat of course provides a particularly striking example of a circadian clock, but there is little doubt that biological timing devices are present in other mammals (see ASCHOFF, 1965).

BERLUCCHI (1970, p. 150) has put forward the hypothesis that the cerebral activities underlying the alternation between sleep and wakefulness depend on an endogenous mechanism, related to the activity of pacemaker neurons which constitute the biological clock. According to BERLUCCHI, however, the same neurons might also be influenced by neural or humoral feed-backs. His conception is therefore different from that of RICHTER (1967), according to whom the biological clock and sleep center are separated structures, with a one-way relationship and without any feed-back control upon the timing device. Rat behavior is well explained by RICHTER's theory, while what we know concerning cat physiology would fit better BERLUCCHI's hypothesis. The accumulation and the dissipation of chemical substances within centers of the brain stem and of the cerebrum, a theme fully developed by JOUVET in these reviews, might provide an example of the endogenous mechanism postulated by BERLUCCHI. Other interesting data come from the study of the rhythmic activity of single nerve cells. The reader is referred to BERLUCCHI's article for a review of the literature.

In our opinion the hypothesis that internal clocks, with their endogenous rhythms, control the sleep-waking cycle and possibly the alternation between the two stages of sleep deserves great attention. The differences between species are of course likely to be important, some clocks being completely independent of any other control, while others are submitted to several kinds of neural or hormonal influences. In either case, accumulation or depletion of local chemical factors might tilt the balance in the reciprocal control of antagonistically organized structures related to sleep and wakefulness.

8. Sleep and Instinctive Behavior

The regulation of the sleep-waking cycle is due to an inborn neural mechanism, which may be modified either by unconditioned impulses or by conditioning. This is of course a general description, which would be acceptable for several neuro-hormonal and behavioral activities concerned with the maintenance of a steady state in the animal organism. CANNON's homeostasis may in fact be attained i) through neural and/or hormonal regulations and ii) via behavioral mechanisms, frequently purely instinctive in character. Thus instinctive behavior is usually associated with classical physiological regulations in the maintenance of a fixed internal environment.

The regulation of the slow processes of recovery occurring during sleep should, in principle, follow this rule. We have seen, however, that there are two important differences between sleep regulation and classical homeostasis. Mechanisms related to a fixed internal environment—be they concerned with the constancy of the water and salt contents or of O_2 and CO_2 tensions—are endowed with extremely precise feed-back controls. Thus water and salt contents of the blood, $[O_2]$, $[CO_2]$ oscillate within very narrow ranges. We have seen, on the other hand, that at least when there is no prolonged sleep deprivation, the sleep-waking cycle is not concerned with the stability of the internal environment, but rather with keeping constant the local physico-chemical properties of neurons, synapses and, possibly, glia cells, all of which are involved in the conscious behavior occurring during wakefulness (see Moruzzi, 1966). We have called the outcome of this process *local homeostasis*. Moreover, it is likely that a feedback regulation from the structures which benefit of the sleep recovery is absent, or at least that any regulation is an indirect one. Thus the control of sleep and wakefulness is bound to be by far less precise than in any classical homeostasis. A consequence of this lack of a direct feed-back control may be either the occasional accumulation of a sleep debt or the tendency of the animal to oversleep with respect to the biochemical needs of its brain. The second alternative is likely to occur more frequently in the domestic cat.

It turns out that *the absence of a precise feed-back regulation and the tendency to overprotect the animal with respect to its needs characterizes also several instinctive behaviors*. This fact appears particularly evident for the instincts which are important for the species rather than for the individual. Sexual and maternal behavior in the vertebrates and several kinds of invertebrate behavior appear related to non-homeostatic drives and may be classified as examples of non-homeostatic behavior. To quote:

"The bee constructs a hive and fills it with honey; the spider weaves an elaborate web; the beaver fells trees and builds a dam. All three activities serve the nutritional needs of the animal in question. But the bee is presumably not hungry all the time that it is storing honey (and, if it is, this particular activity does not satisfy the present hunger). Furthermore many activities do not even indirectly serve the present or future needs of the individual in the ordinary sense of the term. At the proper season the bird builds a nest, mates, lays its eggs, broods, for weeks brings food to the young" (Nissen, 1951, p. 356).

We have seen, moreover, when dealing with Richter's experiments (1967) on the rat, that the biological clocks may also be free of any feedback control.

It would not be advisable, of course, to rely too much on what might be, after all, sheer analogies. It must be conceded, however, that the lack of a feedback control has important consequences, which deserve consideration. The ethological and physiological doctrines on instinctive behavior have been reviewed at length in another article (see Moruzzi, 1969). What matters for a classical homeostatic process is the end effect, e.g. the ingestion of food or the water intake, not the performance of the neural activities underlying the act of eating or drinking. At least for those types of instinctive behavior where a feed-back control is lacking Lorenz's views (1937) may permit useful considerations. He has postulated that some endogeneous process leads to a gradual increase of instability of a specific system, an effect that disappears once the consummatory discharge is over, giving rise to a feeling of "satiation". We should not forget, however, that the "satiation" produced by the consummatory discharge is not exclusively due to dissipation of accumulated endogenous products, as shown by Hinde's important observation (1966, pp. 176 and 234) on the self-inhibitory effect of the "consummatory stimuli" which the animal encounters as a result of its behavior (see Moruzzi, 1969, p. 180). There are undoubtedly several types of instinctive behavior and many possibilities of "satiation". However an attempt to adapt the general theories on instincts to the sleep-waking cycle might be rewarding. This will be our aim in the next pages.

Returning now to the sleep recovery, a distinction should be made between *regulating centers*, which are localized in the brain stem and in the diencephalon, and the *regulated structures*, which during wakefulness are responsible for conscious behavior and which appear to be temporarily disorganized during sleep. The stability of the regulated structures requires a sleep pause, during which consciousness is lost or greatly impaired. There is a centralized control of sleep and wakefulness, an obvious consequence of the fact that a rhythm involving the whole of the animal body, and implying the temporary loss of consciousness, has a great survival value and must occur in an orderly manner.

When we try to work out in detail these general concepts we are confronted with at least three hypotheses. We may i) regard the activities underlying the sleep-waking cycle as a classical example of neural regulation, either reflexive or autochthonous in nature, possibly controlled by a biological clock; or we may ii) concentrate our attention on the behavioral aspects of sleep, and study the sleep-waking cycle as one of the several aspects of instinctive life; finally we may iii) make the hypothesis that a reflexive or autochthonous regulation is associated with a suitable instinctive behavior, as we know to happen for several classical examples of homeostasis.

The control of the constancy of the water and salt contents of the blood, of the glycemia, or of the body temperature is in fact obtained by the combined efforts of neurohormonal mechanisms and of behavioral regulations.

The constancy of the osmotic pressure of blood is controlled i) by a classical mechanism of neuro-hormonal homeostasis, which regulates the water losses through the kidneys, and ii) by a typical instinctive behavior, which is concerned only with water intake. The two processes are closely interrelated, and probably regulated, in fact, by the same osmoreceptors (ANDERSSON, 1966; p. 198). However each of them has its own sphere of action and may be studied independently. The question may be asked whether the same association between physiological and behavioral processes may be found in the slow recovery processes which are usually regarded as the *raison d'être* of sleep.

Ethological observations, reported in another review article (MORUZZI, 1969, pp. 181–183) leave little doubt that sleep is preceded and accompanied by *typical manifestations of instinctive behavior*, occasionally of amazing complexity. This conclusion of course does not imply that sleep itself is purely and exclusively instinctive in nature. Here again purely physiological or neurochemical regulation may be simply associated with instinctive behavior. A preliminary question is to see whether a distinction between purely reflexive regulation and instinctive behavior is justified.

The differences between instincts and purely reflexive regulations have been discussed at length in another article (MORUZZI, 1969, p. 176–181). A conceptual separation is not justified by the different degree of complexity, as was frequently stated in the past. Also purely neuro-hormonal regulations, such as those involved in the control of the body temperature, may be phenomena of tremendous complexity, involving practically the whole animal.

We must start from the fundamental ethological distinction between the two major phases of instinctive behavior (CRAIG, 1918). There is an *appetitive behavior* and an end action, the *consummatory act*, after which the animal appears "satisfied" or "satiated".

An analysis of the general properties of instinctive behavior (MORUZZI, 1969, p. 180) leads to the conclusion that there are, in fact, some patterns of activity which are common to several manifestations of avian and mammalian instincts, and appear to be simple enough to be translated into neurophysiological terms:

1. Large populations of motoneurons are controlled, more or less indirectly, by smaller groups of supraspinal nerve cells which constitute the "center" of instinctive behavior.

2. Enough evidence is at hand (l.c., see p. 178) to postulate that under specific circumstances the excitability of this "center" is enhanced.

3. During the appetitive phase the animal is striving for a situation which permits the accomplishment of a given performance. The animals behavior appears to be dominated by this attempt. The consummatory action is characterized by the occurrence of integrated discharges of several motoneurons following well defined and on the whole stereotyped spatio-temporal patterns.

While the phenomena described under 1) and 2) are not exclusive to instinctive behavior—they may also characterize purely reflexive events such as the response of the respiratory center to anoxia—*the compulsory striving of the animal as a whole—as an individual not as "a mere collection of organs"*[8]*— for a functionally integrated pattern of behavior seems to characterize instinctive life.*

The *period immediately preceding sleep*, and even sleep itself, is characterized by a behavior which is typical of the particular species, and is remarkably constant for any animal. TINBERGEN (1955) has pointed out that both natural and HESS' induced sleep are heralded by a stage which is typical of an appetitive phase: the cat looks around, searches for a corner to go to sleep and finally curls up in the natural sleep position. Obviously *the animal strives for a situation which will permit, or facilitate, the onset of sleep.* If the period immediately preceding sleep—which is characterized in man by a feeling of drowsiness—may be regarded as an appetitive phase, sleep behavior itself should be considered as the corresponding consummatory act, a concept developed by TINBERGEN himself in another paper (1952). Similar concepts have been developed by PLOOG (1953) for human sleep.

That the behavior immediately preceding sleep is typically instinctive in nature is beautifully proved by ethological observations on the tremendous importance to find a "home", or a place for sleep. The "sleep societies"

---

8 See SHERRINGTON (1906a, p. 2).

are an impressive example of instinctive behavior related to sleep (see HOLZ-APFEL, 1940; HEDIGER, 1959, and HASSENBERG, 1965 for ref.). But perhaps the most convincing proof is provided by the observations, quoted by HOLZ-APFEL (1940) and HEDIGER (1959, 1968, 1969), on the behavior of mammals in captivity, when prevented from reaching their usual place of sleep. The appetitive behavior released in this experimental situation is characterized by a drive which is among the strongest in the instinctive life.

In summary, there is little doubt that the sleep-waking cycle involves strong components of typical instinctive behavior. Evidence in support of this statement also comes from other lines of investigation. HESS' experiments on the cat have shown that several types of instinctive behavior occurring during wakefulness are reproduced, with clear signs of the corresponding motivation, by electrical stimulation of the hypothalamus; while sleep with clearly instinctive components may be obtained by stimulating the midline nuclei of the thalamus with the same procedure (see p. 90). In RICHTER's experiments on rats a non-homeostatic circadian clock releases the instinctive behaviors of feeding and drinking during the 12-hour active period, while sleep behavior dominates the 12-hour inactive period. These facts may of course be no more than coincidental similarities between classical instinctive behavior and sleep behavior, but they provide at least suggestive evidence in support of the conclusion reached with ethological studies on the period immediately preceding sleep. The main proof is provided by the observation that whenever sleep appears, either spontaneously or under the influence of an electrical stimulation, it is always heralded by an appetitive phase clearly instinctive in character.

The appetitive phase, or at least the end of it, is characterized in man by a subjective experience called drowsiness. This state is probably due to a fall in the level of reticular activation, which in a proper environment will soon reach the critical point required for the onset of sleep (see MORUZZI, 1969, p. 201).

The distinction between the EEG patterns of drowsiness and those of slow wave and spindle burst sleep is not always easy in the cat (see STERMAN et al., 1965; URSIN, 1968), but there is little doubt that both generalized and localized EEG synchronization may be observed during the waking state, e.g. during eating and drinking behavior (CLEMENTE et al., 1964; STERMAN et al., 1970). However, generalized EEG synchronization combined with waking behavior may be regarded as a sign of reticular deactivation, one that remains below the critical level required for the onset of sleep. In a laboratory cat placed in the usual observation box, provided with sound attenuation properties, this pre-sleep phase is very short, possibly because the reticular deactivation is unhindered and quickly reaches the critical level required for the onset of sleep.

The deactivation process is likely to be much slower in the cat placed in its natural environment. KONORSKI (1967, p. 301) has pointed out that "a sleepy animal looks actively for some place to fall asleep, just as the hungry animal looks for food, and certainly he is very alert to all those stimuli which suit this aim". Hence reticular deactivation would be absent, at least during the initial phase of the appetitive period. Following the same line of thought, BERLUCCHI (1970, p. 170) emphasizes that a good level of integration with the environment—arousal by definition—is necessary for the performance of these pre-sleep activities, whose survival value is very likely that of securing to the animal a proper shelter before it falls asleep. He suggests that the primary effect of midline thalamic stimulation in Hess' experiments would be the production of the species-specific behavior which usually precedes sleep. The hypnogenic effect would simply be the consequence of the onset of a sequence of motor activities which normally anticipate sleep.

The aspects of the appetitive phase of sleep paradoxically implying a state of arousal are of course emphasized in the birds flying towards their sleep-tree (Hediger, 1959), in the animals in captivity which are prevented from reaching their home (l.c.), or in the orang-utan which builds a sleeping platform every day (Portielje, 1939). They are reduced to a minimum, and may even be absent, in the domestic cat or in man in his usual environment. Particularly with animals living in the wild, reticular de-activation is likely to begin only when the animal reaches its shelter, as the last episode of a prolonged appetitive phase.

A separate discussion is required when sleep appears as a displacement activity. The literature has been reviewed by Tinbergen (1952), Hinde (1966) and Delius (1967, 1970). Suffice it to recall here that acts called "irrelevant", because they are entirely out of context with the behavior immediately preceding them, occur in conditions characterized by a high level of arousal, when animals are placed in conflict situations. Hinde has pointed out (1966, p. 429) that displacement activities are causally hetero-genous and has discussed three possible explanations (l.c., p. 279). According to the disinhibition hypothesis "when mutual incompatibility prevents the appearence of those types of behavior which would otherwise have the highest priorities, patterns which would otherwise have been suppressed are permitted to appear" (l.c., p. 279).

Paradoxically, a typical behavior of drowsiness may appear in this conflict situation and sleep may be the final outcome of a state of intense arousal produced by conflict, thwarting, frustation etc. Delius (1970) has reproduced these behavioral syndromes by electrical stimulation of the forebrain and brainstem in gulls. His hypothesis of "arousal homeostasis" implies the activation of an arousal-inhibiting, and ultimately sleep-inducing mechanism, which would counteract the tendency toward excessive arousal. This type of displacement activity would be "the epiphenomenal consequence of this feed-back activation of sleep" (l.c., p. 179).

The conclusion that the phase preceding sleep corresponds to the appetitive phase of an instinctive behavior may lead to a broader view of the problem. Sleep itself might be regarded as the consummatory phase.

An attempt will now be made to translate into neurophysiological terms the behavioral descriptions given by the ethologists.

A behavior or a posture implies a well defined distribution of the activity of several skeletal muscles, and therefore of the corresponding motor units, in both space and time. Thus *any instinctive behavior must be due to a more or less stereotyped spatio-temporal pattern of motoneuronal discharge*. If we now define a given behavior as "instinctive", when it is constantly associated with the animal's striving toward a situation permitting the specific con-summatory discharge, we must recognize that behind this behavior there is a corresponding spatio-temporal pattern of motoneuronal discharge. We might perhaps call this pattern of discharge "neural" instinctive activity. The specific "center" for each type of instinctive behavior appears therefore to be mainly concerned with producing a proper coordination, in both space and time, of the activity of the final common paths which are involved in a given type of appetitive phase or of consummatory action.

We have so far considered only the final common paths belonging to the somatic nervous system. However, the sleep changes of the visceral functions also presuppose stereotyped spatio-temporal patterns of discharge in the final common paths of the autonomic nervous system and in the neurosecretory neurons. There is, therefore, no reason to exclude autonomic and, possibly, neuro-secretory final common paths from the list of the nerve cells which at a given moment may join the motoneurons in producing complex patterns of activity clearly related to instinctive behavior. In fact the typical body postures of a cat during sleep, or the typical movements occurring during the preceding appetitive phase, are the integrated expression of motoneuronal discharges which, from the physiologist's standpoint, can hardly be separated from the increase of the parasympathetic discharge which is responsible for the slowing of heart rate or for the sleep myosis.

It is now possible to make a third, and final step forward in this attempt to find a unitary explanation of the hypnic phenomena. Sleep behavior is concerned not only with final common paths but also, and probably still more, with the control of large populations of *interneurons*. In fact the discharge of most of the cerebral neurons is strongly modified during sleep. This is an introverted type of behavior, and one should not forget that the word "behavior" is utilized, in a somewhat unorthodox way, for describing patterns of neuronal discharges rather than of movements: behavior of nerve cells, rather than of groups of skeletal muscle fibers. It is of course usually easy to recognize the functional significance of the extroverted behavior of the waking animal, but we can do little more than describe the patterns of discharge of cortical neurons during the spindle periods or the desynchronized episodes of sleep (see EVARTS, 1962). We are still relatively far from understanding the functional significance of these discharges in terms of sleep recovery of cortical neurons (see l.c.).

In summary, any attempt to think in terms of neuronal discharges, rather than of muscle fiber contractions, leads to the conclusion that there is no reason, from the physiologist's standpoint, to separate the stereotyped patterns of motoneuronal discharges which produce postures or movements, certainly instinctive in type, from the stereotyped patterns of discharge occurring simultaneously in other neurons, including those of the cerebral cortex. The latter discharges, however, are more likely to be related to the specific task of sleep, the recovery of the structures underlying the higher functions of the brain.

We are thus led to discuss the main issue in any attempt at an interpretation of sleep physiology. We are confronted with two hypotheses.

9*

i) There might be a *causal relationship* between the neural activities underlying the instinctive manifestations of sleep, and the main task of sleep, the maintenance of cerebral homeostasis. The slow processes of recovery would be the final outcome of some kind of neural activity occurring during sleep. For example the typical patterns of discharge of the cerebrocortical neurons observed during the slow and paradoxical phases might represent a necessary step in the sequence of neurophysiological and neurochemical events leading to sleep recovery.

ii) We might also postulate, however, an *association* between manifestations of sleep, partially instinctive in nature, and the neurochemical and neuro-physiological processes of recovery, which by themselves would not be in-stinctive in nature. The situation would be analogous to the association be-tween the instinctive behavior leading to water intake and the purely physiological processes involved in the water absorption or in the regulation of diuresis.

Associations between behavioral and neurohormonal regulations are frequently observed in the classical examples of homeostasis. We have just seen the example of the association of instinctive behavior and of neuroendocrine mechanisms in the regula-tion of water content. DELL (1958) has stressed the importance of these associations in the control of glycemia. RICHTER (1947), finally, has shown that the regulation of sodium content is based upon the appearence of specific appetite for salt in the sodium-deficient animal, besides the well known aldosterone hypersecretion produced by the decrease of plasma sodium concentration.

The hypothesis of sleep as a kind of instinctive behavior may appear unattractive, for the following reason. It is a deeply rooted idea that sleep is a kind of inactivity, or at least of reduced activity, while what we usually call instincts belong to Hess' ergotropic sphere, and are therefore related to a display of activity. It may appear unjustified to call "instinct" a process leading to inactivity or to a decreased activity. However HOLZAPFEL (1940) has introduced the term *triebbedingte Ruhezustände* and has called attention to the fact that appetitive behavior does not necessarily lead to movements, but also to a state of rest. We know now that sleep is related to a different organi-zation of the discharges of the cortical neurons (EVARTS, 1962), not to their cessation, at least for the units that have been sampled so far. During sleep there is simply a different kind of activity, and there is therefore in principle no objection to considering this activity as related to instinctive behavior.

The second difficulty is historical in nature. There is little doubt that several con-siderations put forward by CLAPARÈDE (1905, 1912, 1928, 1929) weakened rather than strengthened his thesis of the instinctive nature of sleep (see MORUZZI, 1969, p. 181). A sharp criticism of his argumentation was made by KLEITMAN (1963, p. 349), who gave all the literature. A modern interpretation of the instinct theory must obviously take into account the ethological and neurophysiological investigations performed since the beginning of the century. A modern theory (see MORUZZI, 1969) would simply start i) from the fact that there are undoubtedly clear-cut instinctive manifestations in the animal behavior related to sleep and ii) from the hypothesis that accumulation and dissipation of endogenous products in the sleep regulating centers—rather than the fatigue and the recovery of the cerebrum itself—are the main causes of the urge to sleep and of the feeling or freshness occurring after it. This situation is very similar to that which has been postulated for the centers of instinctive behavior (see l.c., pp. 176–180). The effects of a few minutes napping undoubtedly recall the "satiation" after an instinctive consummatory discharge (see l.c., p. 181).

Although at present it would be clearly impossible to find a crucial proof in support of either hypothesis, comparative ethology provides us with serious reasons to believe that the instinctive component is of paramount importance and that the sleep-waking cycle is dominated by it, rather than by the neuro-chemical requirements of the brain.

The old actographic techniques (SZYMANSKI, 1918, 1920; NICHOLLS, 1922) had shown that, even among mammals, there were striking differences in the distribution of the periods of rest and of activity (see KLEITMAN, 1963, p. 81–91). Behavioral observations had even led to the extreme view that mammals such as the ruminants (BALCH, 1955, see BELL, 1960, for ref.) or the guinea pig (NICHOLLS, 1922) might never sleep. In fact EEG studies have shown that these differences, striking as they may appear, are purely quantitative in nature. It has been shown that periods of electrocortical synchronization may be observed in the goat (BELL, 1960), that actually both the slow and paradoxical phases of sleep occur in this animal (KLEMM, 1966) and in the sheep (JOUVET and VALATX, 1962; RUCKEBUSCH, 1963); while behavioral signs of sleep may be found in cattle (HEDIGER, 1968), in goats (KLEMM, 1966), and in the deer (BUBENIK, 1960). The periods of sleep, however, are so infrequent and short lasting that they are likely to escape the attention of the observer. The same explanation probably accounts for the fact that no photographs of a sleeping antelope in the wild state have ever been taken (HEDIGER, 1969).

The facts remains nevertheless that the quantitative differences in the organization of the mammalian sleep-waking cycle are really impressive, particularly if one considers the opposite group of animals which spend most of their time sleeping. The opossum (SNYDER ,1967), a primitive mammal, and the gorilla (SCHALLER, 1964) stay awake for only 20% of their time, and the average percentage of the nychthemeron spent in wakefulness by a well-fed domesticated cat is of the order of 28% according to STERMAN et al. (1965) and ANGELERI et al. (1969). Details may be found in a recent review on the phylogeny of sleep (KLEIN, 1963) and in a recent paper dealing specifically with the nychthemeral distribution of sleep in the cat and in the guinea pig (PELLET and BÉRAUD, 1967), taken respectively as representative of good and poor sleepers. It turned out that the cat was awake for 32% of the nychthemeron versus 72% for the guinea pig; the ratios of fast-to-slow sleep were respectively 20% and 3.9%. These four figures may well be taken to epitomize the results of the comparative approach.

The needs with respect to brain recovery are of course likely to be different in different mammals. They can hardly account, however, for quantitative differences of this order of magnitude. It is extremely unlikely that the needs of brain recovery may be about the same in the gorilla and in the opossum, and yet so strikingly lower in the adult man and in the guinea pig. Several explanations of these differences have been suggested, such as the posture required by rumination (BALCH, 1955; BELL, 1960), the long time required by nutrition in the guinea-pig (PELLET and BÉRAUD, 1967), or the fact that an animal like the antelope is always in mortal danger (HEDIGER, 1969). The fact remains, nevertheless, that the recovery requirements may be met in several animals either during periods of somnolence occurring without behavioral signs of sleep (Bell, 1960) or during extremely brief periods of true, deep sleep. It is likely, in fact, that the gorilla, the cat or the opossum oversleep with respect to their neurochemical needs. Thus we are bound to conclude that the needs of brain recovery do not represent the only, or probably even a major factor, in the regulation of the sleep-waking cycle.

The extreme variety of sleep habits, a remarkable fact even if one restricts the study to mammals, fits well the wide variety of instinctive behaviors. Were sleep nothing more than an example of a physiological regulation, it would indeed be a unique kind of regulation, since it implies the temporary loss or the severe impairment of consciousness. All the known examples of instinctive behavior are disrupted during sleep and the animal must adapt its requirements in terms of defense, food, and water supply to the sleep-waking cycle. Hence an appropriate control of sleep has a great survival value for the species, as becomes especially apparent when one considers the implications of a temporary loss of consciousness in the herbivores living in the wild state. It is not surprising, therefore, that the regulation of such an important behavior is taken over by an instinct. A non-instinctive, purely physiological regulation might of course also be present, but in any case the cycle should be dominated by an instinctive mechanism. The requirements of the cerebral homeostasis are likely to be overwhelmed by those of the instinctive behavior.

9. An Attempt at a Synthesis

Let us now start from the hypothesis that an animal's behavior during drowsiness corresponds to the appetitive phase, while sleep itself should be regarded as the consummatory action of a special instinctive behavior. This is of course just a working hypothesis, but it might be rewarding to see its neurophysiological implications.

It should be stated clearly that what we call the consummatory action is represented by a pattern of discharge along final common paths (and, possibly, of interneurons) which is more or less stereotyped for any given species. The neurochemical processes underlying sleep recovery probably *require* a change in the pattern of spontaneous activity of brain neurons, but these processes should be kept conceptually separate from the neurophysiological mechanisms underlying sleep. In fact the relationship is a loose one, and different amounts of recovery may occur during a given period of behavioral sleep. This statement is supported by the fact that the duration of sleep providing an almost complete recovery is out of proportion to the duration of the preceeding sleep deprivation. An imposing syndrome produced in one man by 200 sleepless hours disappeared almost completely after only 13 hour sleep (see Luce, 1965, p. 20). These considerations might incidentally explain why in certain animals a satisfactory brain recovery may be obtained during extremely short episodes of behavioral sleep.

The main point is that *sleep recovery requires a different organization of brain activity, a condition implying absence or impairment of consciousness.* The organization of sleep and wakefulness has therefore a great survival value, and must be adequately controlled with respect to the animal's needs. According to the ethological hypothesis this control would be instinctive in nature. It is closely related, anyway, with the classical instinctive behavior that occurs during the waking state.

The ethological hypothesis is clearly incompatible with any of the passive theories of sleep. It is of course immaterial whether the peripheral effect of a consummatory act is a movement or a cessation of movements. Bodily rest during sleep is a behavior, since during the synchronized phase there is a specific posture, with an increase of the tonic activity of some muscles (see HESS, 1944b, p. 318); while the postural atonia of the desynchronized phase is actively produced by supraspinal inhibitory volleys (see POMPEIANO, 1965, 1967b). Besides these considerations, the ethological hypothesis implies that sleep be *actively produced*, since a group of neurons must start to discharge somewhere in the central nervous system–following specific, more or less stereotyped patterns–whenever the appetitive phase reaches the critical point for the onset of the consummatory action.

Thus our problem is to find out i) whether either or both the synchronized and desynchronized phases of sleep are produced by a specific discharge arising somewhere in the central nervous system, and ii) whether these neural activities may be compared to the patterned discharge responsible for the consummatory action of an instinctive behavior occurring during wakefulness.

We may start by discussing the significance of the slow or synchronized sleep, since normal sleep behavior always starts with this phase. In another article (MORUZZI, 1969, p. 201) and earlier in this review (p. 110) we have pointed out that a sharp conceptual distinction should be made between reticular deactivation and the onset of sleep.

A source of confusion is represented by the tendency to equate a major symptom of sleep, the EEG synchronization, with sleep itself. When low rate electrical pulses are applied to the nucleus of the solitary tract in the *encéphale isolé* cat (MAGNES et al., 1961a, b) EEG synchronization starts immediately (Fig. 31a, b, c). These structures may be called synchronizing, but we have no proof that they are hypnogenic. When stimulation of critical regions of the brain stem reticular formation is carried out in the intact, free moving animal (FAVALE et al., 1961), a sudden onset of EEG synchronization, with practically no aftereffect, may be obtained on a background of drowsiness (Fig. 30a), but behavioral signs of sleep are not observed. Confirming HESS' results (1927,

1944b) on thalamic stimulation, behavioral sleep occurs in the intact cat only after long lasting stimulations, made with several slow rate pulse trains. The results obtained by STERMAN and CLEMENTE (1962a, b) with electrical stimulation of the forebrain area, in the *encéphale isolé* and in the intact unanesthetized cat respectively, have led to the same conclusion (see also MORUZZI, 1969, pp. 198–200 and this review pp. 101–103). *Clearly EEG synchronization reveals only reticular deactivation, a process that must reach a critical level in order to give origin, after an adequate period of time, to genuine manifestations of sleep.* When reticular activation is just above the threshold for the waking state, a situation often observed after spinal section at C_1 (BREMER, 1937) or in the intact but drowsy animal, stimulation of given regions of the brain stem produces a short-lasting fall of reticular activation, as revealed by the EEG synchronization; this fall of reticular tone, however, is apparently not adequate or is too short in duration to permit the onset of true sleep. The two phenomena, EEG synchronization and behavioral sleep, may thus be dissociated.

Summing up, the main idea is to make a sharp conceptual distinction between EEG synchronization and sleep characterized by EEG synchronization. There is true sleep i) when the typical behavioral manifestations are present, ii) when arousal may be obtained with sensory or cortical stimulations and, finally, iii) whenever there is the typical alternation of the synchronized and desynchronized phases. According to the ethological hypothesis these phenomena might be compared with a sequence of consummatory actions, preceded, during the drowsiness stage, by appetitive behavior. *A critical but moderate level of reticular deactivation must be slowly attained in order to permit the onset of the instinctive behavior of sleep.* If the fall of the reticular tone occurs abruptly, and/or is too strong, the behavioral result will not be physiological sleep, but coma. As with any manifestation of instinctive behavior the onset of synchronized sleep must be related to a patterned discharge of a group of neurons, physiologically interrelated but not necessarily localized in a given anatomical space. We may call this group of nerve cells the "center" of the synchronized sleep, if the term "center" is accepted in the loose way defined above. This "center", however, is unlikely to be critically represented by the synchronizing structures of the thalamus, whose random discharge resembles more the released activity of a tonic center than the rigid sequence of patterned discharges that must be responsible for a chain of consummatory actions. Sleep behavior, moreover, has been observed after complete thalamectomy (NAQUET et al., 1965) and following complete chronic decortication (GOLTZ, 1892; KLEITMAN and CAMILLE, 1932) These data by no means imply that in the normal animal the activity of the sleep inducing structures is not modified by the discharges of the cerebral cortex; actually such a modification is quite likely (see JOUVET, 1967a). These findings, however, lead to an interesting conclusion, namely that the cortical control is not of critical importance in the onset of the cycle.

To state in few words our interpretation, the EEG synchronization is merely a consequence of the fact that the onset of sleep requires a critical fall of the ascending reticular tone; a condition that leads, unavoidably, to a release of the synchronizing thalamo-cortical circuits. A widespread synchronization of brain neurons may be the cause, or one of the causes, of the loss of consciousness, a state which certainly characterizes sleep; one, however, that may be produced in other situations, such as coma or barbital anesthesia, two conditions which have very little in common with true physiological sleep.

The evidence relating the episodes of *sleep with desynchronized EEG* to the consummatory actions of instinctive behavior is more impressive, partly as a consequence of the fact that we know the center triggering the paradoxical episodes, possibly also because some behavioral manifestations are phasic in nature and are always particularly striking in character.

During sleep electrophysiological signs of localized discharges, subconvulsive in type, may be recorded from the pons (see JOUVET, 1967a) and from the lateral geniculate body (see MORUZZI, 1963 for ref.) as soon as the synchronized EEG is replaced by desynchronized patterns; these localized discharges suddenly come to an end when the paradoxical episode is over. Convincing evidence is available (see BIZZI and BROOKS, 1963) that there is a triggering structure in the pons, which apparently controls both the electrophysiological and the behavioral manifestations of the desynchronized episodes. On the behavioral side, besides the tonic phenomena of extreme extensor atonia and of postsynaptic inhibition of spinal motoneurons (see POMPEIANO, 1967b), we are confronted with an impressive eruption of phasic effects–rapid eye movements, clonic twitches, visceral manifestations (see BERLUCCHI et al., 1964b; POMPEIANO, 1967)—which suggest that both somatic and autonomic final common paths are driven, directly or indirectly, by the brain stem; possibly by the same pontine center which presents subconvulsive signs of activity.

The sudden onset and the abrupt disappearance of subconvulsive outbursts in a well defined region of the pons, their close relationship with stereotyped manifestations in both motor and autonomic spheres, in short all the electrophysiological and behavioral explosions of activity, are what one would expect were we able to record during a consummatory act the electrical activity of a center for instinctive behavior. It is true that we clearly see a functional significance in the instinctive acts which are carried out during wakefulness, while we are unable to understand the meaning of the strange peripheral manifestations of the paradoxical phase of sleep. However, LORENZ (1937) and in fact all modern ethologists (see TINBERGEN, 1955; THORPE,

1963) have pointed out the fallacy of the finalistic attitude in the study of instinctive behavior. The goal attained by the consummatory act is of course important for the animal itself or for the species. What matters, however, at least from a physiological standpoint, is i) the gradual enhancement in the excitability of a given specific "center", ii) the sudden eruption of the consummatory action, eventually represented by a chain of more or less stereotyped events, finally iii) the "satiation" that follows the consummatory action. All these requirements are met by the structures which are responsible for the triggering of the desynchronized episodes, as shown by the experiments of sleep deprivation (see DEMENT, 1960; DEMENT et al., 1967; JOUVET, 1965c, 1967b) and by recent neurochemical and neuropharmacological studies to be reported by JOUVET in these reviews. The necessity for paradoxical episodes is clear from the experiments of selective sleep deprivation, which have shown that desynchronized sleep is specifically needed, possibly because somerecovery may occur during the paradoxical phase, and during that phase only (see JOUVET, 1967b). We do not know, however, how such a result is achieved, but we know that there is a "satiation" once the paradoxical episode is over, as there is a "satiation" after a central discharge giving rise to a consummatory action.

Summing up the conclusions of this and of an earlier article (MORUZZI, 1969), a review of modern data on the neurophysiology of sleep would lead to the hypothesis that the sequence of episodes with EEG synchronization and desynchronization may be regarded as a chain of consummatory actions, introduced by an appetitive phase, the last part of which corresponds to the period of drowsiness (see p. 128). The reticular deactivation would simply permit the attainment of a critical, moderately low level of the ascending reticular barrage which may be regarded as the optimum for the onset of this particular type of instinctive behavior. Undoubtedly none among these data may be regarded as the proof of the validity of the hypothesis: what we have presented is merely suggestive evidence.

In a previous article an attempt was made to regard sleep merely as an aspect of a wider problem: the study of *the levels of activation which are required for the onset of the different types of instinctive behavior.*

The hypotheses of LINDSLEY (1951), SCHLOSBERG (1954), and HEBB (1955) on the physiological significance of the level of activation and the concept of *activation continuum* have already been discussed (MORUZZI, 1969, pp. 204–206). It has been pointed out (l.c. p. 206) that it would not be advisable to consider only the levels of reticular activation which are associated with conscious behavior. The zero level of reticular activity is not represented by sleep, an active state requiring a low but critical level of reticular activation. The zero level is found during coma. Thus *our interpretation starts from the assumption that the activation continuum should also be considered for those levels of ascending reticular activation which are too low to permit consciousness.*

There are, actually, several levels of coma, as shown by the differences, in the acute *cerveau isolé* (BREMER, 1937, 1938, 1954), between anterior and posterior midbrain transections (ARDUINI and MORUZZI, 1953); as well as by the persistence (BIZZI and SPENCER, 1962) or by the recovery (BATSEL, 1964; VILLABLANCA, 1966) of tonic activating influences, after acute and chronic midbrain transections respectively. These differences have been dealt with at length in this review (see pp. 15–17). For the sake of simplicity they will be ignored at this point and coma will be regarded as representing a unique zero level, characterized either by absence or by a very low level of reticular discharge.

The first step is to analyse the concept of *level of activation for the behavioral manifestations that occur during wakefulness.*

If one accepts the hypothesis that the general aspects of drive are related to the activity of the ascending reticular system, this assumption does not imply that a given range of levels of reticular activation *causes* the discharge of a specific center of instinctive behavior. We simply recognize that the two activities are interrelated or, if we want to avoid physiological hypotheses, that there is a correlation between general and specific aspects of drives. As HINDE (1966) has pointed out "reticular activity must be between certain limits if functionally integrated behavior is to occur. But there is yet no satisfactory evidence that increase in reticular activity within these limits results in an increase in the intensity or frequency with which a particular pattern of behaviour is shown" (l.c. p. 164). The ranges of reticular activation corresponding to the different types of instinctive behavior that occur during wakefulness are probably rather large and certainly overlap (Fig. 39). In fact even the terms "lower and higher levels" are misleading, if one accepts the idea (MORUZZI, 1958) that the differences in the ascending reticular discharge are also qualitative in nature, although the hypothesis that the highest levels of activation are reached during the aggression-defense behavior appears rather likely. These quantitative expressions may still be used, anyway, until we know more about the patterns of ascending reticular discharge during the appetitive and consummatory phases of instinctive behavior.

Summing up, for the triggering of any instinctive behavior during wakefulness two factors should be considered, namely i) the accumulation of endogenous products within a specific center (the "motivation") combined with a proper environmental situation and ii) the level of reticular activation (the "vigilance"). There is of course an interplay between these two factors as shown by the sequence of actions leading from the appetitive phase to the consummatory act.

The second step will be to assume that there are *manifestations of instinctive behavior which appear only when the ascending reticular system is approaching the minimum levels that are needed in order to maintain consciousness. There are several instincts which require alertness, but one which is related to phenomena occurring during unconscious life.* Several electrophysiological and behavioral manifestations of sleep might be regarded as patterns of instinctive behavior which appear for proper levels of reticular deactivation, and possibly also at a critical slope in the temporal course of the reticular deactivation. The level

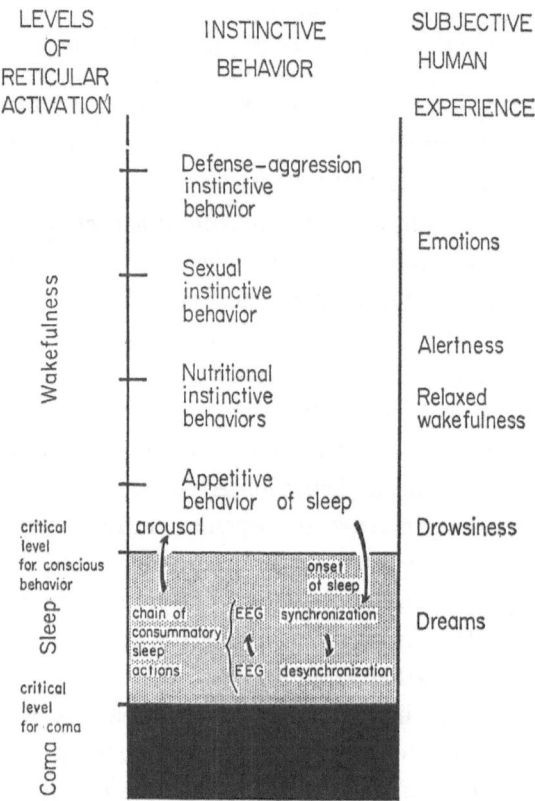

Fig. 39. *Relation between level of reticular activation, instinctive behavior, and subjective human experience.* When the reticular activation is above the critical level required for conscious behavior, i.e. during the waking state, all kinds of instinctive behavior are possible provided there is an adequate excitation of the specific motivational system and a proper environmental situation. Reticular activation has no steering function, but there might be optimal ranges of activation for each type of instinctive behavior, as suggested in the scheme. Viceversa, it is very likely that the level of activation rises during the appetitive phase and declines after the consummatory action is over. The instinctive behavior of sleep starts at levels of reticular activation just above the threshold for consciousness, as an appetitive phase which is typical for each animal. Any further deactivation occurring during, and probably facilitated by, the appetitive phase leads to the chain of sleep consummatory actions, characterized by synchronized or desynchronized EEG. Below a further critical level of reticular deactivation, sleep instinctive behavior is impossible (coma). (From Moruzzi, 1969)

of activation required for sleep is undoubtedly low, but is nonetheless one of critical importance, just as a proper level of vigilance is needed for the manifestations of conscious behavior. When the fall of the reticular tone is too marked, or even (possibly) too abrupt, the hypnic manifestations are missed and the animal precipitates into a state of coma, one that is qualitatively different from sleep (Fig. 39).

One may ask why the tone of the ascending reticular system never falls to comatose levels during physiological sleep. In fact such a fall might well be the outcome of the combined action of the avalanching processes of deactivation (BREMER, 1954) and of the inhibitory influences exerted by the sleep-inducing structures (BREMER, 1970a, b). There must be a reticular homeostasis preventing too marked a fall of the activating barrage. A physiological significance of the paradoxical phase might also be to increase the ascending reticular barrage whenever a dangerous level of reticular deactivation is attained. Phenomena of occlusion (see HUTTENLOCHER, 1961; EVARTS, 1967) combined with brain deefferentation by reticulo-spinal inhibition (POMPEIANO, 1965, 1967b) might give the wrong impression that sleep is deeper during the desynchronized phase. In fact the presence of dreams, implying a rudimentary level of consciousness, would suggest that sleep is lighter, and that the level of reticular activation is higher. Recent experiments by MOLINARI and FOULKES (1969) suggest, however, that mental activity is present also during slow wave sleep. There is simply a relative failure of memory processes which affect the capability of retrieving upon arousal. With respect to mental activity, the distinction is mainly between the phasic episodes of paradoxical sleep on the one hand and the non-phasic periods of the paradoxical sleep, and the slow wave sleep, on the other.

Normal sleep is characterized by i) the presence of a short or long lasting appetitive phase, the end of which is characterized by a feeling of drowsiness; ii) the complete reversibility of all hypnic manifestations, as shown by the arousal produced by sensory (see SHARPLESS and JASPER, 1956) and cortical (BREMER and TERZUOLO, 1952, 1953, 1954; SEGUNDO et al., 1955a, b) stimulations; and finally iii) the alternation between synchronized and desynchronized episodes of sleep, the presence of short lasting paradoxical episodes and of the corresponding hallucinations, the dreams, being an indispensable manifestation of every type of normal mammalian sleep. Once the instinctive behavior of sleep has started, a chain of consummatory actions with complex reverberations between brain stem and cerebrum will prevent a further fall of reticular tone to levels which would lead to coma, a situation that would by definition abolish physiological sleep. A change in the opposite direction occurs at the moment of arousal, since all manifestations of sleep disappear as the reticular tone reaches the critical levels for consciousness (Fig. 39).

It we accept these preliminary assumptions, we may have to change several concepts, which are deeply rooted in both physiology and ethology.

First of all we should give up the idea that sleep is merely absence of wakefulness, a concept that can be traced back to a famous review article of PURKINJE (1846, see MORUZZI, 1964b). Sleep is an active state, one however that requires a level of reticular activation too low to be compatible with conscious behavior. A sleeping animal is unconscious not because it is asleep, but because its level of reticular tone is not adequate for the maintenance of consciousness; it is asleep, and not comatose, because the process of deactivation has been locked at the critical levels which are appropriate for the development of the instinctive behavior of sleep. To take as examples the instinctive manifestations occurring at the two ends of the activation continuum, neither the behavior of defense-aggression nor sleep could occur without an appropriate level of reticular activation.

The validity of the concept of reticular deactivation, indeed of all the "passive" theories (see MORUZZI, 1964b for the literature) which have their origin in BREMER'S classical experiment on the *cerveau isolé* (1935), is by no means shaken by this interpretation. The experimental results are simply viewed from a different angle. Rather than with sleep itself, they would be concerned with the regulation of the tonic activity of the ascending reticular system: a system that controls several other neural activities, besides sleep, during both conscious and unconscious life. The ascending reticular barrage is responsible for the maintenance of a proper background of general activation. It would thus control, indirectly, the perceptual and behavioral manifestations that occur during consciousness, as well as the hallucinations (dreams) and the somatic and autonomic manifestations of sleep which appear only when the level of activation is too low to permit conscious behavior. From sleep to strong emotions most of the neural activities concerned with instinctive behavior, perception, learning, attention, are likely to be indirectly controlled by the ascending reticular system, or at least appear to be correlated with given ranges of ascending reticular barrage. Sleep may be regarded as a special case of instinctive behavior, one requiring *a low but critical* level of reticular activity.

Homeostatic devices control directly the level of tonic activity of the ascending reticular system. This notion has been suggested for the phasic reticular activation (DELL, 1963) and may simply be extended to the tonic activation, both below and above the critical levels required for consciousness. Through cortico-reticulo-cortical loops the onset and the duration of sleep might be regulated, although in a rather loose and certainly indirect way, by the structures which are in need of the slow recovery (see MORUZZI, 1966) occurring during sleep. This however, is only one among several mechanisms which may influence reticular activity, and one can therefore easily understand why it is so easy both to oversleep and to contract sleep debt. The extreme precision of the direct homeostatic devices controlling CO_2 tension is never attained in the regulation of sleep.

A distinction should be made, finally, between the hard core of sleep—the patterned discharge of the neurons constituting the so called "sleep center" and the related behavior—and the condition of the ascending reticular system which makes possible the development of instinctive hypnic behavior. The EEG synchronization is simply the consequence of the withdrawal of the reticular influence exerted upon the thalamo-cortical system. It is merely a manifestation of the reticular deactivation which is required for the onset of sleep, and for this very reason is always associated with the initial phase of physiological sleep, although it should be conceptually separated from it. In fact it is present in neural conditions, such as coma and barbital anesthesia, which share with true physiological sleep only the lack of consciousness.

VI. Summary and Conclusions

This review has been concerned with an analysis of the neural mechanisms underlying the sleep-waking cycle. Our attention has been concentrated upon the experimental attempts to modify, or to abolish, the alternation between sleep and wakefulness, either by lesions or by stimulation in selected parts of the central nervous system. Thus the description and the classification of the phenomena occurring during sleep—the phenomenology of sleep—has remained outside the scope of the present article.

The task of this concluding chapter is made easier by the fact that the main findings have already been discussed at the end of each section. What is now needed, therefore, is an integration of the results obtained along different lines of endeavor.

In our opinion one of the main findings of the lesion experiments is the demonstration that the sleep-waking cycle is present in an isolated cerebrum, provided that the study is made in chronic conditions. Both lesion and stimulation experiments show, in fact, that the structures which appear to be directly and critically responsible for both sleep and wakefulness, and for the alternation between these states, are localized in the diencephalon. Some of these systems have been definitely shown to be tonically active. They are localized, respectively, in the anterior hypothalamus and the basal forebrain, and in the posterior hypothalamus.

The comatose syndrome of the acute *cerveau isolé* is probably due to an imbalance resulting from the sudden withdrawal of the ascending influence of the brain stem. Another imbalance, less strong in intensity and opposite in sign, is produced by a pretrigeminal transection. There is a reciprocal organization between the activating structures of the reticular formation and the deactivating neurons, mainly localized in the lower brain stem.

In summary, the sleep-waking cycle does not arise within, but is strikingly controlled by, the ascending systems of the brain stem. This control—which is both tonic and phasic in nature—is only one aspect of a much wider problem: that of the levels of "activation" which are required for different kinds of behavior. The sleep produced by electrical stimulation of the forebrain and the instinctive behaviors elicited by electrical stimulation of hypothalamic centers are related phenomena, since in both cases a fully integrated pattern of behavior is produced by stimulating specific structures of the central nervous system. A tonic ascending flow of reticular impulses has been shown to be necessary both for the maintenance of wakefulness and for the rage outbursts of the decorticated cat.

While it is generally conceded that an appropriate level of alertness, and therefore of reticular activation, is required for the appearence of the different types of instinctive behavior, it may appear less justified to designate as "instinctive" a behavior which appears only when the ascending reticular system is approaching the minimum level for the maintenance of behavioral responsiveness. The early stages of instinctive activity are usually thought to lead to dissipation of energy, rather than to a prolonged state of quietness such as is required for sleep.

Ethologists had already noted, however, that the period immediately preceding sleep recalls the appetitive phase of instinctive behavior. The physiological significance of these instinctive manifestations is suggested by the fact that sleep recovery is a case of neural regulation which requires the abolition, or a great impairment, of reactivity to the environment. Such a loss, or severe disorganization, of responsiveness is not only dangerous for survival, at least for several animals living in the wild state; it is also clearly incompatible with the usual types of instinctive behavior which require a state of alertness. Consequently, we may hypothesize that the instinctive behavior associated with the onset of sleep serves a dual function: a) it enables the animal to choose an appropriate and safe environment for sleep, and b) it coordinates the period of sleep in the daily cycle of instinctive behaviors.

An introductory appetitive phase implies the eventual onset of a consummatory action. The controversial issue is whether i) all the phenomena occurring during sleep belong to this consummatory phase or ii) the neurochemical and neurophysiological processes underlying sleep recovery are simply associated with instinctive manifestations of sleep. Along the same lines neurophysiological mechanisms regulating water losses are associated with thirst behavior, clearly instinctive in nature, which regulates water intake.

Neurophysiological and behavioral regulations are often associated in maintaining the stability of the internal environment. The recovery of the brain structures which have suffered "wear and tear" may be regarded as a homeostatic problem, although it is one of unusual complexity, since it involves a prolonged loss or a severe impairment of behavioral responsiveness. It is thus not surprising that an instinctive behavior is associated with sleep recovery, and the combined efforts of ethologists and sleep physiologists may be necessary for a full understanding of the phenomenon of sleep.

Acknowledgment. The author is grateful to Drs. Giovanni Berlucchi and Henry Buchtel for stimulating discussions, criticisms, and suggestions. Thanks are also expressed to the Brain Information Service (Los Angeles) and, personally, to Miss Dottie Eakin, for valuable help in gathering and classifying the literature on sleep.

References

ADAMETZ, J. H.: Rate of recovery of functioning in cats with rostral reticular lesions. J. Neurosurg. 16, 85–97 (1959).

ADAMS, D. B., BACCELLI, G., MANCIA, G., ZANCHETTI, A.: Cardiovascular changes during naturally elicited fighting behavior in the cat. Amer. J. Physiol. 216, 1226–1235 (1969).

ADRIAN, E. D.: The electrical activity of the mammalian olfactory bulb. Electroenceph. clin. Neurophysiol. 2, 377–388 (1950).

— BREMER, F., JASPER, H. H. (eds.): Brain mechanisms and consciousness. Oxford: Blackwell 1954. XV + 556 pp.

— MORUZZI, G.: Impulses in the pyramidal tract. J. Physiol. (Lond.) 97, 153–199 (1939).

AFFANNI, J., MARCHIAFAVA, P. L., ZERNICKI, B.: Orientation reactions in the midpontine pretrigeminal cat. Arch. ital. Biol. 100, 297–304 (1962a).

— — — Conditioning in the midpontine pretrigeminal cat. Arch. ital. Biol. 100, 305–310 (1962b).

AKERT, K.: Diencephalon. In: Electrical stimulation of the brain. (D. E. SHEER, ed.), p. 288–310. Austin: University of Texas Press 1961. XV + 641 pp.

— The anatomical substrate of sleep. Progr. Brain Res. 18, 9–19 (1965).

— BALLY, C., SCHADÉ, J. P.: Sleep mechanisms. Progr. Brain Res. 18. Amsterdam: Elsevier 1965. XI + 257 pp.

— KOELLA, W. P., HESS, R., JR.: Sleep produced by electrical stimulation of the thalamus. Amer. J. Physiol. 168, 260–267 (1952).

AKIMOTO, H., YAMAGUCHI, N., OKABE, K., NAKAGAWA, T., NAKAMURA, I., ABE, K., TORII, H., MASAHASHI, K.: On the sleep induced through electrical stimulation on dog thalamus. Folia psychiat. neurol. jap. 10, 117–146 (1956).

ALEKSEYEVA, T. T.: Correlation of nervous and humoral factors in the development of sleep in non-disjointed twins. [Russian.] Zh. vyssh. nerv. Deyat. Pavlova 8, 835–844 (1958).

ALEMÀ, G., PERRIA, L., ROSADINI, G., ROSSI, G. F., ZATTONI, J.: Functional inactivation of the human brain stem related to the level of consciousness. J. Neurosurg. 24, 629–639 (1966).

ANDERSEN, P., ANDERSSON, S. A., LØMO, T.: Some factors involved in the thalamic control of spontaneous barbiturate spindles. J. Physiol. (Lond.) 192, 257–281 (1967a).

— — — Nature of thalamocortical relations during spontaneous barbiturate spindle activity. J. Physiol. (Lond.) 192, 283–307 (1967b).

— McBROOKS, C., ECCLES, J. C., SEARS, T. A.: The ventro-basal nucleus of the thalamus: potential fields, synaptic transmission and excitability of both presynaptic and postsynaptic components. J. Physiol. (Lond.) 174, 348–369 (1964a).

— ECCLES, J. C., SEARS, T. A.: The ventrobasal complex of the thalamus: types of cells, their responses and their functional organization. J. Physiol. (Lond.) 174, 370–399 (1964b).

ANDERSON, E., BATES, R. W., HAWTHORNE, E., HAYMAKER, W., KNOWLTON, K., McK. RIOCH, D., SPENCE, W. T., WILSON, H.: The effects of midbrain and spinal cord transection on endocrine and metabolic functions with postulation of a midbrain hypothalamico-pituitary activating system. Recent. Progr. Hormone Res. 13, 21–66 (1957).

ANDERSSON, B.: The physiology of thirst. Progr. physiol. Psychol. 1, 191–207 (1966).

— WYRWICKA, W.: The elicitation of a drinking motor conditioned reaction by electrical stimulation of the hypothalamic "drinking area" in the goat. Acta physiol. scand. 41, 194–198 (1957).

ANGELERI, F., MARCHESI, G. F., QUATTRINI, A.: Effects of chronic thalamic lesions on the electrical activity of the neocortex and on sleep. Arch. ital. Biol. 107, 633–667 (1969).

Arduini, A., Berlucchi, G., Strata, P.: Pyramidal activity during sleep and wakefulness. Arch. ital. Biol. 101, 530–544 (1963).
— Hirao, T.: On the mechanisms of the EEG sleep patterns elicited by acute visual deafferentation. Arch. ital. Biol. 97, 140–155 (1959).
— — EEG synchronization elicited by light. Arch. ital. Biol. 98, 275–292 (1960).
— Moruzzi, G.: Olfactory arousal reactions in the "cerveau isolé" cat. Electroenceph. clin. Neurophysiol. 5, 243–250 (1953).
Aschoff, J.: Circadian clocks. Amsterdam: North-Holland Publ. Co. 1965. XIX–479 pp.
Baer, A.: Über gleichzeitige elektrische Reizung zweier Großhirnstellen am ungehemmten Hunde. Pflügers Arch. ges. Physiol. 106, 523–567 (1905).
Balch, C. C.: Sleep in ruminants. Nature (Lond.) 175, 940–941 (1955).
Balzano, E., Jeannerod, M.: Activité multi-unitaire de structures sous-corticales pendant le cycle veille-sommeil chez le chat. Electroenceph. clin. Neurophysiol. 28, 136–145 (1970).
Bard, Ph.: A diencephalic mechanism for the expression of rage with special reference to the sympathetic nervous system. Amer. J. Physiol. 84, 490–515 (1928).
— Macht, M. B.: The behaviour of chronically decerebrate cats. In: Neurological basis of behavior (Wolstenholme, G. E. W., and O'Connor, C. M., eds.), p. 55–75. London: Churchill 1958. XII + 400 pp.
Barrett, R., Merritt, H. H., Wolf, A.: Depression of consciousness as a result of cerebral lesions. Res. Publ. Ass. nerv. ment. Dis. 45, 241–276 (1967).
Batini, C., Magni, F., Palestini, M., Rossi, G. F., Zanchetti, A.: Neural mechanisms underlying the enduring EEG and behavioral activation in the mid-pontine pretrigeminal cat. Arch. ital. Biol. 97, 13–25 (1959b).
— Moruzzi, G., Palestini, M., Rossi, G. F., Zanchetti, A.: Persistent patterns of wakefulness in the pretrigeminal midpontine preparation. Science 128, 30–32 (1958).
— — — — — Effects of complete pontine transections on the sleep-wakefulness rhythm: the midpontine pretrigeminal preparation. Arch. ital. Biol. 97, 1–12 (1959a).
— Palestini, M., Rossi, G. F., Zanchetti, A.: EEG activation patterns in the midpontine pretrigeminal cat following sensory deafferentation. Arch. ital. Biol. 97, 26–32 (1959c).
— Pompeiano, O.: Effects of rostro-medial and rostro-lateral fastigial lesions on decerebrate rigidity. Arch. ital. Biol. 96, 315–329 (1958).
Batsel, H. L.: Electroencephalographic synchronization and desynchronization in the chronic "cerveau isolé" of the dog. Electroenceph. clin. Neurophysiol. 12, 421–430(1960).
— Spontaneous desynchronization in the chronic cat "cerveau isolé". Arch. ital. Biol. 102, 547–566 (1964).
Baumgarten, R. von, Mollica, A., Moruzzi, G.: Modulierung der Entladungsfrequenz einzelner Zellen der Substantia reticularis durch corticofugale und cerebelläre Impulse. Pflügers Arch. ges. Physiol. 259, 56–78 (1954).
Baust, W.: Ermüdung, Schlaf und Traum. Stuttgart: Wissenschaftliche Verlags G.m.b.H. 1970a. 314 pp.
— Die Phänomenologie des Schlafes. In: Ermüdung, Schlaf und Traum (Baust, W., ed.), p. 99–144. Stuttgart: Wissenschaftliche Verlags G.m.b.H. 1970b. 314 pp.
— Autonomic functions during sleep. Physiol. Rev. (1971) (in preparation).
— Niemczyk, H., Vieth, J.: The action of blood pressure on the ascending reticular activating system with special reference to adrenaline induced EEG arousal. Electroenceph. clin. Neurophysiol. 15, 63–72 (1963).
Bazett, H. C., Penfield, W. G.: A study of the Sherrington decerebrate animal in the chronic as well as the acute condition. Brain 45, 185–265 (1922).
Bell, F. R.: The electroencephalogram of goats during somnolence and rumination. Anim. Behav. 8, 39–42 (1960).
Belugou, J. L., Benoit, O., Leygonie, F.: Décharges neuronales de l'hippocampe au cours de la veille et du sommeil. J. Physiol. (Paris) 60 (Suppl. 2), 399–399 (1968).

BERLUCCHI, G.: Callosal activity in unrestrained, unanesthetized cats. Arch. ital. Biol. **103**, 623–634 (1965).
— Electroencephalographic activity of the isolated hemicerebrum of the cat. Exp. Neurol. **15**, 220–228 (1966).
— Mechanismen von Schlafen und Wachen. In: Ermüdung, Schlaf und Traum (BAUST, W., ed.), S. 145–203. Stuttgart: Wissenschaftliche Verlags G.m.b.H. 1970. 314 pp.
— MAFFEI, L., MORUZZI, G., STRATA, P.: EEG and behavioral effects elicited by cooling of medulla and pons. Arch. ital. Biol. **102**, 372–392 (1964a).
— — — — Mécanismes hypnogènes du tronc de l'encéphale antagonistes du système réticulaire activateur. In: Aspects anatomo-fonctionnels de la physiologie du sommeil (JOUVET, M., ed.), p. 89–105. Paris: Centre National de la Recherche Scientifique 1965.
— MORUZZI, G., SALVI, G., STRATA, P.: Pupil behavior and ocular movements during synchronized and desynchronized sleep. Arch. ital. Biol. **102**, 230–244 (1964b).
BERNARD, CL.: Introduction à l'étude de la médecine expérimentale. Paris: J. B. Baillière & fils 1865. 400 pp.
BIZZI, E.: Discharge patterns of single geniculate neurons during the rapid eye movements of sleep. J. Neurophysiol. **29**, 1087–1095 (1966).
— BROOKS, D. C.: Functional connections between pontine reticular formation and lateral geniculate nucleus during deep sleep. Arch. ital. Biol. **101**, 666–680 (1963).
— POMPEIANO, O., SOMOGYI, I.: Spontaneous activity of single vestibular neurons of unrestrained cats during sleep and wakefulness. Arch. ital. Biol. **102**, 308–330 (1964).
— SPENCER, W. A.: Enhancement of EEG synchrony in the acute "cerveau isolé". Arch. ital. Biol. **100**, 234–247 (1962).
BONVALLET, M.: Système nerveux et vigilance. Paris: Presses Universitaires de France 1966. VIII + 131 pp.
— ALLEN, M. B., JR.: Prolonged spontaneous and evoked reticular activation following discrete bulbar lesions. Electroenceph. clin. Neurophysiol. **15**, 969–988 (1963).
— BLOCH, V.: Bulbar control of cortical arousal. Science **133**, 1133–1134 (1961).
— DELL, P., HIEBEL, G.: Tonus sympathique et activité électrique corticale. Electroencephal. clin. Neurophysiol. **6**, 119–144 (1954).
— SIGG, B.: Etude électrophysiologique des afférences vagales au niveau de leur pénétration dans le bulbe. J. Physiol. (Paris) **50**, 63–74 (1958).
BREMER, F.: Contribution à l'étude de la physiologie du cervelet. La fonction inhibitrice du paléocerebellum. Arch. int. Physiol. **19**, 189–226 (1922).
— See discussion of Hess' report, p. 1363 (HESS, 1931).
— Cerveau "isolé" et physiologie du sommeil. C. R. Soc. Biol. (Paris) **118**, 1235–1241 (1935).
— L'activité cérébrale au cours du sommeil et de la narcose. Contribution à l'étude du mécanisme du sommeil. Bull. Acad. roy. Méd. Belg. **4**, 68–86 (1937).
— L'activité électrique de l'écorce cérébrale et le problème physiologique du sommeil. Boll. Soc. ital. Biol. sper. **13**, 271–290 (1938).
— Le problème physiologique du sommeil. Medicina (Parma) **1**, 589–611 (1951).
— Some problems in neurophysiology. London: The Athlone Press 1953. 79 pp.
— The neurophysiological problem of sleep. In: Brain mechanisms and consciousness. (ADRIAN, E. D., BREMER, F., and JASPER, H. H. eds.), p. 137–162. Oxford: Blackwell 1954. XV + 556 pp.
— See discussion of Hess' report, p. 126–127 (HESS, 1954).
— Analyse des processus corticaux de l'éveil. Electroencephal. clin. Neurophysiol., Suppl. **13**, 125–136 (1960).
— Processus d'inhibition et physiologie du sommeil. In: Problemi di Neurologia e Psichiatria. (ALEMÀ, G. et al., eds.), p. 120–125. Roma: Il Pensiero Scientifico 1968a. XXXVI + 1453 pp.
— Veille, sommeil et rêve. Bull. Acad. roy. Belg. **54**, 1580–1586 (1968b).

Bremer, F.: Inhibitions intrathalamiques récurrentielles et physiologie du sommeil. Electroenceph. clin. Neurophysiol. **28**, 1–16 (1970a).
— Preoptic hypnogenic focus and mesencephalic reticular formation. Brain Res. **21**, 132–134 (1970b).
— Terzuolo, C.: Rôle de l'écorce cérébrale dans le processus du réveil. Arch. int. Physiol. **60**, 228–231 (1952).
— — Nouvelles recherches sur le processus du réveil. Arch. int. Physiol. **61**, 86–90 (1953).
— — Contribution à l'étude des mécanismes physiologiques du maintien de l'activité vigile du cerveau. Interaction de la formation réticulée et de l'écorce cérébrale dans le processus du réveil. Arch. int. Physiol. **62**, 157–178 (1954).
Brodal, A.: Neurological anatomy, 2nd ed. New York-London-Toronto: Oxford University Press 1969. XX + 807.
— Rossi, G. F.: Ascending fibers in brain stem reticular formation of cat. Arch. Neurol. Psychiat. (Chic.) **74**, 68–87 (1955).
— Taber, E., Walberg, F.: The raphe nuclei of the brain stem in the cat. II. Efferent connections. J. comp. Neurol. **114**, 239–260 (1960).
Brown, T. Graham: The intrinsic factors in the act of progression in the mammal. Proc. roy. Soc. B **84**, 308–319 (1911).
— On the nature of the fundamental activity of the nervous centres, together with an analysis of the conditioning of rhythmic activity in progression, and a theory of the evolution of function in the nervous system. J. Physiol. (Lond.) **48**, 18–46 (1914).
— Studies in the physiology of the nervous system. XXII: On the phenomenon of facilitation. 1: Its occurrence in reactions induced by stimulation of the "motor" cortex of the cerebrum in monkeys. Quart. J. exp. Physiol. **9**, 81–99 (1915a).
— Studies in the physiology of the nervous system. XXIII: On the phenomenon of facilitation. 2: Its occurrence in response to subliminal cortical stimuli in monkeys. Quart. J. exp. Physiol. **9**, 101–116 (1915b).
— Studies in the physiology of the nervous system. XXV: On the phenomenon of facilitation. 4: Its occurrence in the subcortical mechanism by the activation of which motor effects are produced on artificial stimulation of the "motor" cortex. Quart. J. exp. Physiol. **9**, 131–145 (1915c).
— Die Reflexfunktionen des Zentralnervensystems mit besonderer Berücksichtigung der rhythmischen Tätigkeiten beim Säugetier. Ergebn. Physiol. **15**, 480–790 (1916).
— Die Großhirnhemisphären. In: Handbuch der normalen pathologischen Physiologie (herausgeg. von A. Bethe, G. v. Bergmann, G. Embden u. A. Ellinger), Bd. 10, S. 418–524, see 439–447. Berlin: Springer 1927.
Bubenik, A.: Wild u. Hund **63**, 3 (1960). (Quoted by Hediger, H., 1969).
Bürger-Prinz, H., Fischer, P. A. (eds.): Schlaf, Schlafverhalten, Schlafstörungen. Stuttgart: F. Enke 1967. VIII + 152 pp.
Bureš, J., Burešová, O.: The use of Leão's spreading cortical depression in research on conditioned reflexes. Electroenceph. clin. Neurophysiol. Suppl. **13**, 359–373 (1960).
Burešová, O., Bureš, J.: Can the brain be improved? Endeavour **28**, 139–145 (1969).
Burgi, S.: Le sommeil et le rêves. Quelques données récentes sur un problème millénaire. Schweiz. med. Wschr. **96**, 1620–1626 (1966).
Calvet, J., Calvet, M. C.: Étude quantitative et organisation en fonction de la vigilance de l'activité unitaire des diverses régions du cortex cérébral. Brain Res. **10**, 183–199 (1968).
Camis, M.: Recherches sur le mécanisme central des mouvements de déambulation. Arch. int. Physiol. **20**, 340–370 (1923).
Candia, O., Favale, E., Giussani, A., Rossi, G. F.: Blood pressure during natural sleep and during sleep induced by electrical stimulation of the brain stem reticular formation. Arch. ital. Biol. **100**, 216–233 (1962).
— Rossi, G. F., Sekino, T.: Brain stem structures responsible for the electroencephalographic patterns of desynchronized sleep. Science **155**, 720–722 (1967).

CANNON, W. B.: The wisdom of the body. New York: The Norton Library 1932. 312 pp.
— BRITTON, S. W.: Studies on the conditions of activity in endocrine glands. XV. Pseudo-affective medulliadrenal secretion. Amer. J. Physiol. **72**, 283–313 (1925).
CARLI, G.: Dissociation of electrocortical activity and somatic reflexes during rabbit hypnosis. Arch. ital. Biol. **107**, 219–234 (1969).
— ARMENGOL, V., ZANCHETTI, A.: Brain stem-limbic connections and the electrographic aspects of deep sleep in the cat. Arch. ital. Biol. **103**, 725–750 (1965).
— ZANCHETTI, A.: A study of pontine lesions suppressing deep sleep in the cat. Arch. ital. Biol. **103**, 751–788 (1965).
CASPERS, H.: Changes of cortical D.C. potentials in the sleep-wakefulness cycle. In: The nature of sleep (WOLSTENHOLME, G. E. W., and O'CONNOR, C. M., eds.), p. 237–259. London: Churchill 1961. XII–416 pp.
— WINKEL, K.: Die Beeinflussung der Großhirnrindenrhythmik durch Reizungen im Zwischen- und Mittelhirn bei der Ratte. Pflügers Arch. ges. Physiol. **259**, 334–356 (1954).
CHATRIAN, G. E., WHITE, L. E., SHAW, C. M.: EEG patterns resembling wakefulness in unresponsive decerebrate state following traumatic brain-stem infarct. Electroenceph. clin. Neurophysiol. **16**, 285–289 (1964).
CLAPARÈDE, E.: Esquisse d'une théorie biologique du sommeil. Arch. Psychol. **4**, 246–349 (1905).
— La question du sommeil. Ann. psychol. **18**, 419–459 (1912).
— Opinions et travaux divers relatifs à la théorie biologique du sommeil et de l'hystérie. Arch. Psychol. **21**, 113–172 (1928).
— Le sommeil et la veille. J. Psychol. norm. path. **26**, 433–492 (1929).
CLEMENTE, C. D., STERMAN, M. B.: Cortical synchronization and sleep patterns in acute restrained and chronic behaving cats induced by basal forebrain stimulation. Electroenceph. clin. Neurophysiol., Suppl. **24**, 172–187 (1963).
— — Basal forebrain mechanisms for internal inhibition and sleep. Res. Publ. Ass. nerv. ment. Dis. **45**, 127–147 (1967a).
— — Limbic and other forebrain mechanisms in sleep induction and behavioral inhibition. Progr. Brain Res. **27**, 34–47 (1967b).
— — WYRWICKA, W.: Forebrain inhibitory mechanisms: conditioning of basal forebrain induced EEG synchronization and sleep. Exp. Neurol. **7**, 404–417 (1963).
— — — Post-reinforcement EEG synchronization during alimentary behavior. Electroenceph. clin. Neurophysiol. **16**, 355–365 (1964).
CORDEAU, J. P.: Functional organization of the brain stem reticular formation in relation to sleep and wakefulness. Rev. canad. Biol. **21**, 113–125 (1962).
— MANCIA, M.: Effect of unilateral chronic lesions of the midbrain on the electrocortical activity of the cat. Arch. ital. Biol. **96**, 374–399 (1958).
— — Evidence for the existence of an EEG synchronization mechanism originating in the lower brain stem. Electroenceph. clin. Neurophysiol. **11**, 551–564 (1959).
— WALSH, J., MAHUT, H.: Variations dans la transmission des messages sensoriels en fonction des différents états d'éveil et de sommeil. In: Aspects Anatomo-fonctionnels de la Physiologie du sommeil (JOUVET, M., ed.), p. 477–507. Paris: Centre National de la Recherche Scientifique 1965. 657 pp.
CRAIG, W.: Appetites and aversions as constituents of instincts. Biol. Bull. **34**, 91–107 (1918).
CREUTZFELDT, O., JUNG, R.: Neuronal discharge in the cat's motor cortex during sleep and arousal. In: The nature of sleep. (WOLSTENHOLME, G. E. W., and O'CONNOR, C. M., eds.), p. 131–170. London: Churchill 1961. XII + 416 pp.
DELIUS, J. D.: Displacement activities and arousal. Nature (Lond.) **214**, 1259–1260 (1967).
— Irrelevant behaviour, information processing and arousal homeostasis. Psychol. Forsch. **33**, 165–188 (1970).

Dell, P.: Reticular homeostasis and critical reactivity. In: Brain Mechanisms (Moruzzi, G., Fessard, A., and Jasper, H. H., eds.). Amsterdam: Elsevier 1963. Progr. Brain Res. 1, 82–114 (1963). See also Hugelin's discussion: p. 105–111.

— Bonvallet, M., Hugelin, A.: Mechanisms of reticular deactivation. In: The nature of sleep (Wolstenholme, G. E. W., and O'Connor, C. M., eds.), p. 86–107. London: Churchill 1961. XII + 416.

Dell, P. C.: Some basic mechanisms of the translation of bodily needs into behaviour. In: Neurological basis of behaviour. (Wolstenholme, G. E. W., and O'Connor, C. M., eds.), p. 187–203. London: Churchill 1958. XII + 400 pp.

Dement, W.: The effect of dream deprivation. Science 131, 1705–1707 (1960).

— Henry, P., Cohen, H., Ferguson, J.: Studies on the effect of REM deprivation in humans and in animals. Res. Publ. Ass. nerv. ment. Dis. 45, 456–468 (1967).

— Kales, A., Rechtschaffen, A. (eds.): Sleep reviews. Started in 1970 under the auspices of UCLA Brain Information Service.

— Rechtschaffen, A.: Narcolepsy: polygraphic aspects, experimental and theoretical considerations. In: The abnormalities of sleep in man (Gastaut, H., Lugaresi, E., Berti Ceroni, G. and Coccagna, G., eds.), p. 147–164. Bologna: Aulo Gaggi 1968. 326 pp.

Deutsch, J. A., Deutsch, D.: Arousal, sleep and attention. In: Physiological psychology (Deutsch, J. A., and Deutsch, D., eds.), p. 145–176. Homewood, Ill.: Dorsey Press 1966. 553 pp.

Doty, R. W.: Conditioned reflexes formed and evoked by brain stimulation. In: Electrical stimulation of the brain (Sheer, D. E., ed.), p. 397–412. Austin: University of Texas Press 1961. XV + 641 pp.

— Electrical stimulation of the brain in behavioral context. Ann. Rev. Psychol. 20, 289–320 (1969).

Dow, R. S., Moruzzi, G.: The physiology and pathology of the cerebellum. Minneapolis: The University of Minnesota Press 1958. XV + 675 pp.

Dreher, B., Zernicki, B.: Visual fixation reflex: behavioral properties and neural mechanism. Acta Biol. exp. 29, 359–383 (1969).

Eakin, D. (ed.): Sleep bulletin, a bimontly publication published since 1968 by the Brain Information Center of the University of California at Los Angeles.

Eccles, J. C. (ed.): Brain and conscious experience. Berlin-Heidelberg-New York: Springer 1966. XXI + 591.

Economo, C. von: Die Pathologie des Schlafes. In: Handbuch der normalen und pathologischen Physiologie (herausgeg. von A. Bethe, G. v. Bergmann, G. Embden u. A. Ellinger), Bd. 17, S. 591–610. Berlin: Springer 1926.

— Schlaftheorie. Ergebn. Physiol. 28, 312–339 (1929).

Ellison, G. D., Flynn, J. P.: Organized aggressive behavior in cats after surgical isolation of the hypothalamus. Arch. ital. Biol. 106, 1–20 (1968).

Elul, R., Marchiafava, P. L.: Accommodation of the eye as related to behavior in the cat. Arch. ital. Biol. 102, 616–644 (1964).

Ephron, H. S., Carrington, P.: Rapid eye movement sleep and cortical homeostasis. Psychol. Rev. 73, 500–526 (1966).

Evarts, E. V.: Activity of neurons in visual cortex of the cat during sleep with low voltage fast EEG activity. J. Neurophysiol. 25, 812–816 (1962).

— Temporal patterns of discharge of pyramidal tract neurons during sleep and waking in the monkey. J. Neurophysiol. 27, 152–171 (1964).

— Relation of discharge frequency to conduction velocity in pyramidal tract neurons. J. Neurophysiol. 28, 215–228 (1965).

— Activity of individual cerebral neurons during sleep and arousal. Res. Publ. Ass. nerv. ment. Dis. 45, 319–337 (1967).

— Magoun, H. W.: Some characteristics of cortical recruiting responses in unanesthetized cats. Science 125, 1147–1148 (1957).

EWALD, J. RICH.: Über künstliche Reizung der Großhirnrinde. Dtsch. med. Wschr. **24**, 180–181 (1898).

FADIGA, E., MANZONI, T., SAPIENZA, S., URBANO, A.: Synchronizing and desynchronizing fastigial influences on the electrocortical activity of the cat, in acute experiments. Electroenceph. clin. Neurophysiol. **24**, 330–342 (1968).

— — URBANO, A.: The tonic action of cerebellar efferents on the level of electrocortical activity as appearing from acute ablation experiments in the cat. Arch. Sci. Biol. **51**, 24–40 (1967).

FAVALE, E., LOEB, C., PARMA, M., ROSSI, G. F., SACCO, G.: Effets de la stimulation de structures du tronc cérebral sur le comportement du chat. Neurochirurgie **6**, 89–91 (1960).

— — ROSSI, G. F., SACCO, G.: EEG synchronization and behavioral signs of sleep following low frequency stimulation of the brain stem reticular formation. Arch. ital. Biol. **99**, 1–22 (1961).

FEINBERG, I., EVARTS, E. V.: Changing concepts of the function of sleep: discovery of intense brain activity during sleep calls for revision of hypotheses as to its function. Biol. Psychiat. **1**, 331–348 (1969).

FELDMAN, S. M., WALLER, H. J.: Dissociation of electrocortical activation and behavioural arousal. Nature (Lond.) **196**, 1320–1322 (1962).

FINDLAY, A. L. R., HAYWARD, J. N.: Spontaneous activity of single neurones in the hypothalamus of rabbits during sleep and waking. J. Physiol. (Lond.) **201**, 237–258 (1969).

FISCHGOLD, H. (ed.): Le sommeil de nuit normal et pathologique. Paris: Masson 1965. 391 pp.

FOULKES, D.: The psychology of sleep. New York: C. Scribner's Sons 1966. XII + 265 pp.

— HOBSON, J. A. (eds.): Abstracts of papers presented to the ninth annual Meeting of the Association for the psychophysiological study of sleep. Psychophysiology **6**, 214–272 (1969); see p. 227–231.

FREDERICKSON, C. J., HOBSON, J. A.: Electrical stimulation of the brain stem and subsequent sleep. Arch. ital. Biol. **108** 564–576 (1970).

FRENCH, J. D., MAGOUN, H. W.: Effects of chronic lesions in central cephalic brain stem of monkeys. Arch. Neurol. Psychiat. (Chic.) **68**, 591–604 (1952).

FULTON, J. F., McCOUCH, G. P.: The relation of the motor area of primates to the hypo-reflexia ("spinal shock") of spinal transection. J. nerv. ment. Dis. **86**, 125–146 (1937).

FUSTER, J. M.: Effects of stimulation of brain stem on tachistoscopic perception. Science **127**, 150 (1958).

GAMPER, E.: Bau und Leistungen eines menschlichen Mittelhirnwesens (Arhinencephalie mit Encephalocele) zugleich ein Beitrag zur Teratologie und Fasersystematik. Z. ges. Neurol. Psychiat. **104**, 67–120 (1926).

GARCÍA AUSTT, E.: Mecanismos neurofisiológicos de la vigilia y sus implicaciones psico-lógicas. Acta neurol. lat.-amer. **11**, 11–40 (1965).

GASTAUT, H., BERT, J.: Electroencephalographic detection of sleep induced by repetitive sensory stimuli. In: On the nature of sleep (WOLSTENHOLME, G. E. W., and O'CONNOR, C. M., eds.), p. 260–283. London: Churchill 1961. XII + 416 pp.

— LUGARESI, E., BERTI CERONI, G., COCCAGNA, G. (eds.): The abnormalities of sleep in man. Proc. XVth Europ. Meeting of electroencephalography, Bologna 1967. Bologna: Aulo Gaggi, 326 pp., 1968.

GAUTHIER, C., MOLLICA, A., MORUZZI, G.: Physiological evidence of localized cerebellar projections to bulbar reticular formation. J. Neurophysiol. **19**, 468–483 (1956).

GENOVESI, U., MORUZZI, G., PALESTINI, M., ROSSI, G. F., ZANCHETTI, A.: EEG and behavioral patterns following lesions of the mesencephalic reticular formation in chronic cats with implanted electrodes. Abstr. Comm. 20th Int. Physiol. Congr., Bruxelles, 335–336 (1956).

GENTILOMO, A., ROSADINI, G., ROSSI, G. F., ZATTONI, J.: Modificazioni respiratorie, cardiocircolatorie e pupillari da inattivazione barbiturica del tronco encefalico. Boll. Soc. ital. Biol. sper. **40**, 843–845 (1964).

GEREBTZOFF, M. A.: État fonctionnel de l'écorce cérébrale au cours de l'hypnose animale. Arch. int. Physiol. **51**, 365–378 (1941).

GIANNAZZO, E., MANZONI, T., RAFFAELE, R., SAPIENZA, S., URBANO, A.: Effects of chronic fastigial lesions on the sleep-wakefulness rhythm in the cat. Arch. ital. Biol. **107**, 1–18 (1969).

GIAQUINTO, S., POMPEIANO, O., SOMOGYI, I.: Supraspinal modulation of heteronymous monosynaptic and of polysynaptic reflexes during natural sleep and wakefulness. Arch. ital. Biol. **102**, 245–281 (1964a).

— — — Descending inhibitory influences on spinal reflexes during natural sleep. Arch. ital. Biol. **102**, 282–307 (1964b).

GILMAN, T. T., MARCUSE, F. L.: Animal hypnosis. Psychol. Bull. **46**, 151–165 (1949).

GODDARD, G. V.: Amygdaloid stimulation and learning in the rat. J. comp. physiol. Psychol. **58**, 23–30 (1964).

GOLTZ, F.: Der Hund ohne Großhirn—Siebente Abhandlung über die Verrichtungen des Großhirns. Pflügers Arch. ges. Physiol. **51**, 570–614 (1892).

GOODMAN, S. J., MANN, P. E. G.: Reticular and thalamic multiple unit activity during wakefulness, sleep and anesthesia. Fxp. Neurol. **19**, 11–24 (1967).

GRANIT, R.: Recurrent inhibition as a mechanism of control. In: Brain mechanisms (MORUZZI, G., FESSARD, A. and JASPER, H. H., eds.). Amsterdam: Elsevier 1963. Progr. Brain Res. **1**, 23–37 (1963).

— (ed.): Muscular afferents and motor control. First Nobel Symposium. Stockholm: Almqvist & Wiksell 1966. 466 pp.

GUGLIELMINO, S., STRATA, P.: Cerebellum and atonia of the desynchronized phase of sleep. Arch. ital. Biol. **109** (1971) (in press).

HALÁSZ, B., PUPP, L.: Hormone secretion of the anterior pituitary gland after physical interruption of all nervous pathways to the hypophysiotropic area. Endocrinology **77**, 553–562 (1965).

HARDY, J. D.: Physiology of temperature regulation. Physiol. Rev. **41**, 521–606 (1961).

HARMS, E. (ed.): Problems of sleep and dreaming in children. Oxford: Pergamon Press 1964. 147 pp.

HARRISON, F.: The hypothalamus and sleep. Res. Publ. Ass. nerv. ment. Dis. **20**, 635–656 (1939).

— An attempt to produce sleep by diencephalic stimulation. J. Neurophysiol. **3**, 156–165 (1940).

HARTMANN, E.: The biology of deraming. Springfield, Ill.: Ch. C. Thomas 1967. XIII–206 pp.

HASSENBERG, L.: Ruhe und Schlaf bei Säugetieren. Wittenberg-Lutherstadt: A. Ziemsen 1965. 160 pp.

HASSLER, R.: Motorische und sensible Effekte umschriebener Reizungen und Ausschaltungen im menschlichen Zwischenhirn. Dtsch. Z. Nervenheilk. **183**, 148–171 (1961).

— Funktionelle Neuroanatomie und Psychiatrie. In: Psychiatrie der Gegenwart (herausgeg. v. H. W. GRUHLE, R. JUNG, W. MAYER-GROSS u. M. MÜLLER), Bd. I/1 A, S. 152–285; see 189–203, 243–270. Berlin-Heidelberg-New York: Springer 1967.

— Personal communication (1970).

— RIECHERT, T.: Wirkungen der Reizungen und Koagulationen in den Stammganglien bei stereotaktischen Hirnoperationen. Nervenarzt **32**, 97–109 (1961).

HEAD, H.: The conception of nervous and mental energy (II). ("Vigilance": a physiological state of the nervous system). Brit. J. Psychol. **14**, 126–147 (1923).

HEBB, D. O.: The organization of behavior. A neuropsychological theory. See p. 211. New York: J. Wiley & Sons 1949. XIX + 335 pp.

HEBB, D.O.: Drives and the C.N.S. (conceptual nervous system). Psychol. Rev. **62**, 243–254 (1955).

HEDIGER, H.: Wie Tiere schlafen. Med. Klin. **54**, 938–946, 965–968 (1959).

— The psychology of animals in zoos and circuses. NewYork: Dover Publ. 1968. VII + 166 pp.

— Comparative observations on sleep. Proc. roy. Soc. Med. **62** (2), 153–156 (1969).

HENSEL, H.: Physiologie der Thermoreception. Ergebn. Physiol. **47**, 166–368 (1952).

HERNÁNDEZ-PEÓN, R.: Sleep induced by localized electrical or chemical stimulation of the forebrain. Electroenceph. clin. Neurophysiol. **14**, 423–424 (1962).

— (ed.): The physiological basis of mental activity. Electroenceph. clin. Neurophysiol., Suppl. **24** (1963). X + 283 pp.

— Die neuralen Grundlagen des Schlafes. Arzneimittel-Forsch. **15**, 1099–1118 (1965).

HESS, R., Jr.: The EEG in sleep. Electroenceph. clin. Neurophysiol. **16**, 44–55 (1964).

— AKERT, K., KOELLA, W.: Die hirnelektrische Aktivität des Cortex und des Thalamus der Katze bei künstlich modifizierter Reaktionsbereitschaft des Zentralnervensystems. Helv. physiol. pharmacol. Acta **8**, 60–62 (1950).

— KOELLA, W.P., AKERT, K.: Cortical and subcortical recordings in natural and artificially induced sleep in cats. Electroenceph. clin. Neurophysiol. **5**, 75–90 (1953).

HESS, W.R.: Über die Wechselbeziehungen zwischen psychischen und vegetativen Funktionen. Neurol. Psychiatr. Abhandl. aus dem Schweiz. Arch. Neurol. Psychiat. H. 2. Zürich-Leipzig-Berlin: Orell Füssli Verlag 1925. 60 pp.

— Stammganglien Reizversuche. Tagg. Dtsch. Physiol. Ges. in Frankfurt a.M., 27. bis 30. Sept. 1927. Ber. ges. Physiol. **42**, 554–555 (1927).

— Le sommeil. C.R. Soc. Biol. (Paris) **107**, 1333–1364 (1931).

— Der Schlaf. Klin. Wschr. **12**, 129–134 (1933).

— Hypothalamische Adynamie. Helv. physiol. pharmacol. Acta **2**, 137–147 (1944a).

— Das Schlafsyndrom als Folge dienzephaler Reizung. Helv. physiol. pharmacol. Acta **2**, 305–344 (1944b).

— Le sommeil comme fonction physiologique. J. Physiol. (Paris) **41**, 61 A–67 A (1949a).

— Das Zwischenhirn — Syndrome, Lokalisationen, Funktionen. Basel: Benno Schwabe & Co. 1949b. 187 pp.

— Symposium über das Zwischenhirn. Helv. physiol. pharmacol. Acta, Suppl. **6**, 1–80 (1950).

— The diencephalic sleep centre. In: Brain mechanisms and consciousness (ADRIAN, E.D., BREMER, F. and JASPER, H.H., eds.), p. 117–136. Oxford: Blackwell 1954. XV + 556 pp.

— Hypothalamus und Thalamus, 2. Aufl. Stuttgart: G. Thieme 1968a. 77 pp.

— Psychologie in biologischer Sicht, 2. Aufl. Stuttgart: G. Thieme 1968b. XI + 132 pp.

HILL, D.: Normal and pathological sleep. Proc. roy. Soc. Med. **55**, 905–907 (1962).

HINDE, R.A.: Animal behaviour. NewYork: McGraw-Hill Book Co. 1966. X + 534 pp.

HISHIKAWA, Y.: Neurophysiological nature of narcoleptic symptoms. In: The abnormalities of sleep in man (GASTAUT, H., LUGARESI, E. BERTI CERONI, G., and COCCAGNA, G.. eds.), p. 165–175. Bologna: Aulo Gaggi 1968. 326 pp.

HO, T., WANG, Y.R., LIN, T.A.N., CHENG, Y.F.: Predominance of electrocortical sleep patterns in the "encéphale isolé" cat and new evidence for a sleep center. Physiol. bohemoslov. **9**, 85–92 (1960).

HOBSON, J.A.: The effects of chronic brain-stem lesions on cortical and muscular activity during sleep and waking in the cat. Electroenceph. clin. Neurophysiol. **19**, 41–62 (1965).

HODES, R.: Electrocortical synchronization resulting from reduced proprioceptive drive course by neuromuscular blocking agents. Electroenceph. clin. Neurophysiol. **14**, 220–232 (1962).

— Electrocortical desynchronization resulting from spinal block: evidence from synchronizing influences in the cervical cord. Arch. ital. Biol. **102**, 183–196 (1964).

Hösli, L., Monnier, M.: Schlaf- und Weckwirkungen des intralaminären Thalamus. Reizparameter und Koagulationsbefunde beim Kaninchen. Pflügers Arch. ges. Physiol. **275**, 439–451 (1962).

— — Die funktionelle Dualität des mesencephalen Retikularsystems. I. Aktivierende und dämpfende Reizwirkungen auf die elektrische Hirnaktivität und das Verhalten. Pflügers Arch. ges. Physiol. **278**, 241–249 (1963).

Hoffmeister, F.: Über den physiologischen und den medikamentösen Schlaf. Dtsch. med. Wschr. **93**, 85–92 (1968).

Holzapfel, M.: Triebbedingte Ruhezustände als Ziel von Appetenzhandlungen. Naturwissenschaften **28**, 273–280 (1940).

Hoofdakker, R. H., van den: Behaviour and EEG of drowsy and sleeping cats. Groningen: Drukkerij van Denderen 1966. III + 111 pp.

Housepian, E. M., Purpura, D. P.: Electrophysiological studies of subcortical-cortical relations in man. Electroenceph. clin. Neurophysiol. **15**, 20–28 (1963).

Hunsperger, R. W., Brown, J. L., Rosvold, H. E.: Combined stimulation in areas governing threat and flight behaviour in the brain stem of the cat. Progr. brain Res. **6**, 191–197 (1964).

Hunter, J., Jasper, H. H.: Effect of thalamic stimulation in unanesthetized animals. Electroenceph. clin. Neurophysiol. **1**, 305–324 (1949).

Huttenlocher, P. R.: Evoked and spontaneous activity in single units of medial brain stem during natural sleep and waking. J. Neurophysiol. **24**, 451–468 (1961).

Ingram, W. R., Barris, R. W., Ranson, S. W.: Catalepsy. An experimental study. Arch. Neurol. Psychiat. (Chic.) **35**, 1175–1197 (1936).

Iwamura, Y., Sterman, M. B., McGinty, D. J.: Spinal and brain stem reflex activity during paradoxical sleep in brain transected cats. Physiologist **12**, 260 (1969).

Jasper, H. H.: Diffuse projection systems: the integrative action of the thalamic reticular system. Electroenceph. clin. Neurophysiol. **1**, 405–419 (1949).

— Thalamic reticular system. In: Electrical stimulation of the brain (Sheer, D. E., ed.), p. 277–287. Austin: University of Texas Press 1961. XV + 641 pp.

— Proctor, L. D., Knighton, R. S., Nushay, W. C., Costello, R. T. (eds.): Reticular formation of the brain. Boston: Little, Brown & Co. 1958. XIV + 766.

— Smirnov, G. D. (eds.): The Moscow Colloquium on electroencephalography of higher nervous activity. Electroencephal. clin. Neurophysiol. Suppl. **13** (1960). XVI + 420 pp.

Jeannerod, M.: Sommeil, éveil et formation réticulée. Rev. lyon. Méd. **18**, 9–28 (1969).

Jouvet, D., Valatx, J. L.: Etude polygraphique du sommeil chez l'agneau. C.R. Soc. Biol. (Paris) **156**, 1411–1414 (1962).

Jouvet, M.: Recherches sur les structures nerveuses et les mécanismes responsables des différentes phases du sommeil physiologique. Arch. ital. Biol. **100**, 125–206 (1962).

— (ed.): Aspects anatomo-fonctionnels de la physiologie du sommeil. Paris: Centre National de la Recherche Scientifique 1965a. 657 pp.

— Etude de la dualité des états de sommeil et des mécanismes de la phase paradoxale. In: Aspects anatomo-fonctionnels de la physiologie du sommeil (Jouvet, M., ed.), p. 397–449. Paris: Centre National de la Recherche Scientifique 1965b. 657 pp.

— Behavioural and EEG effects of paradoxical sleep deprivation in the cat. Proc. XXIII int. Congr. physiol. Sci. Tokyo 1965. Lectures and Symposia, p. 344–353. Amsterdam: Excerpta Medica Foundation 1965c. 644 pp.

— Neurophysiology of the states of sleep. Physiol. Rev. **47**, 117–177 (1967a).

— Mechanisms of the state of sleep: a neuropharmacological approach. Res. Publ. Ass. nerv. ment. Dis. **45**, 86–126 (1967b).

— The states of sleep. Sci. Amer. **216**, 62–72 (1967c).

— Jouvet, D.: Le sommeil et les rêves chez l'animal. In: Psychiatrie animale (Brion, A., et H. Ey, eds.), p. 149–167. Paris: Desclée de Brouwer 1964. 605 pp.

— Renault, J.: Insomnie persistante après lésions des noyaux du raphé chez le chat. C.R. Soc. Biol. (Paris) **160**, 1461–1465 (1966).

JOUVET-MOUNIER, D.: Ontogenèse des états de vigilance chez quelques mammifères. Thèse Université de Lyon, 231 pp., 1968.

JOVANOVIĆ, U. J. (ed.): Der Schlaf. Neurophysiologische Aspekte. München: J. A. Barth 1969. 252 pp.

JUNG, R.: Der Schlaf: In: Physiologie und Pathologie des vegetativen Nervensystems (MONNIER, M., ed.), Bd. 2, S. 651–684. Stuttgart: Hippokrates Verlag 1963. 2 Vols.

— Physiologie und Pathophysiologie des Schlafes. Verh. dtsch. Ges.inn. Med. 71, 788–797 (1965).

— Neurophysiologie und Psychiatrie. In: Psychiatrie der Gegenwart (herausgeg. v. H. W. GRUHLE, R. JUNG, W. MAYER-GROSS u. M. MÜLLER), Bd. I/1 A, S. 325–928; see p. 556–611. Berlin-Heidelberg-New York: Springer 1967.

KANZOW, E., KRAUSE, D., KÜHNEL, H.: Die Vasomotorik der Hirnrinde in den Phasen desynchronisierter EEG-Aktivität im natürlichen Schlaf der Katze. Pflügers Arch. ges. Physiol. 274, 593–607 (1962).

KARPLUS, J. P., KREIDL, A.: Gehirn und Sympathicus. VII. Mitt. Über Beziehungen der Hypothalamuszentren zu Blutdruck und innerer Sekretion. Pflügers Arch. ges. Physiol. 215, 667–670 (1927).

KAWAMURA, H., POMPEIANO, O.: Phasic D. C. potentials shifts in the sensorimotor and visual cortices during desynchronized sleep. Pflügers Arch. ges. Pyhsiol. 317, 10–19(1970).

— SAWYER, C. H.: D-C potential changes in rabbit brain during slow wave and paradoxical sleep. Amer. J. Physiol. 207, 1379–1386 (1964).

KELLER, A. D.: Autonomic discharges elicited by physiological stimuli in midbrain preparations. Amer. J. Physiol. 100, 576–586 (1932).

— The striking inherent tonus of the deafferented central pupillo-constrictor neurons. Fed. Proc. 5, 55–57 (1946).

KETY, S. S.: Relationship between energy metabolism of the brain and functional activity. Res. Publ. Ass. nerv. ment. Dis. 45, 39–47 (1967).

— EVARTS, E. V., WILLIAMS, H. L. (eds.): Sleep and altered states of consciousness. Res. Publ. Ass. nerv. ment. Dis. 45 (1967). XII + 591 pp.

KING, F. A., MARCHIAFAVA, P. L.: Ocular movements in the midpontine pretrigeminal preparation. Arch. ital. Biol. 101, 149–160 (1963).

— — MORUZZI, G.: The time course of the mydriatic response to darkness in the midpontine pretrigeminal cat. Arch. ital. Biol. 101, 545–551 (1963).

KLAUE, R.: Die bioelektrische Tätigkeit der Großhirnrinde im normalen Schlaf und in der Narkose durch Schlafmittel. J. Psychol. Neurol. (Lpz.) 47, 510–531 (1937), see p. 513–515.

KLEE, M. R.: Different effects on the membrane potential of motor cortex units after thalamic and reticular stimulation. In: The thalamus (PURPURA, D. P., and YAHR, M. D. eds.), p. 287–321. New York: Columbia University Press 1966. IX + 438 pp.

KLEIN, M.: Étude polygraphique et phylogénique des états de sommeil. Thèse Universitè de Lyon, 101 pp., 1963.

KLEITMAN, N.: Sleep. Physiol. Rev. 9, 624–665 (1929). See pp. 655–658.

— Sleep, wakefulness and consciousness. Psychol. Bull. 54, 354–359 (1957).

— Sleep and wakefulness, 2nd ed. Chicago: Chicago University Press 1963. X + 552 pp.

— Phylogenetic, ontogenetic and environmental determinants in the evolution of sleep-wakefulness cycles. Res. Publ. Ass. nerv. ment. Dis. 45, 30–38 (1967). See pp. 32, 36.

— CAMILLE, N.: Studies on the physiology of sleep. VI. The behavior of decorticated dogs. Amer. J. Physiol. 100, 474–480 (1932).

KLEMM, W. R.: Electroencephalographic-behavioral dissociations during animal hypnosis. Electroenceph. clin. Neurophysiol. 21, 365–372 (1966).

— Sleep and paradoxical sleep in ruminants. Proc. Soc. exp. Biol. (N.Y.) 121, 635–638 (1966).

KNOTT, J. R., INGRAM, W. R., CHILES, W. D.: Effects of subcortical lesions on cortical electroencephalogram in cats. Arch. Neurol. Psychiat. (Chic.) 73, 203–215 (1955).

Koch, E.: Die Irradiation der pressor-receptorischen Kreislaufreflexe. Klin. Wschr. **11**, 225–227 (1932).

Koella, W. A.: Sleep. Its nature and physiological organization. Springfield: Ch. C. Thomas 1967. XIV + 199 pp.

Konorski, J.: Conditioned reflexes and neuron organization. Cambridge: Cambridge University Press 1948. XIV + 267 pp.

— Introduction to the Symposium "The functional properties of hypothalamus". Acta Biol. exp. (Warszawa) **27**, 265–267 (1967a).

— Integrative activity of the brain. Chicago and London: Chicago University Press 1967b. XII + 531 pp.

Kornmüller, A. E., Lux, H. D., Winkel, K., Klee, M.: Neurohumoral ausgelöste Schlafzustände an Tieren mit gekreuztem Kreislauf unter der Kontrolle von EEG-Ableitungen. Naturwissenschaften **48**, 503–505 (1961).

Kruta, V.: J. E. Purkyně's conception of the physiological basis of wakefulness and sleep. Scr. med. Fac. Med. Brun. **40**, 281–290 (1967).

Kumazawa, T.: Deactivation of the rabbit's brain by pressure application to the skin. Electroenceph. clin. Neurophysiol. **15**, 660–671 (1963).

— Baccelli, G., Guazzi, M., Mancia, G., Zanchetti, A.: Hemodynamic patterns during desynchronized sleep in intact cats and in cats with sinoaortic deafferentation. Circulat. Res. **24**, 923–937 (1969).

Legendre, R., Piéron, H.: Recherches sur le besoin de sommeil consécutif à une veille prolongée. Z. allg. Physiol. **14**, 235–262 (1913).

Liddell, E. G. T.: Spinal shock and some features in isolation alteration of the spinal cord in cats. Brain **57**, 386–400 (1934).

Lindsley, D. B.: Emotion. In: Handbook of exp. psychol. (Stevens, S. E., ed.), p. 473–516. New York: Wiley & Sons 1951. XI + 1436 pp.

— The reticular activating system and perceptual integration. In: Electrical stimulation of the brain (Sheer, D. E., ed.), p. 331–349. Austin: University of Texas Press 1961. XV + 641 pp.

— Bowden, J. W., Magoun, H. W.: Effect upon the EEG of acute injury to the brain stem activating system. Electroenceph. clin. Neurophysiol. **1**, 475–486 (1949).

— Schreiner, L. H., Knowles, W. B., Magoun, H. W.: Behavioral and EEG changes following chronic brain stem lesions in the cat. Electroenceph. clin. Neurophysiol. **2**, 483–498 (1950).

Loeb, C., Poggio, G.: Electroencephalograms in a case with ponto-mesencephalic haemorrage. Electroenceph. clin. Neurophysiol. **5**, 295–296 (1953).

Lorenz, K.: Über die Bildung des Instinktbegriffes. Naturwissenschaften **25**, 289–300, 307–318, 324–331 (1937).

Luce, G. G.: Current research on sleep and dreams. Public Health Service Publ. No 1389 (1965). VII + 125 pp.

Macchi, G.: Introductory statement about the thalamo-cortical connections. Arch. ital. Biol. **107**, 547–569 (1969).

Madoz Jáuregui, P.: Influencia de la región preóptica en la regulación de la actividad eléctrica cerebral. An. Anat. **18**, 477–537 (1969).

Maffei, L., Moruzzi, G., Rizzolatti, G.: Influence of sleep and wakefulness on the response of lateral geniculate units to sinewave photic stimulation. Arch. ital. Biol. **103**, 596–608 (1965).

— Rizzolatti, G.: Effect of synchronized sleep on the response of lateral geniculate units to flashes of light. Arch. ital. Biol. **103**, 609–622 (1965).

Magnes, J., Moruzzi, G., Pompeiano, O.: Synchronization of the EEG produced by low-frequency electrical stimulation of the region of the solitary tract. Arch. ital. Biol. **99**, 33–67 (1961a).

— — — Electroencephalogram-synchronizing structures in the lower brain stem. In: The nature of sleep (Wolstenholme, G. F. W., and O'Connor, M. C., eds.), p. 57–85. London: Churchill 1961b. XII + 416 pp.

Magni, F., Moruzzi, G., Rossi, G. F., Zanchetti, A.: EEG arousal following inactivation of the lower brain stem by selective injection of barbiturate into the vertebral circulation. Arch. ital. Biol. 97, 33–46 (1959).

— Willis, W. D.: Identification of reticular formation neurons by intracellular recording. Arch. ital. Biol. 101, 681–702 (1963).

— — Cortical control of brain stem reticular neurons. Arch. ital. Biol. 102, 418–433 (1964).

Magnus, R.: Körperstellung. See p. 4. Berlin: Springer 1924. XIII + 740 pp.

Magoun, H. W.: Bulbar inhibition and facilitation of motor activity. Science 100, 549–550 (1944).

— The waking brain, 2nd ed. Springfield: Ch. C. Thomas 1963. VIII + 188 pp.

— Harrison, F., Brobeck, J. R., Ranson, S. W.: Activation of heat loss mechanisms by local heating of the brain. J. Neurophysiol. 1, 101–114 (1938).

— Rhines, R.: Spasticity. The stretch-reflex and extrapyramidal systems. Springfield: Ch. C. Thomas 1947. VII + 59 pp.

Malcom, L. J., Bruce, I. S. C., Burke, W.: Excitability of the lateral geniculate nucleus in the alert, non-alert, and sleeping cat. Exp. Brain Res. 10, 283–297 (1970).

Malliani, A., Bizzi, E., Apelbaum, J., Zanchetti, A.: Ascending afferent mechanisms maintaining sham rage behavior in the acute thalamic cat. Arch. ital. Biol. 101, 632–647 (1963).

Mancia, M.: Electrophysiological and behavioural changes owing to splitting of the brain stem in cats. Electroenceph. clin. Neurophysiol. 27, 487–502 (1969).

— Desiraju, T., Chhina, G. S.: The Monkey split brain-stem: effects on the sleep-wakefulness cycle. Electroenceph. clin. Neurophysiol. 24, 409–416 (1968).

— Mechelse, K., Mollica, A.: Microelectrode recording from midbrain reticular formation in the decerebrate cat. Arch. ital. Biol. 95, 110–119 (1957).

— Meulders, M., Santibañez, H. G.: Synchronisation de l'électroencéphalogramme provoquée par la stimulation visuelle répétitive chez le chat "médiopontin prétrigéminal". Arch. int. Physiol. 67, 661–670 (1959).

Mantegazzini, P., Poeck, K., Santibañez, G.: The action of adrenaline and noradrenaline on the cortical electrical activity of the "encéphale isolé" cat. Arch. ital. Biol. 97, 222–242 (1959).

Manzoni, T., Sapienza, S., Urbano, A.: EEG and behavioural sleep-like effects induced by the fastigial nucleus in unrestrained, unanaesthetized cats. Arch. ital. Biol. 106, 61–72 (1968).

Marchesi, G. F., Strata, P.: Climbing fibers of cat cerebellum: modulation of activity during sleep. Brain Res. 17, 145–148 (1970).

Marchiafava, P. L., Pompeiano, O.: Pyramidal influences on spinal cord during desynchronized sleep. Arch. ital. Biol. 102, 500–529 (1964).

Martini, L., Ganong, W. F.: Neuroendocrinology. 2 vols. New York and London: Academic Press 1966 and 1967.

Massion, J., Angaut, P., Albe-Fessard, D.: Étude des facteurs impliqués dans l'accroissement des activités corticales observé après cérébellectomie. Electroenceph. clin. Neurophysiol. 18, 455–463 (1965).

Mazzella, H., García Austt, E., García Mullin, R.: Carotid sinus and EEG. Electroenceph. clin. Neurophysiol. 8, 155 (1956).

— García Mullin, R., García Austt, E.: Effect of carotid sinus stimulation on the EEG. Acta neurol. lat.-amer. 3, 361–364 (1957).

McCarley, R. W., Hobson, J. A.: Cortical unit activity in desynchronized sleep. Science 167, 901–903 (1970).

McGinty, D. J.: Somnolence, recovery, and hyposomnia following ventro-medial diencephalic lesions in the rat. Electroenceph. clin. Neurophysiol. 26, 70–79 (1969).

— Sterman, M. B.: Sleep suppression after basal forebrain lesions in the cat. Science 160, 1253–1255 (1968).

MEULDERS, M.: Approche neurophysiologique des mécanismes du sommeil. J. Neurol. Sci. **2**, 459–473 (1965).

MICHEL, F.: L'encéphale dédoublé chez le Chat. Communications Association des Physiologistes, Bordeaux 1967. J. Physiol. (Paris) **59**, 266 (1967).

— ROFFWARG, H. P.: Chronic split brain stem preparation: effect on the sleep-waking cycle. Experientia (Basel) **23**, 126–128 (1967).

MINK, W. D., BEST, P. J.: Neurons in paradoxical sleep and motivated behavior. Science **158**, 1335–1337 (1967).

MOLINARI, S., FOULKES, D.: Tonic and phasic events during sleep: psychological correlates and implications. Perceptual and motor skills **29**, 343–368 (1969).

MOLLICA, A.: Absence de réaction électroencéphalographique d'éveil par stimulation électrique corticale chez le chat "cerveau isolé". Arch. ital. Biol. **96**, 216–230 (1958).

— MORUZZI, G., NAQUET, R.: Décharges réticulaires induites par la polarisation du cervelet: leur rapports avec le tonus postural et la réaction d'éveil. Electroenceph. clin. Neurophysiol. **5**, 571–584 (1953).

MONAKOW, C. VON: Über den gegenwärtigen Stand der Frage nach der Lokalisation im Großhirn. Ergebn. Physiol. **1** (II. Abt.), 534–665 (1902). See p. 569–570.

MONNIER, M., FALLERT, M.: Neurophysiologische und biochemische Regulationsmechanismen der Wachfunktion. Wien. klin. Wschr. **79**, 509–515 (1967a).

— — Neurophysiologische und biochemische Mechanismen der Schlafsteuerung. Schweiz. med. Wschr. **97**, 866–875 (1967b).

MORISON, R. S., DEMPSEY, E. W.: A study of thalamo-cortical relations. Amer. J. Physiol. **135**, 281–292 (1942).

MORRISON, A. R., POMPEIANO, O.: Vestibular influences during sleep. II. Effects of vestibular lesions on the pyramidal discharge during desynchronized sleep. Arch. ital. Biol. **104**, 214–230 (1966a).

— — Vestibular influences during sleep. IV. Functional relations between vestibular nuclei and lateral geniculate nucleus during desynchronized sleep. Arch. ital. Biol. **104**, 425–458 (1966b).

MORUZZI, G.: The physiological properties of the brain stem reticular system. In: Brain mechanisms and consciousness. (ADRIAN, E. D., BREMER, F., and JASPER, H. H., eds.) p. 21–53. Oxford: Blackwell 1954. XV + 556 pp.

— The functional significance of the ascending reticular system. Arch. ital. Biol. **96**, 17–28 (1958).

— Synchronizing influences of the brain stem and the inhibitory mechanisms underlying the production of sleep by sensory stimulation. Electroenceph. clin. Neurophysiol. Suppl. **13**, 231–256 (1960).

— The midpontine pretrigeminal cat. Arch. int. Pharmacodyn. **140**, 227–230 (1962).

— The physiology of sleep. Endeavour **22**, 31–36 (1963a).

— Active processes in the brain stem during sleep. Harvey Lect. **58**, 233–297 (1963b).

— Reticular influences on the EEG. Electroenceph. clin. Neurophysiol. **16**, 2–17 (1964a).

— The historical development of the deafferentation hypothesis of sleep. Proc. Amer. Phil. Soc. **108**, 19–28 (1964b).

— Summary statement. Progr. Brain Res. **18**, 241–243 (1965).

— The functional significance of sleep with particular regard to the brain mechanisms underlying consciousness. In: Brain and conscious experience (ECCLES, J. C., ed.), p. 345–388. Berlin-Heidelberg-New York: Springer 1966. XXI + 591 pp.

— Sleep and instinctive behavior. Arch. ital. Biol. **107**, 175–216 (1969).

— FESSARD, A., JASPER, H. H.: Brain mechanisms. Progr. Brain Res. **1**. Amsterdam: Elsevier 1963. XIX + 493 pp.

— MAGOUN, H. W.: Brain stem reticular formation and activation of the EEG. Electroenceph. clin. Neurophysiol. **1**, 455–473 (1949).

— POMPEIANO, O.: Effects of vermal stimulation after fastigial lesions. Arch. ital. Biol. **95**, 31–55 (1957).

Mukhametov, L. M., Rizzolatti, G.: The responses of lateral geniculate neurons to flashes of light during the sleep-waking cycle. Arch. ital. Biol. **108**, 348–368 (1970).

— — Tradardi, V.: Spontaneous activity of neurones of nucleus reticularis thalami in freely moving cats. J. Physiol. (Lond.), **210**, 651–667 (1970).

Naquet, R., Denavit, M., Albe-Fessard, D.: Comparaison entre le rôle du subthalamus et celui des différentes structures bulbomésencéphaliques dans le maintien de la vigilance. Electroenceph. clin. Neurophysiol. **20**, 149–164 (1966).

— — Lanoir, J., Albe-Fessard, D.: Altérations transitoires ou définitives de zones diencéphaliques chez le chat. Leurs effets sur l'activité électrique corticale et le sommeil. In: Aspects Anatomo-fonctionnels de la physiologie du sommeil. (Jouvet, M., ed.), p. 107–131. Paris: Centre National de la Recherche Scientifique 1965. 657 pp.

Nauta, W. J. H.: Hypothalamic regulation of sleep in rats. Experimental study. J. Neurophysiol. **9**, 285–316 (1946).

— Koella, W. P., Quarton, G. C. (eds.): Sleep, wakefulness, dreams, and memory. Neurosciences Research Symposium Summaries, **2**, p. 1–90. Boston: M.I.T. Press 1967. XIII + 642 pp.

Nicholls, E. E.: A study of the spontaneous activity of the guinea pig. J. comp. Psychol. **2**, 303–330 (1922).

Nissen, H. W.: Phylogenetic comparison. In: Handbook of experimental psychology (Stevens, S. S., ed.), p. 347–386. New York: J. Wiley & Sons 1951. XI + 1436 pp.

Noda, H., Adey, W. R.: Firing variability in cat association cortex during sleep and wakefulness. Brain Res. **18**, 513–526 (1970).

— — Changes in neuronal activity in association cortex of the cat in relation to sleep and wakefulness. Brain Res. **19**, 263–275 (1970).

— Manohar, S., Adey, W. R.: Spontaneous activity of cat hyppocampal neurons in sleep and wakefulness. Exp. Neurol. **24**, 217–231 (1969).

Orthner, H.: Neuroanatomische Gesichtspunkte der Schlaf-Wach-Regelung. In: Der Schlaf (Jovanović, U. J., ed.), p. 49–84. München: J. A. Barth 1969. 252 pp.

Oswald, I.: Sleeping and waking. Physiology and psychology. Amsterdam: Elsevier 1962. 232 pp.

— Sleep. Baltimore: Penguin Books 1966. 141 pp.

Palestini, M.: Mecanismos neurofisiológicos del sueño y de la actividad onírica. Acta neurol. lat.-amer. **11**, 41–54 (1965).

Pappenheimer, J. R., Miller, T. B., Goodrich, C. A.: Sleep-promoting effects of cerebrospinal fluid from sleep-deprived goats. Proc. nat. Acad. Sci. (Wash.) **58**, 513–517 (1967).

Parmeggiani, P. L.: Sleep behaviour elicited by electrical stimulation of cortical and subcortical structures in the cat. Helv. physiol. pharmacol. Acta **20**, 347–367 (1962).

— A study on the central representation of sleep behaviour. Progr. Brain Res. **6**, 180–190 (1964).

— Telencephalo-diencephalic aspects of sleep mechanisms. Brain Res. **7**, 350–359 (1968).

— Rabin, C.: Sleep and envinronmental temperature. Arch. ital. Biol. **108**, 369–388 (1970).

Passouant, P. (ed.): Physiologie de l'hippocampe. Paris: Centre National de la Recherche Scientifique 1962. 512 pp.

— Problèmes physiopathologiques de la narcolepsie et périodicité du "Sommeil Rapide" au cours du nycthèmere. In: The abnormalities of sleep in man (Gastaut, H., Lugaresi, E., Berti Ceroni, E., and Coccagna, G., eds.), p. 177–189. Bologna: Aulo Gaggi 1968. 326 pp.

— Cadilhac, J.: Les rythmes theta hippocampiques au cours du sommeil. In: Physiologie de l'Hippocampe (Passouant, P., ed.), p. 331–347. Paris: Centre National de la Recherche Scientifique 1962. 512 pp.

— — Baldy-Moulinier, M.: Physiopathologie des hypersomnies. Rev. neurol. **116**, 585–629 (1967).

Pavlov, I. P.: „Innere Hemmung" der bedingten Reflexe und der Schlaf — ein und derselbe Prozeß. Skand. Arch. Physiol. **44**, 42–58 (1923).
— Conditioned reflexes. An investigation of the physiological activity of the cerebral cortex (Anrep, G. V., ed.). Oxford: Oxford University press 1927. XV–430 pp.
— Sämtliche Werke, 2. Aufl. Berlin: Akademie-Verlag 1953–1955. 6 Bände.
— In: Pawlowsche Mittwochkolloquien. Berlin: Akademie Verlag 1955–1956. 3 Bände.
Pellet, J., Béraud, G.: Organisation nycthémérale de la veille et du sommeil chez la cobaye (*Cavia Porcellus*). Physiol. Behav. **2**, 131–137 (1967).
Penfield, W.: Speech, perception, and the uncommitted cortex. In: Brain and conscious experience (Eccles, J. C., ed.), p. 217–237. Berlin-Heidelberg-New York: Springer 1966. XXI + 591 pp.
— Perot, P.: The brain's record of auditory and visual experience. Brain **86**, 595–702 (1963).
Penfield, W. G.: Group discussion in brain mechanisms and consciousness (Adrian, E. D., Bremer, F., and Jasper, H. H., eds.), p. 125–136. Oxford: Blackwell 1954. XV +556 pp.
Piéron, H.: Le problème physiologique du sommeil. Paris: Masson 1913. XV +520 pp.
Ploog, D.: Über den Schlaf und seine Beziehungen zu endogenen Psychosen. Münch. med. Wschr. **95**, 897–900 (1953).
— Verhaltensforschung und Psychiatrie. In: Psychiatrie der Gegenwart (herausgeg. H. W. v. Gruhle, R. Jung, W. Mayer-Gross und M. Müller), Bd. I, S. 291–443; see 368–375. Berlin-Göttingen-Heidelberg-New York: Springer 1964.
Podvoll, E. M., Goodman, S. J.: Averaged neural electrical activity and arousal. Science **155**, 223–225 (1967).
Poeck, K.: Die Formatio reticularis des Hirnstamms. Nervenarzt **30**, 289–298 (1955).
Pompeiano, O.: Ascending and descending influences of somatic afferent volleys in unrestrained cats; supraspinal inhibitory control of spinal reflexes during natural and reflexly induced sleep. In: Aspects Anatomo-fonctionnels de la Physiologie du Sommeil. (Jouvet, M., ed.), p. 309–395. Paris: Centre National de la Recherche Scientifique 1965. 657 pp.
— Muscular afferents and motor control during sleep. In: Muscular afferents and motor control (Granit, R., ed.), First Nobel Symposium, p. 415–436. Stockholm: Almqvist & Wiksell 1966. 466 pp.
— Sensory inhibition during motor activity in sleep. In: Neurophysiological basis of normal and abnormal motor activities (Yahr, M. D. and Purpura, D. P., eds.), p. 323–375. Hewlett, N.Y.: Raven Press 1967a. XI +5000 pp.
— The neurophysiological mechanisms of the postural and motor events during desynchronized sleep. Res. Publ. Ass. nerv. ment. Dis. **45**, 351–423 (1967b).
— Morrison, A. R.: Vestibular influences during sleep. I. Abolition of the rapid eye movements of desynchronized sleep following vestibular lesions. Arch. ital. Biol. **103**, 569–595 (1965).
— — Vestibular influences during sleep. III. Dissociation of the tonic and phasic inhibition of spinal reflexes during desynchronized sleep following vestibular lesions. Arch. ital. Biol. **104**, 231–246 (1966).
— Swett, J. E.: EEG and behavioral manifestations of sleep induced by cutaneous nerve stimulation in normal cats. Arch. ital. Biol. **100**, 311–342 (1962a).
— — Identification of cutaneous and muscular afferent fibers producing EEG synchronization or arousal in normal cats. Arch. ital. Biol. **100**, 343–380 (1962b).
— — Actions of graded cutaneous and muscular afferent volleys on brain stem units in the decerebrate cerebellectomized cat. Arch. ital. Biol. **101**, 552–583 (1963).
Portielje, A. F. J.: Triebleben bzw. intelligente Äußerungen beim Orang-Utan (*Pongo pigmaeus* Hoppius). Bijdr. Dierk. **27**, 61–114 (1939). (Quoted by Tinbergen, 1952.)
Purkinje, J. E.: Wachen, Schlaf, Traum und verwandte Zustände. In: Handwörterbuch der Physiologie (herausgeg. von R. Wagner,). 9 Bände. Bd. 3 (Abt. 2), S. 412–480. Braunschweig: Friedr. Vieweg & Sohn 1846.

PURPURA, D. P., FRIGYESI, T. L., McMURTRY, J. G., SCARFF, T.: Synaptic mechanisms in thalamic regulation of cerebello-cortical projection activity. In: The thalamus (PURPURA, D. P., and M. D. YAHR, eds.), p. 153–172. New York and London: Columbia University Press 1966. IX + 438 pp.

— YAHR, M. D. (eds.): The thalamus. New York and London: Columbia University Press 1966. IX + 438 pp.

QUARTON, G. C., MELNECHUK, TH., SCHMITT, F. O. (eds.): The neurosciences, see p. 499–633. New York: The Rockefeller University Press 1967. XIII + 962 pp.

RANSON, S. W.: Somnolence caused by hypothalamic lesions in the monkey. Arch. Neurol. Psychiat. (Chic.) 41, 1–23 (1939).

— MAGOUN, H. W.: The hypothalamus: Ergebn. Physiol. 41, 56–163 (1939).

RANSTROM, S.: The hypothalamus and sleep regulation. An experimental and morphological study. Acta path. microbiol. scand., Suppl. 70, 1–90 (1947).

RECHTSCHAFFEN, A., EAKIN, D.: Sleep and dream research. A bibliography. Los Angeles: Published by Brain Information Service, University of California 1968.

REETH, P. C. VAN, CAPON, A.: Sommeil provoqué chez le Lapin par des stimulations profondes, céphaliques et cervicales. C.R. Acad. Sci. (Paris) 255, 3050–3052 (1962).

RICHTER, C. P.: Biology of drives. J. comp. physiol. Psychol. 40, 129–134 (1947).

— Sleep and activity: their relation to the 24 hour clock. Res. Publ. Ass. nerv. ment. Dis. 45, 8–29 (1967).

RINALDI, F., HIMWICH, H. E.: Cholinergic mechanisms involved in function of meso-diencephalic activating system. Arch. Neurol. Psychiat. (Chic.) 73, 396–402 (1955).

RIOCH, D. Mc. K.: Group discussion in Brain Mechanisms and Consciousness (ADRIAN, E. D., BREMER, F., and JASPER, H. H., eds.), p. 125–136. Oxford: Blackwell 1954. XV + 556 pp.

ROBERTS, W. W., BERGUIST, E. H., ROBINSON, T. C. L.: Thermoregulatory grooming and sleep-like relaxation induced by local warming of preoptic area and anterior hypothalamus in opossum. J. comp. physiol. Psychol. 67, 182–188 (1969).

— ROBINSON, T. C. L.: Relaxation and sleep induced by warming of preoptic region and anterior hypothalamus in cats. Exp. Neurol. 25, 282–294 (1969).

ROBINSON, K. W., LEE, H. K.: Animal behaviour and heat regulation in hot atmospheres. University of Queensland Papers, Department of Physiology 1, 1–8 (1946). (Quoted by ROBERTS and ROBINSON, 1969).

ROELOFS, G. A., HOOFDAKKER, R. H. VAN DER, PRECHTL, H. F. R.: Sleep effects of subliminal brain stimulation in cats. Exp. Neurol. 8, 84–92 (1963).

ROFFWARG, H. P., MUZIO, J. N., DEMENT, W. C.: Ontogenetic development of the human sleep–dream cycle. Science 152, 604–619 (1966).

ROGER, A., ROSSI, G. F., ZIRONDOLI, A.: Le rôle des afférences des nerfs craniens dans le maintien de l'état vigile de la préparation "encéphale isolé". Electroenceph. Clin. Neurophysiol. 8, 1–13 (1956).

ROITBAK, A. I.: Electrical phenomena in the cerebral cortex during the extinction of orientation and conditioned reflexes. Electroenceph. clin. Neurophysiol., Suppl. 13, 91–100 (1960).

— ERISTAVI, N.: EEG and behavior reactions upon stimulation of nonspecific thalamic nuclei in unanesthetized cats. Acta Biol. exp. (Warszawa) 26, 463–482 (1966).

ROSADINI, G., GENTILOMO, A., ALEMÀ, G., ROSSI, G. F.: Effetti neurologici ed elettro-encefalografici dell'iniezione diretta di un barbiturico nel circolo vertebro-basilare del coniglio. Boll. Soc. ital. Biol. sper. 40, 838–841 (1964).

ROSINA, A., MANCIA, M.: Electrophysiological and behavioural changes following selective and reversible inactivation of lower brain-stem structures in chronic cats. Electroenceph. clin. Neurophysiol. 21, 157–167 (1966).

— SOTGIU, M. L.: Effects of intravertebral injection of a barbiturate on unit activity in lower brain stem. Brain Res. 6, 510–522 (1967).

ROSSI, G. F.: Ricerche sulla natura della miosi nel sonno e nella narcosi barbiturica. Arch. Sci. Biol. (Bologna) 41, 46–56 (1957).

Rossi, G. F.: A hypothesis on the neural basis of consciousness. Acta neurochir. (Wien) 12, 187–197 (1964).
— Brain stem facilitating influences on EEG synchronization. Experimental findings and observations in man. Acta neurochir. (Wien) 13, 257–288 (1965).
— Minobe, K., Candia, O.: An experimental study of the hypnogenic mechanisms of the brain stem. Arch. ital. Biol. 101, 470–492 (1963).
— Palestini, M., Pisano, M., Rosadini, G.: An experimental study of the cortical reactivity during sleep and wakefulness. In: Aspects Anatomo-fonctionnels de la Physiologie du Sommeil. (Jouvet, M., ed.), p. 509–532. Paris: Centre National de la Recherche Scientifique 1965. 657 pp.
— Steffanon, L.: Effetti della stimolazione olfattiva sulla miosi del preparato "cervello isolato". Arch. Fisiol. 52, 468–474 (1953).
— Zanchetti, A.: Brain stem reticular formation. Anatomy and physiology. Arch. ital. Biol. 95, 199–435 (1957).
Rothballer, A. B.: Studies on the adrenaline-sensitive component of the reticular activating system. Electroenceph. clin. Neurophysiol. 8, 603–621 (1956).
Rougeul, A., Le Yaouana, A., Buser, P.: Activités neuroniques spontanées dans le tractus pyramidal et certains structures sous-corticales au cours du sommeil naturel chez le chat libre. Exp. Brain Res. 2, 129–150 (1966).
Routtenberg, A.: Neural mechanisms of sleep: changing view of reticular formation function. Psychol. Rev. 73, 481–499 (1966).
— The two-arousal hypothesis: reticular formation and limbic system. Psychol. Rev. 75, 51–80 (1968).
Ruckebusch, Y.: Etude poligraphique et comportementale de l'évolution post-natale du sommeil physiologique chez l'agneau. Arch. ital. Biol. 101, 111–132 (1963).
— Le sommeil et les rêves chez les animaux. In: Psychiatrie animale (Brion, A., and H. Ey, eds.), p. 139–148. Paris: Desclée de Brouwer 1964. 605 pp.
Sakahura, H.: Spontaneous and evoked unitary activities of cat lateral geniculate neurons in sleep and wakefulness. Jap. J. Physiol. 18, 23–42 (1968).
Schaller, G. B.: The year of the gorilla. Chicago: Chicago Press University 1964. (Quoted by Kleitman, N., 1967.)
Scheibel, M. E., Scheibel, A. B.: Structural substrates for integrative patterns in the brain stem reticular core. In: Reticular formation of the brain (Jasper, H. H., et al., eds.), p. 31–55. Boston and Toronto: Little, Brown & Co. 1958.
— — Structural organization of nonspecific thalamic nuclei and their projection toward cortex. Brain Res. 6, 60–94 (1967).
Schlag, J. D., Balvin, R.: Background activity in the cerebral cortex and reticular formation in relation with the electroencephalogram. Exp. Neurol. 8, 203–219 (1963).
— Chaillet, F.: Thalamic mechanisms involved in cortical desynchronization and recruiting responses. Electroenceph. clin. Neurophysiol. 15, 39–62 (1963).
— Faidherbe, J.: Recruiting responses in the brain stem reticular formation. Arch. ital. Biol. 99, 135–162 (1961).
— Scheibel, A. B. (eds.): Forebrain inhibitory mechanisms. Brain Res. 6, 1–200 (1967).
Schlosberg, H.: Three dimensions of emotion. Psychol. Rev. 61, 81–88 (1954).
Segundo, J. P., Arana, R., French, J. D.: Behavioral arousal by stimulation of the brain in the monkey. J. Neurosurg. 12, 601–613 (1955a).
— Naquet, R., Buser, P.: Effects of cortical stimulation on electrocortical activity in monkeys. J. Neurophysiol. 18, 236–245 (1955b).
Serkov, F. N., Makulkin, R. F., Tychina, D. N.: Electrical activity of the brain following mesencephalic transection under chronic experiment (Russian). Sechenov physiol. J. U.S.S.R. 52, 837–846 (1966).
Sharpless, S. K.: Reorganization of function in the nervous system—use and disuse. Ann. Rev. Physiol. 26, 357–388 (1964).
— Jasper, H. H.: Habituation of the arousal reaction. Brain 79, 655–680 (1956).

SHERRINGTON, C. S.: The integrative action of the nervous system. New York: Charles Sribner's Sons 1906a. XVI + 411 pp.
— Observations on the scratch-reflex in the spinal dog. J. Physiol. (Lond.) **34**, 1–50 (1906b).
— Flexion-reflex of the limb, crossed extension reflex, and reflex stepping and standing. J. Physiol. (Lond.) **40**, 28–121 (1910).
— Man on his nature. Cambridge: Cambridge University Press 1946. 413 pp.
— SOWTON, S. C. M.: Observations on reflex responses to single break-shocks. J. Physiol. (Lond.) **49**, 331–348 (1915).
SIMONOFF, L. N.: Die Hemmungsmechanismen der Säugethiere experimentell bewiesen. Arch. Anat. Physiol. wiss. Med. **33**, 545–564 (1866).
SLÓSARSKA, M., ZERNICKI, B.: Synchronized sleep in the chronic pretrigeminal cat. Acta Biol. exp. (Warszawa) **29**, 175–184 (1969).
— — Wakefulness and sleep in the isolated cerebrum of the pretrigeminal cat. Arch. ital. Biol. **109** (1971) (in press).
SNYDER, F.: Discussion of the report by KLEITMAN, N. (1967).
SOKOLOV, YE. N.: Perception and the conditioned reflex. Oxford: Pergamon Press 1963. X + 309 pp.
SPERRY, R. W.: Cerebral organization and behavior. Science **133**, 1749–1757 (1961).
SPRAGUE, J. M.: The effects of chronic brainstem lesions on wakefulness, sleep and behavior. Res. Publ. Ass. nerv. ment. Dis. **45**, 148–188 (1967).
— BERLUCCHI, G., RIZZOLATTI, G.: The role of the superior colliculus and pretectum in vision and visually guided behavior. In: Handbook of sensory physiology, vol. VII/2 B: Central processing of visual information. Berlin-Heidelberg-New York: Springer 1971. In press.
STAVRAKY, G. W.: Supersensitivity following lesions of the nervous system. Toronto: University of Toronto Press 1961. X + 210 pp.
STEINIGER, F.: Die Biologie der sog. "tierischen Hypnose". Ergebn. Biol. **13**, 348–451 (1936).
STERIADE, M.: Ascending control of thalamic and cortical responsiveness. Int. Rev. Neurobiol. **12**, 87–144 (1970).
STERMAN, M. B., CLEMENTE, C. D.: Cortical recruitment and behavioral sleep induced by basal forebrain stimulation. Fed. Proc. **20**, 334 (1961).
— — Forebrain inhibitory mechanisms: cortical synchronization induced by basal forebrain stimulation. Exp. Neurol. **6**, 91–102 (1962a).
— — Forebrain inhibitory mechanisms: sleep patterns induced by basal forebrain stimulation in the behaving cat. Exp. Neurol. **6**, 103–117 (1962b).
— — Basal forebrain structures and sleep. Acta physiol. lat.-amer. **14**, 228–244 (1968).
— KNAUSS, T., LEHMANN, D., CLEMENTE, C. D.: Circadian sleep and waking patterns in the laboratory cat. Electroenceph. clin. Neurophysiol. **19**, 509–517 (1965).
SWETT, C. P., HOBSON, J. A.: The effects of posterior hypothalamic lesions on behavioral and electrographic manifestations of sleep and waking in cats. Arch. ital. Biol. **106**, 283–293 (1968).
SZENTÁGOTHAI, J., FLERKÓ, B., MESS, B., HALÁSZ, B.: Hypothalamic control of the anterior pituitary. Budapest: Akadémiai Kiadó 1968. 400 pp.
SZYMANSKI, J. S.: Die Verteilung der Ruhe- und Aktivitätsperioden bei weißen Ratten und Tanzmäusen. Pflügers Arch. ges. Physiol. **171**, 324–347 (1918).
— Aktivität und Ruhe bei Tieren und Menschen. Z. allg. Physiol. **18**, 105–162 (1920).
TAKAGI, K.: Über den Einfluß des mechanischen Hautdruckes auf die vegetativen Funktionen. Acta neuroveg. (Wien) **16**, 439–446 (1957).
— NAKAYAMA, T., NAGASAKA, T.: Skin pressure reflex and the EEG inhibitory response. Electroenceph. clin. Neurophysiol. Suppl. **18**, 3–5 (1959).
TALBERT, G. A.: Über Rindenreizung am freilaufenden Hunde nach J. R. EWALD. Arch. (Anat.) Physiol. **24**, 195–208 (1960).

Thauer, R.: Der Mechanismus der Wärmeregulation. Ergebn. Physiol. **41**, 607–805 (1939).
— Homoiothermie als Fortschritt und Schicksal des Menschen. Jb. Max Planck Ges. **17**, 39–75 (1967).
Thorpe, W. H.: Learning and instinct in animals. London: Methuen 1963. X + 558.
Tiberin, P., Rhodes, J., Naquet, R.: Les hémi-synchronisations corticales à point de départ cortical. C. R. Soc. Biol. (Paris) **155**, 1346–1349 (1961).
Tinbergen, N.: "Derived" activities; their causation, biological significance, origin. and emancipation during evolution. Quart. Rev. Biol. **27**, 1–32 (1952).
— The study of instinct, 2nd ed. Oxford: Clarendon Press 1955. XII + 228 pp.
Tissot, R.: Aspects récents de l'étude du sommeil chez l'homme. Schweiz. Arch. Neurol. Psychiat. **96**, 115–141 (1965).
Tönnies, J. F.: Automatische EEG-Intervall-Spektrumanalyse (EISA) zur Langzeit-darstellung der Schlafperiodik und Narkose. Arch. Psychiat. Nervenkr. **212**, 423–445 (1969).
Tournay, A.: Sémiologie du sommeil, p. 36–54. Paris: Doin 1934. 131 pp.
Trömner, E.: Das Problem des Schlafs. Biologisch und psychophysiologisch betrachtet. See p. 24–28. Wiesbaden: Bergmann 1912. 89 pp.
Ursin, R.: The two stages of slow wave sleep in the cat and their relation to REM sleep, Brain Res. **11**, 347–356 (1968).
Valatx, J. L.: Ontogenèse des différents états de sommeil. Thèse Université de Lyon. 1963. 81 pp.
Valleala, P.: The temporal relation of unit discharge in visual cortex and activity of the extraocular muscles during sleep. Arch. ital. Biol. **105**, 1–14 (1967).
Verzeano, M., Negishi, K.: Neuronal activity in wakefulness and in sleep. In: The nature of sleep (Wolstenholme, G. E. W., and O'Connor, C. M., eds.), p. 108–130. London: J. A. Churchill 1961. XII + 416 pp.
Villablanca, J.: Electroencephalogram in the permanently isolated forebrain of the cat. Science **138**, 44–46 (1962).
— The electrocorticogram in the chronic *cerveau isolé* cat. Electroenceph. clin. Neuro-physiol. **19**, 576–586 (1965).
— Behavioral and polygraphic study of "sleep" and "wakefulness" in chronic decere-brate cats. Electroenceph. clin. Neurophysiol. **21**, 562–577 (1966a).
— Ocular behavior in the chronic *cerveau isolé* cat. Brain Res. **2**, 99–102 (1966b).
— Electrocorticogram in the chronic "isolated hemisphere" of the cat. Effect of atropine and eserine. Brain Res. **3**, 287–291 (1967).
Ward, A. A.: Decerebrate rigidity. J. Neurophysiol. **10**, 89–103 (1947).
Webb, W. B.: Sleep: an experimental approach. New York: MacMillan 1968. VI + 250 pp.
Weinberger, N. M., Velasco, M., Lindsley, D. B.: Effects of lesions upon thalamically induced electrocortical desynchronization and recruiting. Electroenceph. clin. Neuro-physiol. **18**, 369–377 (1965).
Weyn, A., Latash, L.: Physiologie und Pathologie des Schlafes. Ideen des exakten Wiss. **7**, 409–416 (1969).
Whitlock, D. G., Arduini, A., Moruzzi, G.: Microelectrode analysis of pyramidal system during transition from sleep to wakefulness. J. Neurophysiol. **16**, 414–429 (1953).
Wolstenholme, G. E. W., O'Connor, C. M., (eds.): Neurological basis of behaviour. Ciba Foundation Symposium. London: J. A. Churchill 1958. XII + 400 pp.
— — The nature of sleep. Ciba Foundation Symposium. London: J. A. Churchill 1961. XII + 416 pp.
Woods, J. W.: Behavior of chronic decerebrate rats. J. Neurophysiol. **27**, 635–644 (1964).
— Bard, P.: Thyroid activity in the chronic decerebrate cat with an isolated "island" of hypothalamus. Fed. Proc. **18**, 173 (1959).

Wyss, O. A. M.: Die nervöse Steuerung der Atmung. Ergebn. Physiol. **54**, 1–479 (1964).

Yamaguchi, N., Ling, G. M., Marczynski, T. J.: Recruiting responses observed during wakefulness and sleep in unanesthetized chronic cats. Electroenceph. clin. Neurophysiol. **17**, 246–254 (1964).

Zanchetti, A.: Neurale Mechanismen emotionalen Verhaltens. Verh. Ges. Kreislaufkr. **32**, 46–57 (1966).

— Subcortical and cortical mechanisms in arousal and emotional behavior. In: The Neurosciences (Quarton, G. C., Melnechuk, T., and Schmitt, F. O., eds.), p. 602–614. New York: The Rockefeller University Press 1967a. XIII + 962 pp.

— Brain stem mechanisms of sleep. Anesthesiology **28**, 81–99 (1967b).

— Guazzi, M., Baccelli, G.: Influence of sleep on circulation in normal and hypertensive animals. In: Antihypertensive therapy (Gross, F., ed.). Berlin-Heidelberg-New York: Springer 1966. XII + 632 pp.

Zattoni, J., Rossi, G. F.: A study of the hypnogenic action of "mogadon" by selective injection into carotid and vertebral circulation. Physiol. Behav. **2**, 277–282 (1967).

Zbrozina, A., Bonvallet, M.: Influence tonique inibitrice du bulbe sur l'activité du noyau d'Edinger-Westphal. Arch. ital. Biol. **101**, 208–222 (1963).

Zernicki, B.: Isolated cerebrum of midpontine pretrigeminal preparation: a review. Acta Biol. exp. (Warszawa) **24**, 247–284 (1964).

— Pretrigeminal cat. Brain Res. **9**, 1–14 (1968).

— Doty, R. W., Santibañez-H., G.: Isolated midbrain in cats. Electroenceph. clin. Neurophysiol. **28**, 221–235 (1970).

— Dreher, B.: Studies on the visual fixation reflex. I. General properties of the orientation fixation reflex in pretrigeminal and intact cats. Acta Biol. exp. (Warszawa) **25**, 187–205 (1965).

— — Krzywosinski, L., Sychowa, B.: Some properites of the acute midpontine pretrigeminal cat. Acta Biol. exp. (Warszawa) **27**, 123–139 (1967).

— Osetowska, E.: Conditioning and differentation in the chronic midpontine pretrigeminal cat. Acta Biol. exp. (Warszawa) **23**, 25–32 (1963).

The Role of Monoamines and Acetylcholine-Containing Neurons in the Regulation of the Sleep-Waking Cycle

MICHEL JOUVET***

With 24 Figures

Table of Contents

* Department of Experimental Medicine — School of Medicine — Lyon — France.
** The experiments reported from this laboratory have been supported by DRME (Grant 69—047), INSERM and CNRS.

I. Introduction

Any approach to the mechanisms controlling sleep and waking should take into consideration the well-established fact that the central nervous system is a super-organization comprising many sub-systems. Since limited lesions may totally suppress states of either sleep or waking in their central EEG or behavioral manifestations, it might be postulated that specific systems of neurons play a determinant role in actively triggering sleep or waking. Thus, any explanation of sleep and waking should first postulate the *anatomical specificity* of the sleep and waking mechanisms.

Over many years, indeed, localized brain stem lesions or transections of the brain stem have enabled us to map out rather accurately the smallest lesion able to suppress either slow-wave sleep (SWS), paradoxical sleep (PS) or waking. Most of these lesions were located in the complicated network of the reticular formation. But it was impossible, by classical histological techniques, to determine whether these postulated "sleep or waking neurons" had any histochemical specificity. The pathways originating from these cells were very difficult to trace with standard neuroanatomical techniques, and the projection of these systems as established by means of neurophysiological procedures (stimulation or lesion) gave disappointing results. In fact, nothing in the cells themselves, nor in their axonal ramifications, yielded any information concerning the *mechanisms* which would explain the onset of sleep, or the sustained tonic arousal following the presentation of a stimulus.

At the same time, it became evident that a purely electrophysiological theory of the sleep-waking cycle was untenable. According to this theory, waking was induced by the continuous bombardment of reticular or cortical cells by peripheral or central impulses. Sleep, in turn, was explained as a passive resting state of these neurons, provoked by the cessation of afferent impulses (deafferentation theory) or by neuronal fatigue (see BREMER, 1954) (see MORUZZI, this review). The active theory of sleep, which goes back to the observation that insomnia may be created by setting a lesion (see MORUZZI, 1963, 1964 and this review) or to the demonstration of a state of sleep with low-voltage fast activity (KLAUE, 1937; DEMENT, 1958; JOUVET et al., 1959), could not be explained either by classical electrophysiological mechanisms. How could electrophysiology alone account for the striking periodicity of paradoxical sleep in the pontile cat (JOUVET, 1962) and its long-lasting rebound (up to several days) after its previous selective suppression? (VIMONT-VICARY et al., 1966).

Hence the re-emergence of the rather naive concept that "something" might "accumulate" in the cerebro-spinal fluid, in the blood or in a group of neurons during waking, and that this substance might be "eliminated" during sleep. This was the revival of the old theory of "hypnotoxin" (PIÉRON, 1913). Naive as it was, this concept paved the way for the introduction of "wet neurophysiology" (SCHMITT, 1962) into the field of sleep research.

"Wet neurophysiology" could easily enter this area, since it had acquired an "anatomical dimension", thanks to modern histochemistry and neurochemistry. Through the application of histofluorescence, a new map of the brain emerged in which systems of neurons with a neurotransmitter function appeared in bright yellow or green colors. This map was not unlike a modern painting, especially where the colors overlapped the strict boundary of the nuclei of the reticular formation, as in the dorsolateral pontine tegmentum. New systems appeared. In short, this was a revolution. Numerous "terrae incognitae" in the reticular formation disappeared and acquired histochemical specificity. Since many of these systems were located in places where lesions had been shown to impair sleep or waking, a new dimension became accessible to the study of the sleep-waking cycle. It was now possible to approach the sleep mechanism by systematically studying its histochemical systems through lesion, stimulation, recording, neuropharmacological alteration, and biochemical analysis.

In this review, we shall first summarize the problem of the humoral control of sleep. Although of historical importance, research in this field is now less active since it has become clear that a purely humoral theory of sleep cannot satisfactorily explain the central mechanisms involved in the regulation of sleep and waking. Thus, we shall be mainly concerned with the neurohumoral mechanisms underlying the sleep-waking cycle. Since a knowledge of the anatomical organization of the neurons specialized in neurohumoral function is a prerequisite for any approach to their role in the sleep-waking cycle, we shall limit ourselves to the study of those systems whose organization has been mapped out, at least schematically: the monoaminergic and cholinergic systems. This does not imply that other putative neurotransmitters do not play an important role. But at the present time nothing, or not enough, is known concerning the topography of neurons containing GABA, histamine, glycine, prostaglandins etc. to permit any valuable correlation with neurophysiological data.

We shall therefore outline the topography of neurons containing monoamines (MA) and acetylcholine (ACH) before reviewing some aspects of their functioning which have permitted us to postulate with some certainty that they can be considered as central neurotransmitters. Then we shall discuss theoretically the possible strategies and tactics for the study of MA- or ACH-containing neurons in the sleep-waking cycle. This chapter will oblige us to delineate briefly, but with the utmost precision, the electrical and behavioral frontiers of the object of our study: the alternation between slow wave sleep (SWS), paradoxical sleep (PS) and waking.

In the subsequent chapters, we shall successively review the indirect and direct evidence which suggests the intervention of serotonin (5-Hydroxy-tryptamine) (5-HT), catecholamines (CA), dopamine (DA) or noradrenaline (NA) and ACH-containing neurons in the sleep-waking cycle. Paradoxical sleep will be discussed in a special chapter for several reasons. First, the neurophysiological mechanisms of PS were only briefly outlined in MORUZZI's review;

second, PS appears to involve a succession of several mechanisms belonging to different monoaminergic and cholinergic systems. Thus, it is easier to consider their action in a synthesizing chapter; third, the relationship between the central phasic electrical activity of PS [ponto-geniculo-occipital waves (PGO)] and the alteration of monoamine metabolism is so intimate that it deserves a special chapter.

Finally, we shall synthesize our data in a monoaminergic theory of the sleep-waking cycle. This theory will be critically correlated with the main neurophysiological data presented in MORUZZI's review.

MA was discovered in the brain relatively recently, but the literature on this topic is already quite extensive. The following books, symposia or review articles which deal partially or wholly with the central aspect of MA (but not in relation to the sleep-waking cycle) have been published during recent years:

— *Books and Symposia:* HIMWICH and HIMWICH (1964), HIMWICH and SCHADE (1965), ACHESON (1966), ERSPAMER (1966), KETY and SAMSON (1967), AKERT and WASER (1969), BOURNE (1969), HOOPER (1969), LAJTHA (1969), COSTA and GIACOBINI (1970), McLENNAN (1970), RALL and GILMAN (1970), SMYTHIES (1970).

— *Reviews:* GIARMAN (1959), CARLSSON (1964), COSTA and BRODIE (1964), HORNYKIEWICZ (1966), SALMOIRAGHI et al. (1965), WERMAN (1966), VOTAVA (1967), BLOOM and GIARMAN (1968), HIMWICH and ALPERS (1970), HEBB (1970). The literature devoted to the neurophysiological aspects of sleep has been reviewed by MORUZZI.

These 2 fields of research have generated an enormous amount of literature and to review it would be beyond the scope of a single reviewer. But for many reasons, these 2 fields have only recently overlapped. Although brain MA were thought from the beginning to act upon "trophotropic" or "ergotropic" function (BRODIE and SHORE, 1957), it was only quite recently that the chronic polygraphic recordings of the sleep state (on the one hand), and the anatomical delineation of MA systems and modern biochemical techniques (on the other hand) have permitted a thorough study of the relationships between MA, ACH and the sleep-waking cycle.

The reader will find recent references under the heading: biochemistry, pharmacology and endocrinology in Sleep and Dream Research, a bibliography, edited by RECHTSCHAFFEN and EAKIN (from 1962 to 1968), further in the Sleep Bulletin, edited by EAKIN since 1968, and in Sleep Reviews edited by DEMENT, KALES and RECHTSCHAFFEN. Since 1970, the weekly bulletin "Biogenic Amines and Transmitters in the Nervous System", edited by E. YEAGLIN, has published numerous references concerning the sleep-waking cycle.

The following books, symposia, or reviews also contain references pertinent to our subject: *Books* edited by AJURIAGUERRA (1962), NAUTA et al. (1966),

LUCE (1966), KETY et al. (1967), KOELLA (1967), KALES (1969), SCHILDKRAUT (1970).

— *Reviews* written by ROTHBALLER (1959), COSTA (1960), JUNG (1963), FREEDMAN and GIARMAN (1963), MONNIER (1964), MANDEL and GODIN (1965), HERNANDEZ-PÉON (1965), JOUVET (1965b, c, 1966, 1967a–c, 1968, 1969a, b, 1971), JEANNEROD (1966), MONNIER and FALLERT (1967a, b), EVANS (1968), HARTMANN (1968a, 1969a, b, 1970), OSWALD (1968), KOELLA (1969a, b), JASPER (1969), HAVLICEK (1969), CORDEAU (1970), SJOERDSMA et al. (1970), TISSOT (1970), VAN PRAAG (1970).

II. Humoral Control of Sleep-Waking

Two basically different approaches have been used in the search for some humoral control of sleep: cerebrospinal-fluid factor and blood factor.

1. Cerebrospinal-Fluid (CSF) Sleep Factor

At the beginning of this century, PIÉRON (1913), LEGENDRE and PIÉRON (1913) furnished the first experimental evidence for the "wet" nature of sleep control (as opposed to PAVLOV's contemporary "dry" internal inhibition theory of sleep) (see PAVLOV, 1927). In PIÉRON and LEGENDRE's experiments, dogs were kept awake by devoted assistants who took night walks in the streets of Paris. Otherwise the dogs were attached to the wall by a short collar so that they could not lower their heads. This procedure probably did not totally suppress SWS but presumably deprived the dog of PS. Injection of CSF from these sleep-deprived dogs into the cisterna magna of normal animals induced sleep in the recipient for 2–6 hours following the injection. Recipients of CSF from normal dogs remained alert. The "hypnotoxic" factor was non-dialyzable and thermolabile. SCHNEDORF and IVY (1939) reinvestigated this phenomenon in 1939 and reported positive results in 9 out 20 trials. Their animals were deprived of sleep for 7–16 days and the injection of CSF was made (as in PIÉRON's technique) without anesthesia. SCHNEDORF and IVY tested the CSF for acetylcholine but were not successful and they emphasized that the central "depressing" effects of the injection could also be due to the concomitant elevation of intracranial pressure and to hyperthermia, which usually followed the intracisternal injection. Recently PAPPENHEIMER et al. (1967) have carried out a new investigation of the PIÉRON phenomenon. They obtained perfusate from sleep-deprived goats whose ventricular system was cannulated. When injected into the ventricular system of cats and rats, the perfusate induced clinical signs of sleep (inactivity) lasting up to 18 hours, whereas there was no change in the activity of animals receiving CSF from non-sleep-deprived goats. No EEG studies of the recipient animals have yet been reported. The factor responsible for the inactivity has not yet been identified but has been reported to be dialyzable (molecular weight under 1000–2000). Finally RINGLE and HERNDON (1969) have made some unsuccessful attempts to induce sleep in rats from the cerebrospinal fluid of sleep-deprived rabbits.

2. Plasma-Dialyzable Sleep-Promoting Factor

In order to test the hypothesis that sleep-inducing factors occur in the blood, KORNMÜLLER et al. (1961) and MONNIER et al. (1963, 1965), MONNIER (1964), MONNIER and FALLERT (1967a, b), HÖSLI and MONNIER (1965) used the technique of crossed blood circulation in cats and rabbits, respectively. The stimulation of the medio-central intralaminary thalamus in the donor induces a significant increase in cortical slow waves. Synchronously, the recipient, crossed with the donor, shows a statistically significant increase in cortical slow activity (usually after the fourth stimulation of the thalamus). On the other hand, stimulation of the midbrain reticular system which induces arousal in the donor, significantly reduces cortical slow waves in the recipient. Subsequent experiments (MONNIER and HÖSLI, 1964, 1965) indicated that dialysis of cerebral blood, performed upon rabbits during sleep induced by electrical stimulation of the "somnogenic" thalamus, provided a dialysate which, when later injected into recipient rabbits, elicited EEG manifestations of sleep. On the other hand, recipients receiving dialysate from non-stimulated control rabbits had wakeful EEG patterns. Numerous checks of visceral activities (blood pressure, heart rate, respiration and rectal temperature) failed to reveal any significant alteration (SCHNEIDERMAN et al., 1966). Finally, comparisons between "sleep dialysate" and other pharmacological substances indicated that its effects generally resembled those of chlorpromazine and substance P. Apparently the sleep dialysate factor obtained from the rabbit is active only after electrically induced sleep, since RINGLE and HERNDON (1968) have failed to show any sleep-promoting effects from the reconstituted plasma dialysates obtained from 72-hours sleep-deprived rabbits upon recipient rabbits, rats or mice.

3. Neurohumoral Induction of Sleep

Although this method is concerned with neurohumoral control, it has been included in this chapter since the possible sleep-inducing perfusate has not been identified. Using the technique developed by MYERS (1967) for the study of thermoregulatory mechanisms in the hypothalamus of monkeys, DRUCKER-COLIN et al. (1970) perfused sleep-deprived cats with push-pull cannulae inserted in the midbrain reticular formation. The outgoing perfusate of cats, during rebound of PS, induced sleep when injected into push-pull cannulae inserted the midbrain reticular formation of recipient cats.

4. Discussion

From these three different approaches it is difficult to draw any conclusion, since no sleep-promoting factor has yet been chemically identified in the CSF, the blood or the brain perfusate. The low molecular weight of the CSF factor isolated by Pappenheimer et al. (1967) is not incompatible with a MA metabolite.

It is possible that "hypnogenic" central stimulation or sleep deprivation may release a greater quantity of transmitter or metabolites than physiological sleep. Since the work of Söderberg (1962), Sharpless and Rothballer (1961) and Rothballer (1956, 1957), it has been shown that the stimulation of the reticular formation may release a blood factor whose activity has been compared with vasopressin. On the other hand, stimulation of a MA bundle or terminals may release either MA (McLennan, 1964; Stein and Wise, 1969; Arbuthnott et al., 1970) or MA metabolites which may be eliminated either in the blood or in the CSF. If these metabolites are still active when injected into the blood or CSF, it remains to be shown at what postsynaptic level they act. These results should be kept in mind when we review the problem of the chronic "cerveau isolé" preparation. It is quite possible that any arousal induced by external or internal stimulation may activate the reticular activating system and that some substance may be released into the circulation and through this humoral link activate postsynaptic receptors situated in the deafferented brain.

The fact that blood perfusate can induce sleep only after stimulation of the thalamus, but not when taken from sleep-deprived rabbits, shows that we must also differentiate between electrically induced "sleep" and normal sleep. On the other hand, the thalamus per se is not involved in the sleep mechanism since its total destruction does not significantly alter the sleep-waking cycle (Angeleri et al., 1969; see the Discussion in Moruzzi's review). For these reasons, one should accept with caution the results obtained after stimulation of the thalamus. Finally, the well-known fact that siamese twins, sharing the same blood circulation, may sleep independently (Alekseyeva, 1958) makes it unlikely that sleep depends *exclusively* upon a blood factor.

The experimental approach of Drucker-Colin et al. would have considerably more value if the perfusate from a sleep-deprived cat (or from a cat recuperating from sleep deprivation) were to induce natural sleep in an otherwise totally insomniac preparation (i.e. after raphe lesion), whereas its value is lessened in normal cats due to their spontaneous tendency to sleep. Such an approach could theoretically lead to the dissection of the chain of transmitters which may be involved in the triggering of sleep. But chronic push-pull cannulae generally induce rather large brain lesions and it would be difficult to prove that the perfusate which is released comes from presynaptic "ergic" terminals of the donor and acts upon postsynaptic "homoceptive" receptors of the recipient.

III. Neurohumoral Control of Sleep-Waking

Before reviewing the indirect and direct approaches to the relationship between central monoamine (MA) metabolism and the sleep-waking states, we should first try to answer two main questions: 1) Where are the mono-amine and acetylcholine-containing neurons located? 2) Are monoamines central neurotransmitters?

A. Anatomical Distribution of Monoamines-Containing Neurons

The concept that MA-containing neurons could exist appeared when bio-chemical analysis revealed that the distribution of 5-HT (AMIN et al., 1954), NA (VOGT, 1954) and DA (CARLSSON et al., 1958) is not identical in different parts of the brain. Later it was shown that limited lesions in the ventral hypothalamus could decrease in 5-HT in the telencephalon (HELLER et al., 1962; HARVEY et al., 1963; MOORE et al., 1965; MOORE and HELLER, 1967). Thus, it appeared possible that an ascending system of 5-HT neurons might exist. At about the same time, various histochemical techniques became available and it was possible to map out the monoamine oxidase (MAO) which is involved in the catabolism of MA. This map corresponds almost exactly to the distribution of MA-containing perikarya in the brain stem (SHIMIZU et al., 1959; MAEDA et al., 1960; HASHIMOTO et al., 1962).

However, thanks to the work of FALK et al. (1962) (see ERANKO, 1967), fluorescence histochemistry has proven invaluable in establishing one of the crucial pieces of evidence required for the identification of the MA's as neuro-transmitters, namely their occurrence within particular perikarya and nerve endings. The histofluorescence method can be combined with lesion or stimula-tion experiments to delineate the organization of the MA system. Indeed, after a lesion involving MA-containing perikarya or a bundle of MA neurons, the specific fluorescence in the terminals first decreases and, after 8–10 days, disappears, whereas there is some increase in the fluorescence in the axons situated above the lesion (due to the axoplasmic flow of the amine from the perikarya) (ANDÉN et al., 1964; HAGGENDAL and DAHLSTRÖM, 1969). On the other hand, if some perikarya or axons belonging to MA-containing neurons are stimulated, there is depletion of the terminals, provided that the synthesis of MA is inhibited by some pharmacological agent (ARBUTHNOTT et al., 1970).

All these techniques are rather new and much uncertainty remains con-cerning the exact organization of the MA systems and their possible inter-actions. However, since the anatomical organization is of paramount importance for understanding the possible role of MA in the regulation of the sleep-waking cycle, we shall describe the topography of these systems in some details.

1. Organization of the 5-HT System

In the Rat. Most of the perikarya containing 5-HT are located in the nuclei of the raphe system (ANDÉN et al., 1965, 1966d; DAHLSTRÖM and FUXE, 1964, 1965; FUXE, 1965). (See BRODAL et al., 1960a, b and TABER et al., 1960, for the cytoarchitectonics of those nuclei.) The 5-HT-containing cell bodies extend from N. raphe pallidus in the caudal medulla (group B 1), to N. raphe dorsalis in the caudal mesencephalon (group B 9). Some groups are located more laterally in the N. paraganto cellularis and in the ventral part of the area postrema. The pathways ascending from the brain stem are not yet known in detail but it is certain that a large part of the ascending fibers enters the medial forebrain bundle. 5-HT terminals (shown as very bright yellow terminals whose fluorescence increase after block of MAO with nialamide and diminishes after reserpine) have been located in the ponto-mesencephalic reticular formation, hypothalamus, lateral geniculate nuclei, the amygdala, pallidum system, hippocampus, anterior hypothalamus and preoptic area, neocortex and in most nuclei of the brain stem. The raphe nuclei also send descending fibers to the spinal cord and 5-HT terminals may be found in the grey matter, the posterior, lateral and anterior horns. It is likely that there are some synapses with moto-neurons.

In the cat the topography of 5-HT cell bodies and terminals is almost similar to that of the rat, yet with some minor differences (PIN et al., 1968; JONES, 1968) (Fig. 1). A large parasagittal bundle of axons, ascending from the region of nucleus raphe pontis and running parasagittaly rostrally and dorsally has been mapped out after rostro-pontine transection. These axons enter, as in the rat, the medial forebrain bundle and contribute to the innervation of the lateral hypothalamic and preoptic area, the septal area, the amygdaloid complex (MORGANE, 1971). It is also probable that other fibers might ascend more dorsally in a mesencephalic central 5-HT bundle, since a lesion of the central mesencephalic tegmentum sparing the medial forebrain bundle decreases 5-HT levels in the telediencephalon (JONES, 1968).

2. Organization of the Catecholamine System

i) Dopamine

Most of the DA-containing cell bodies are located in the pars compacta of the substantia nigra (group A 9) and in an area dorsal to the N. interpeduncularis (group A 10). The DA-terminals are accumulated mainly in the neostriatum and the tuberculum olfactorium. They are also found in the N. accumbens, the median eminence, the amygdaloid nucleus and N. paraventricularis. The existence of a dopaminergic nigro-striatal system is now

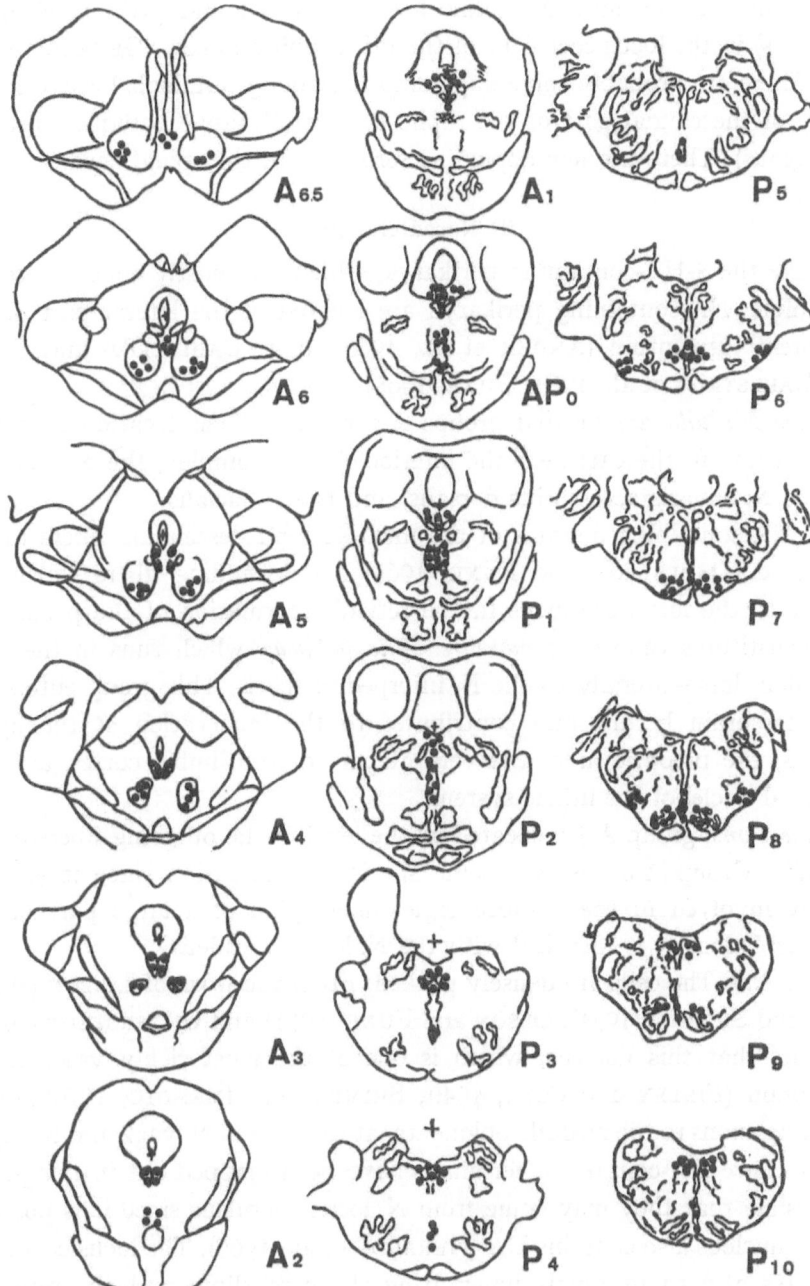

Fig. 1. *Topography of 5-HT-containing cell bodies (dots) in the brain stem of the cat.* Frontal section according to the Horsley Clark coordinates. (PIN et al., 1968)

well established (ANDÉN et al., 1964a, 1966b, e, g; SOURKES and POIRIER, 1965; PARENT and POIRIER, 1969; BEDARD et al., 1969a, b; YORK, 1970). In addition, there is another system from group A 10 to the limbic system (tuberculum olfactorium). Dopamine has also been found (with biochemical techniques) in the locus coeruleus of the rat, rabbit and cow (GERARDY et al., 1969). There are probably some dopamine-containing terminals located in the vicinity of the caudal part of the raphe system (N. raphe pallidus, obscurus and magnus). There are few dopamine terminals in the spinal cord.

ii) Noradrenaline

Unlike the 5-HT-containing perikarya, which are mostly concentrated in the midline, NA-containing perikarya are located in the lateral part of the brain stem tegmentum (ANDÉN et al., 1966c, d, g; DAHLSTRÖM and FUXE, 1964; DAHLSTRÖM et al., 1964; FUXE, 1965).

In the Medulla are located groups A 1 to A 4. Their location is similar in the rat and in the cat: near the inferior olivary complex, the N. motorius vagi, the N. olivaris accessorius dorsalis, and the N. facialis.

The NA-containing neurons of the medulla send descending fibers to the spinal cord (DAHLSTRÖM and FUXE, 1965) and ascending fibers which run medially to the locus coeruleus in the reticular formation of the pons. This group constitutes the *ventral noradrenaline pathway* which runs in the mesencephalon dorso-laterally to the N. interpeduncularis. This group enters the medial forebrain bundle and contributes to the innervation of the hypothalamus, the preoptic area, the ventral part of the limbic cortex and the subcortical nuclei of the limbic system.

In the Pons, group A 5 is located at the level of the outgoing fibers of the N. facialis. Group A 6 deserves some special consideration since it is most probably involved in the noradrenergic innervation of a large part of the telediencephalon. It is identical with the N. locus coeruleus.

In the Rat. The cells are densely packed. From the work of LOIZOU (1969), MAEDA and SHIMIZU (1971), OLSON and FUXE (1971) and UNGERSTEDT (1971), it appears that this nucleus, which is one of the most richly vascularized in the brain (FINLEY and COBB, 1940; SHIMIZU and IMAMOTO, 1970), sends descending axons to the medulla oblongata, at the level of N. vagi, and N. raphe pallidus. Since DA-containing terminals have been mapped out in this place, it is possible that they may come from N. locus coeruleus since it is possible that this nucleus also contains DA (GERARDY et al., 1969). The locus coeruleus contributes also to terminals innervating the cerebellum and the reticular formation of the pons. Ascending NA projections from the locus coeruleus ascend through a dorsal NA bundle just lateral to the central grey substance (OLSON and FUXE, 1971; UNGERSTEDT, 1971) and also through an intermediate bundle (MAEDA and SHIMIZU, 1971). They turn abruptly ventrally at the border between the mesencephalon and the diencephalon and approach the dorso-lateral hypothalamic area after passing though the subthalamus. The dorsal NA bundle innervates mainly the cortex cerebri and the hippo-

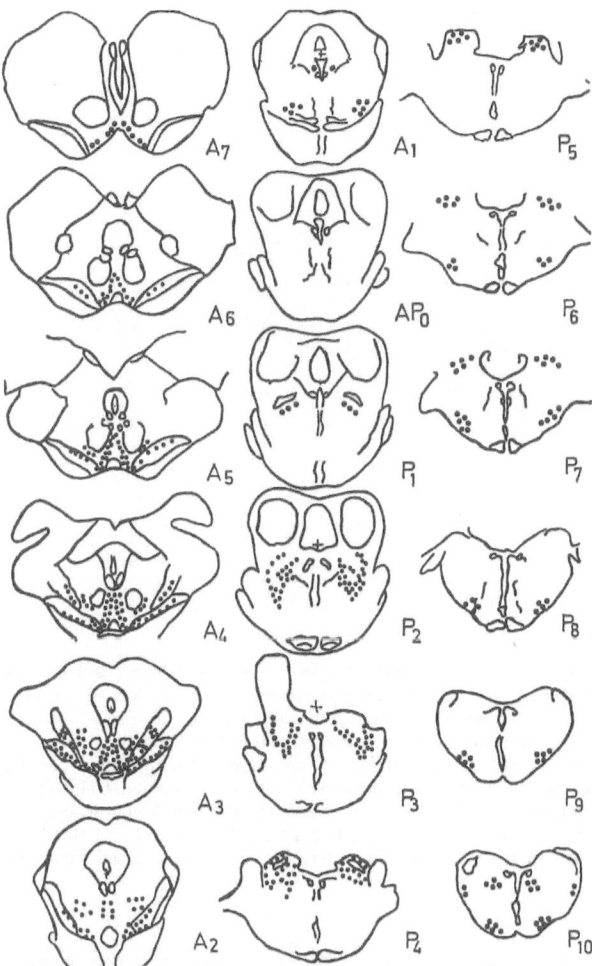

Fig. 2. *Topography of catecholamine-containing cell bodies (dots) in the brain stem of the cat.* Frontal section according to the Horsley Clark coordinates. (PIN et al., 1968)

campal formation. Most of the terminals found in the lateral geniculate bodies also come from N. locus coeruleus. There is still some discussion concerning the ascending MA pathways. For MOORE et al. (1965, 1967) some polysynaptic pathways must exist and the decrease of amine after a lesion of the medial forebrain bundle is a result of transynaptic metabolic changes in morphologically intact neurons. However, the histochemical fluorescence technique has failed to provide evidence of MA cell bodies situated between the lesion and the telencephalon. This is why OLSON and FUXE (1971) and UNGERSTEDT (1971) are in favor of direct uninterrupted ascending monosynaptic fibers from the brain stem to the telencephalon.

In the Cat, the topography of NA-containing perikarya in the pons is more diffuse than in the rat and corresponds to N. locus coeruleus, subcoeruleus, N. parabrachialis medialis, lateralis and group K (Fig. 2). Through

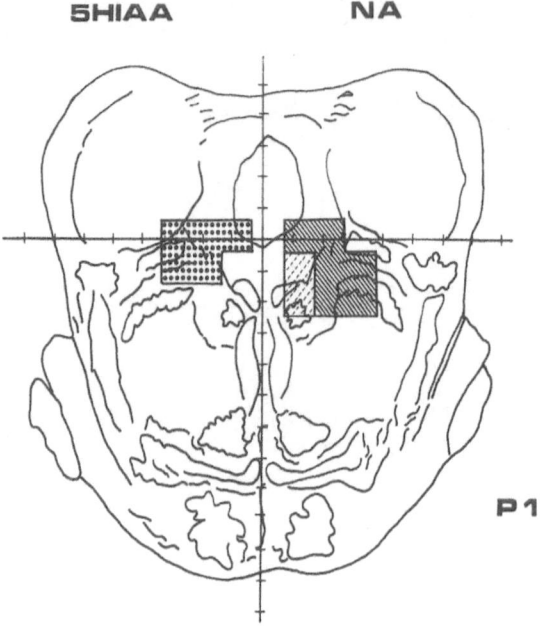

Fig. 3. *On the right* (oblique hatching) *unilateral projection of a bilateral lesion common to 10 cats which decreases the tele-diencephalic level of noradrenaline* (10 days after the lesion) by more than 50% as compared with intact cats. Dotted lesion common to 8 out of 10 cats. On the left (dots) unilateral projection of a bilateral lesion common to the same cats, which also increases the tele-diencephalic level of tryptophan and 5-hydroxyindolacetic acid by more than 30% as compared with intact cats. There is no significant alteration of telediencephalic 5-HT. All these lesioned cats presented true hypersomnia with increase of both slow-wave sleep and paradoxical sleep. It is likely that the lesion destroys the dorsal NA bundle ascending from the locus coeruleus and that interruption of this bundle at this level may interfere with the metabolism of 5-HT in the rostral raphe system (see p. 221). (Modified from Nemoz et al., 1970, and Petitjean, 1970)

the lesion technique, associated with biochemical evaluation of NA in the terminals located in the telediencephalon, the dorsal NA bundle has also been mapped out just rostrally to the rostral part of the locus coeruleus immediatly lateral to the grey matter (Nemoz et al., 1970) (Fig. 3). It is also likely that the caudal portion of the locus coeruleus sends descending NA axons to the reticular formation of the pons and the cervical spinal cord, since its destruction is followed by a decrease of NA in these regions (Buguet, 1969).

In the Mesencephalon. The localization is the same in rat and cat. Group A 10, located in the area dorsal to the N. interpeduncularis, is mainly composed of DA-containing cell bodies innervating the limbic system. Group A 8 is also almost exclusively composed of DA-containing neurons and possibly NA-containing neurons in the cat (Jones, 1969). It is located in the mesencephalic reticular formation lateral and dorsal to the red nucleus. Its rostral pole descends ventrally to mix with the anterior part of group A 9. The cells of group A 8 are situated ventrally to axons with numerous varicosities ascending into the adjacent mesencephalic reticular formation.

In the Diencephalon. Whereas there are no 5-HT-containing perikarya rostral to the caudal part of the mesencephalon, both in rat and cat there are 2 groups (A 11 and A 12) of noradrenaline-containing perikarya located near the third ventricle within the substantia grisea periventricularis and within N. arcuatus and N. periventricularis anterior.

iii) Adrenaline

There is some adrenaline in the mammalian brain (4 to 17% of combined NA and adrenaline) (BARCHAS et al., 1969). Most of it is concentrated in the pineal and the hypophysis. No adrenaline-containing system of neurons has yet been mapped out.

3. Monoamine Catabolizing Enzymes

Only monoamine oxidase (MAO) which catabolizes most (if not all) 5-HT and some part of CA, is visualized with the formazan technique. No methods are yet available to visualize catecho-O-methyl transferase (COMT) which also catabolizes CA (ALBERICI et al., 1965).

The correlation between the localization of MAO-containing structures and the MA system is not a simple one (MAEDA et al., 1960; MAEDA, 1970). Thus, in the rat and rabbit the nucleus interpeduncularis shows a very strong MAO activity, whereas it is entirely devoid of MA-containing neurons. However most of the MAO-containing structures coincides with the topography of MA-containing perikarya. Thus, some MAO may be found in the raphe nuclei. MAO is also concentrated in the caudal part of CA-containing cell bodies (group A 1 to A 7). Thus there is an exact coincidence between the topography of NA perikarya of the bulbar and pontile tegmentum and that of MAO-containing cells.

Interestingly enough, neither in the cat nor in the rat, is MAO located in the mesencephalic group of CA-containing neurons. No, or very light, MAO staining is seen in substantia nigra (group A 9) or in the group of CA-containing neurons of the mesencephalic tegmentum (group A 8). Thus there is a possibility that two different groups might exist in the CA system: the first one (principally located in the pontine tegmentum) would be principally (or selectively) concerned with MAO catabolism; the second (located in the mesencephalon) would be principally (or selectively) catabolized through COMT.

4. Ontogenesis of the Monoaminergic System

As an architectural principle of the central nervous system, it has been known that the sulcus limitans of Hiss runs rostro-caudally on both lateral walls of the neural tube. MASAI et al. (1965) have demonstrated that as early as 5 days (in the chick embryo) MAO staining could be demonstrated along the septal area, preoptic area, hypothalamus, central grey area along the sulcus limitans and the raphe of the pons and medulla. Thus, apparently both the 5-HT system and the CA system originate very early from the so-called visceral columns. Shimizu and Morikawa (1959), Robinson (1968), Maeda and Gerebtzoff (1969), Maeda and Dresse (1969) have studied the development of the locus coeruleus in rat fetus. Both MAO and CA fluorescence appears around the 14th day. There are two sudden augmentations of activity, at 17 days, and at birth.

In the newborn kitten, only CA-containing terminals can be found in the brain stem, whereas there are few 5-HT terminals. No MA terminals are found in the cortex. The NA terminals precede the appearance of 5- HT terminals, which look mature only after the second month (Loup and Cadilhac, 1970). Biochemical studies evaluating the regional changes in MA during postnatal development (Glowinski et al., 1964; Loizou and Salt, 1970; Loizou, 1970) suggest also that the increase in the level of brain MA is a consequence of the progressive proliferation of axon terminals and also possibly of an increase of synthesizing enzymes. Much work remains to be done concerning the ontogeny and phylogeny of the MA system, but a picture is emerging of two fundamental systems originating either from the floor of the ventricle (5-HT system) or from the sulcus limitans (CA system). The fact that these systems are operating in utero (as shown by MAO staining) or immediately after birth will be discussed later when some ontogenetic aspects of the sleep-waking cycle and the possible function of PS are considered.

5. Cholinergic System

Thanks to the histochemical method of Koelle (1954) the distribution of cholinesterase in the central nervous system can be studied. There is still some discussion concerning the question whether acetylcholinesterase can be considered a reliable marker of cholinergic neurons (see Karczmar, 1969; Krnjevic, 1969). However, lesion techniques have shown that it was possible to map out probable cholinergic systems by observing the decrease of acetylcholinesterase in the distal part of the axons and the increase in the proximal part. In the rat brain, Shute and Lewis (1963, 1966, 1967), Lewis et al. (1967), Krnjevic and Silver (1963), Lewis and Shute (1967), Shute (1969) have mapped out two main ascending cholinergic pathways whose topography can be summarized as follows (Fig. 4): the *dorsal tegmental pathway* which runs rostrally from the mid-brain tegmentum (N. cuneiformis) and supplies the tectum, pretectal area, geniculate bodies and thalamus. This system receives afferents from the nucleus locus coeruleus: the *ventral tegmental pathway* arises from the pars compacta of the substantia nigra and runs rostrally to the zona incerta, through the mamillary region and the lateral hypothalamic area. More rostrally, this pathway appears to connect with the ento-peduncular

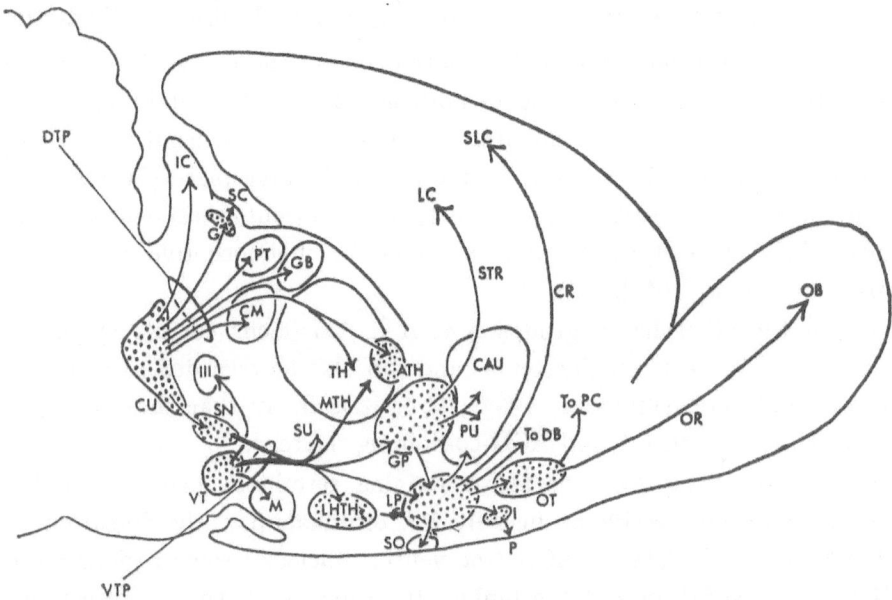

Fig. 4. *Diagram showing the constituent nuclei* (stippled) *of the ascending cholinergic reticular system in the midbrain and forebrain*, with projections to the cerebellum, tectum, thalamus, hypothalamus, striatum, lateral cortex and olfactory bulb. Abbreviations: *ATH* antero-ventral and antero-dorsal thalamic nuclei, *CAU* caudate, *CM* centromedian (parafascicular) nucleus, *CR* cingulate radiation, *CU* nucleus cuneiformis, *DB* diagonal band, *DTP* dorsal tegmental pathway, *G* stratum griseum intermediale of superior colliculus, *GB* medial and lateral geniculate bodies, *GP* globus pallidus and entopeduncular nucleus, *I* islets of Calleja, *IC* inferior colliculus, *III* oculomotor nucleus, *LC* lateral cortex, *LHTH* lateral hypothalamic area, *LP* lateral preoptic area, *M* mammillary body, *MTH* mammillo-thalamic tract, *OB* olfactory bulb, *OR* olfactory radiation, *OT* olfactory tubercle, *P* plexiform layer of olfactory tubercle, *PC* precallosal cells, *PT* pretectal nuclei, *PU* putamen, *SC* superior colliculus, *SLC* supero-lateral cortex, *SN* substantia nigra pars compacta, *SO* supraoptic nucleus, *STR* striatal radiation, *SU* subthalamus, *TH* thalamus, *TP* nucleus reticularis tegmenti pontis (of Bechterew), *VT* ventral tegmental area and nucleus of basal optic root, *VTP* ventral tegmental pathway. (Shute and Lewis, 1967)

nucleus, globus pallidus and lateral preoptic area. This pathway receives afferents from dorsal and median raphe nuclei.

These two pathways compose the *ascending cholinergic* reticular system of Shute and Lewis. The hippocampal formation also receives a possible cholinergic innervation from the medial septum and diagonal band. Hippocampal efferents travelling by way of the fornix project on to cholinesterase-containing neurons in the hippocampal commissure, anterior thalamus, interpeduncular nuclei and the midbrain tegmentum. This is the *cholinergic limbic system*, which has also some connection with the previous ascending cholinergic reticular system. Another system, very rich in cholinesterase, has recently been mapped out by Olivier et al. (1970). It corresponds to striato-pallidal and striato-nigral fibers. These systems may contribute to a striato-nigro-striatal loop in which both cholinergic and dopamine-containing neurons could interact. There are, in fact, many other structures where MA and ACH may interact.

6. Relationship between Monoaminergic and Cholinergic Systems

In view of the widespread distribution of cholinesterase, it is not surprising that some regions belong to both MA and ACH-containing systems. Since it is possible that ACH might act upon MA mechanisms, it is interesting to point out some regions where interaction might take place (SHUTE, 1969).

a) Nucleus raphe pallidus (group B 1) of 5-HT-containing neurons, which stains lightly for acetylcholinesterase, would be cholinoceptive according to SHUTE and LEWIS (1967).

b) On the other hand, groups A 4, A 5, A 6 (locus coeruleus) and A 9 stain heavily for acetylcholinesterase and can also be regarded as cholinergic (HOLMES and WOLSTENCROFT, 1964). Thus it is likely that those groups of neurons contain both CA and ACH-containing neurons.

c) In considering ACH or MA-containing terminals, many possibilities arise. Some regions receive exclusively MA terminals (i.e., the dorsal nucleus of the lateral geniculate) or ACH (the ventral nucleus); some regions receive both, e.g. the striatum and the limbic structures. However, no direct proof of MA-ACH synapses in the central nervous system of mammals has yet been provided, due to the difficulty of recognizing MA or ACH terminals by electron microscopy.

7. Subcellular Correlates of the Storage Process of Monoamines and Acetylcholine

The histofluorescence or histochemical technique is not precise enough to give a clear picture of the subcellular storage of MA or ACH in the presynaptic terminals. However, electron microscopy has provided considerable information (DE ROBERTIS et al., 1963, 1965; WOOD, 1966; AGHAJANIAN and BLOOM, 1967a, b; BAK, 1967; HÖKFELT, 1967). The large granular vesicles (800 to 1200 Å diameter) are the only consistent form of granular synaptic vesicles observed in the brain. However, there remains the question concerning the relationship between these large granular vesicles and storage of MA and ACH, since the electron density of the granular content of vesicles has been reported either to change (BAK, 1967; see BLOOM and GIARMAN, 1968) or to resist change (BLOOM and AGHAJANIAN, 1968) after a variety of pharmacological manipulations. Recently, with more sophisticated techniques, however, small-dense-core granular vesicles (400–600 Å diameter) have been observed and may represent the storage form of MA. On the other hand, it is possible that synaptic vesicles which stain with zinc iodide-osmium impregnation may be the storage form of ACH (see AKERT et al., 1969).

The enzymes responsible for the synthesis or catabolism of MA are synthesized in the perikarya. The enzymes are transported to the terminals through axoplasmic flow (HAGGENDAL and DAHLSTRÖM, 1969). MAO is mostly associated with mitochondria. COMT is also concentrated in the nerve endings but this enzyme is so highly soluble that it is difficult to determine its exact localization within the synaptic complex (ALBERICI et al., 1965). The role of axoplasmic flow in transporting the enzymes from the perikarya to the terminals may be less important than was formerly believed, since it has recently been demonstrated that (in the peripheral nervous system) some induction of synthesis of tyrosine hydroxylase could take place in terminals disconnected from the perikarya (THOENEN et al., 1970). If this is true of the central nervous system, there would be some possibility of short-term regulation of synthesis or of catabolism directly at the terminal level without the intervention of axoplasmic flow.

8. Summary

Four new systems of neurons have recently been mapped out: 5-HT, NA, DA and ACH-containing neurons. They are most probably closely interconnected with each other, but much work remains to be done concerning their exact organization. Since one of the main strategies of neurophysiologists is to attack (destroy or stimulate) these systems as selectively as possible, the obvious tactic is to approach these systems at the place where most of the cell bodies (or ascending pathways) are concentrated. Since 5-HT-containing perikarya are concentrated in the midline raphe, the target is obviously the raphe system. NA-containing neurons of the medulla being widespread, the best target for studying their ascending influences is the ventral NA bundle. Direct attack upon the locus coeruleus is easier in the rat than in the cat, where the "coeruleus complex" is rather diffuse. However, some ascending axons from this complex can be destroyed in the region of the dorsal NA bundle. Unfortunately, cholinergic connections from the locus coeruleus to the N. cuneiformis may also ascend through this region. The DA system can be approached directly in the substantia nigra (group A 9) but this will also affect the ventral tegmental cholinergic pathway.

Finally, the midline DA neurons (group A 10) are also a difficult target, since the ventral NA bundle runs very closely. The development of drugs, like 6 hydroxydopamine, which can "selectively" destroy CA neurons is a step forward in the selective approach to the systems of CA-containing neurons.

B. Are Monoamines and Acetylcholine Neurotransmitters?

The histofluorescence method has met some of the most important criteria. There are specific systems of neurons where the amines are synthesized by specific enzymes, and are transported through axonal flow to the presynaptic terminals where they are stored. But before it can be said to be a neurotransmitter, some other criteria must also be fulfilled: MA must be released at the presynaptic junction into the synaptic cleft in response to depolarization and stimulation, must trigger some response in the postsynaptic neurons, and must be inactivated (either by enzymatic destruction or by reuptake). "Release, response, reuptake, these 'three R's' underlie the initiation, elicitation and termination of the amine response", say BLOOM and GIARMAN (1968). Since excellent recent reviews are available on this subject (AXELROD, 1962, 1966; KOPIN, 1964, 1966; ACHESON, 1966; GLOWINSKI and BALDESSARINI, 1966; BLOOM and GIARMAN, 1968; GLOWINSKI, 1970; WEINER, 1970; AXELSSON, 1971; WURTMAN, 1971), we shall emphasize only that aspect of MA metabolism which may be relevant to the problem of the control of the sleep-waking cycle.

1. Depolarization—Release

Some indirect data concerning a possible relationship between the firing of presumed 5-HT-containing perikarya and their functional aspect have been provided by AGHAJANIAN and WEISS (1968) (AGHAJANIAN et al., 1970a, b). Extracellular micro-electrode recordings in the dorsal raphe nucleus have shown a characteristic 2/sec. activity which may be altered by drugs acting upon MA. Thus LSD, which is believed to decrease 5-HT turnover, totally blocks this activity (Fig. 5). MAO inhibitors have the same action. Since MAO inhibitors may inhibit the synthesis of 5-HT (MACON et al., 1971), this may be interpreted as a possible feedback inhibition (the increase of 5-HT in the synaptic cleft would lead to a compensatory reduction in the firing). However, this interpretation is not supported by the fact that 5-HTP has no action upon the firing, nor has p-chlorophenylalanine, which decreases 5-HT synthesis. In contrast, amphetamine increases the firing of these neurons (like that of many other neurons in the brain stem) (BRADLEY and WOLSTENCROFT, 1965). In fact, these findings are difficult to interpret since there is no *absolute* proof (through histofluorescence or electron microscopy) that this characteristic electrical activity comes from 5-HT containing perikarya. Although the NA-containing neurons are quite densely packed in the locus coeruleus of the rat, there has not yet been any study of the correlation between their extra- or intracellular electrical activity and their metabolism.

2. Stimulation—Release

Both in vitro and in vivo MA can be released by central stimulation: 5-HT and NA can be recovered in vitro from the incubation fluid surrounding isolated nervous structures after tetanic stimulation (ANDÉN et al., 1965a; KAWAI, 1970). Stimulation of the dorsal raphe nucleus, in the rat, results in a slight decrease of 5-HT in the forebrain and in a long lasting increase of 5-hydroxyindolacetic acid (AGHAJANIAN et al., 1967c; SHEARD and AGHAJANIAN, 1968; KOSTOWSKI et al., 1969; ECCLESTON et al., 1969, 1970; GUMULKA et al., 1971). It is not certain, however, whether the catabolism of 5-HT by

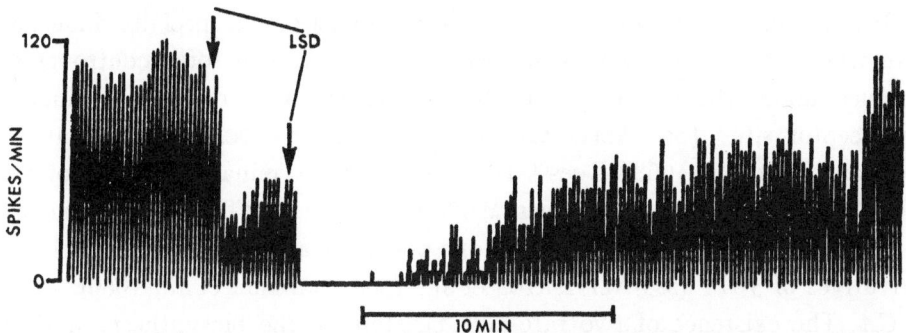

Fig. 5. *Typical response of a raphe unit to LSD*. This unit, which was situated in the dorsal raphe, had a spontaneous rate of about 120 spikes/min. Within 30 seconds after an i.v. injection of 10 μg/kg of LSD, this unit slowed; after a second dose of 5 μg/kg it ceased firing entirely. Recovery took place gradually over a period of about 20 minutes. The record consists of consecutive 10-second samples of the analog output of a counter triggered by the unit spikes. (AGHAJANIAN et al., 1970)

MAO is always effected before or after neuronal release. However, the fact that hyperthermia and an altered rate of habituation to sensory stimulation or sleep follow stimulation of the dorsal raphe may be a good argument in favor of release of 5-HT or its metabolites (if they play a physiological role; see p. 199). The stimulation of NA-containing cell bodies of the caudal medulla or brain stem results in a decrease of NA level and fluorescence in the spinal cord (DAHLSTRÖM et al., 1965). Similarly, stimulation or self-stimulation of the ascending dorsal and ventral NA bundles results in a disappearance of the fluorescence of the terminals, provided that the synthesis of NA is inhibited (ARBUTHNOTT et al., 1970, 1971). On the other hand, stimulation of the amygdala results in a decreased telencephalic NA level. Since there are no NA-containing cell bodies in the amygdala, there is indeed some evidence that this is a polysynaptic effect of the stimulation (FUXE and GUNNE, 1964).

The use of push-pull cannulae implanted in the brain has made it possible to prove the release of 5-HT, NA, DA and ACH either spontaneously or after stimulation (MCLENNAN, 1964; PORTIG et al., 1968; STEIN and WISE, 1969; CARR and MOORE, 1970; BELESLIN and MYERS, 1970; MYERS and BELESLIN, 1970). This technique is, however, not without its critics (CHASE and KOPIN, 1968), since the implantation of rather large cannulae might lead to tissue damage and to nonspecific release of substances unrelated to transmitter activity. On the other hand, the release observed through a push pull cannula is not an absolute proof that the release is onto receptor sites, which should be the case for a physiological action.

3. Monoamines and their "Catabolites"

5-HT. More than 90% of the 5-HT formed in the brain is eventually eliminated from the brain as 5-hydroxyindoleacetic acid (5-HIAA). MAO converts 5-HT to 5-hydroxyindolacetaldehyde and the aldehyde can be converted to either 5-HIAA or to 5-hydroxytryptophol. 5-HIAA is transported through the blood-brain barrier into the blood or the ventricles by a specialized mechanism which can be blocked by probenecid. There is still some controversy, however, about the fate of 5-HT after its release: how does 5-HT bind to the receptor site? (See ALIVISATOS et al., 1970a, b; ROCABOY et al., 1969; WISE and RUELIUS, 1968.) Does 5-HT re-enter the terminal to be catabolized by MAO intraneuronally, or does MAO catabolize 5-HT extraneuronally? This opens up the interesting question of a possible transmitter role of aldehyde derivatives of 5-HT (see ALIVISATOS et al., 1968; SABELLI et al., 1969).

CA. The existence of two rate-limiting steps in the biosynthesis of NA: tyrosine hydroxylase and dopamine-β-hydroxylase and of two catabolic enzymes [MAO and catechol-O-methyl transferase (COMT)] make the model of CA release from the terminal much more complex than for 5-HT neurons. Furthermore, the methods of studying the metabolism of CA, either by intraventricular injection of labelled NA or by intravenous injection of the different precursors (tyrosine or DOPA), have sometimes given different results. The currently accepted hypothesis of noradrenergic release and reuptake is schematized in Fig. 6. Most of the NA contained in presynaptic terminals is probably stored in two pools or compartments (at least). i) A so-called labile or functional pool. This pool of newly synthesized NA is preferentially released by nerve stimulation or by some pharmacological agents. This pool may regulate its own synthesis by feedback inhibition of tyrosine hydroxylase and is very sensitive to inhibition of the different steps of NA synthesis. ii) A storage pool, in which NA is stored in membrane-bound granules where it is protected from metabolic degradation.

Fig. 6. *The different steps of synthesis and inactivation of central monoamines.* (PUJOL, 1970)

I. Biosynthesis and inactivation of NA
  1. Tyrosine hydroxylase, the limiting enzyme for tyrosine (Tyr)
  2. Dopa decarboxylase (probably non-specific)
     DOPA dihydroxyphenylalanine
  3. Dopamine beta hydroxylase
  MAO   monoamine oxidase
  m   mitochondria
  v   synaptic vesicle—R: post synaptic receptor
  Comt   catechol-O-methyl transferase
  DHPG   3-4.dihydroxyphenylglycol
  DHMA   3-4.dihydroxymandelic acid
  VMA   vanillmandelic acid
  NMA   normethoxyadrenalin

II. Biosynthesis and inactivation of dopamine (DA)
  DHPE   3-4.dihydroxyphenylethanol
  DHPAA 3-4.dihydroxyphenylacetic acid
  HVA   homovanillic acid
  30MeDA 3.0-methyldopamine

III. Biosynthesis and inactivation of 5-HT
  I. Tryptophan hydroxylase
     TRY   tryptophan

  II. 5-HTP decarboxylase (probably non-specific)
      5-HTP   5 hydroxytryptophan
      5-HIAA 5 hydroxyindolacetic acid

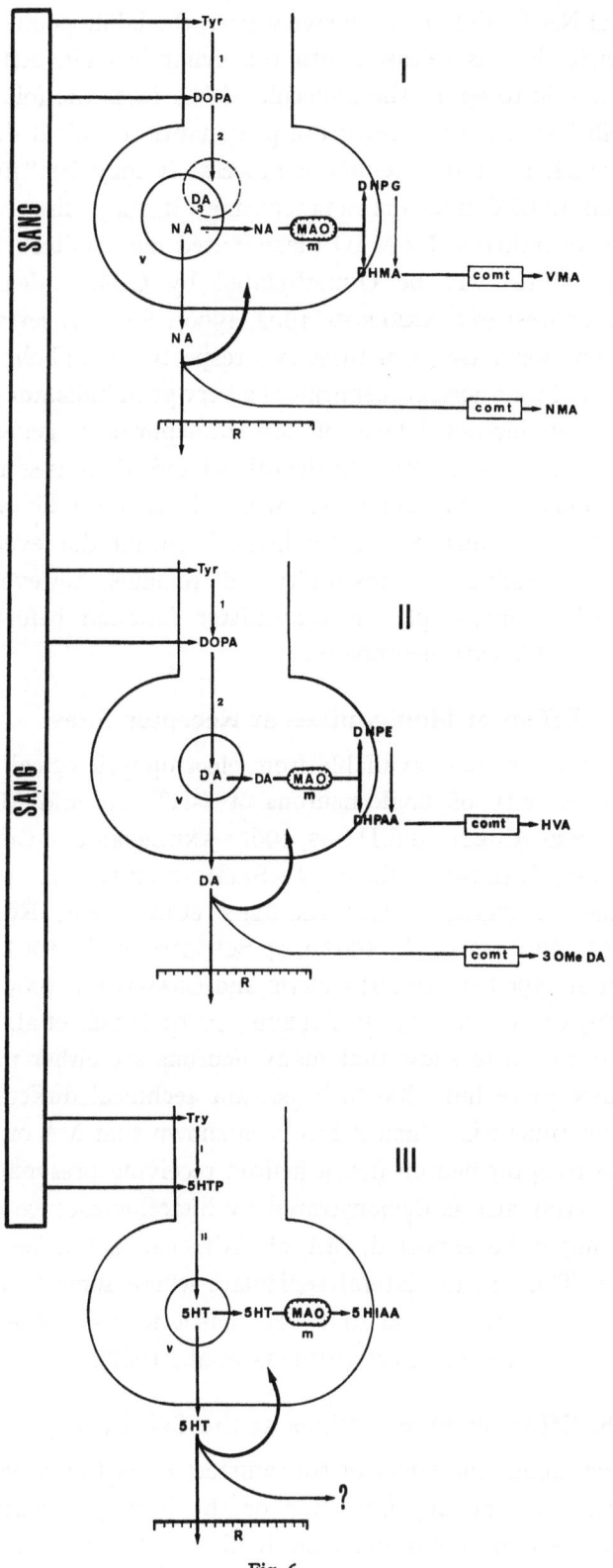

Fig. 6

Intraneuronal NA (mainly or exclusively from the labile pool), in response to nerve depolarization, is released into the synaptic cleft. After reacting with the postsynaptic receptor, the molecule of NA faces the following three different possibilities: it may re-enter the presynaptic terminal (in the functional pool) through an active reuptake process; it may be "inactivated" by extraneuronal COMT into normetanephrine; it may also re-enter the terminals and be deaminated by MAO (deaminated metabolites). Later these deaminated metabolites will be O-methylated by COMT (deaminated-O-methylated metabolites) (see AXELROD, 1962, 1966). Some uncertainty exists concerning the functional aspect of these two respective metabolic pathways: it is usually accepted that normetanephrine is a very good indicator of neuronal NA release. Thus an increased level of normetanephrine is generally taken as proof of increased release of NA. On the other hand, there is some evidence that MAO acts only on catecholamines which have been liberated intraneuronally. MAO would thus exert a "policing" role for the level of amines in the presynaptic terminals. A possibility still remains, however, that deaminated metabolites might play a transmitter function before they are O-methylated by COMT extraneuronally.

4. Effect of Monoamines at Receptor Sites

Considerable data are now available from electrophysiological analysis of the effect on the activity of single neurons of 5-HT, CA and ACH applied by microiontophoresis (CURTIS and DAVIS, 1962; ANDERSEN and CURTIS, 1964; BLOOM et al., 1964; RANDIC et al., 1964; SALMOIRAGHI et al., 1964, 1965; SALMOIRAGHI and STEPHANIS, 1965; BRADLEY et al., 1966; ROBERTS and STRAUGHAN, 1967; PHILLIS et al., 1967b, c; SCHMIDT et al., 1967; STEINER, 1968; BOAKES et al., 1969, 1970a, b; CURTIS and CRAWFORD, 1969; BRADLEY and CANDY, 1970; CRAWFORD, 1970; TEBECIS, 1970; HÖSLI et al., 1971). All these data are in accord to show that many neurons are either monoaminoceptive or cholinoceptive but, due to important technical difficulties, there is not a single experiment in which it has been shown that MA or ACH were applied to a true receptor neuron (i.e. a neuron receiving presynaptic MA or ACH-containing terminals) as demonstrated by histofluorescence or electron microscopy. As might be expected, MA or ACH can act differently upon different neurons. Thus, in the lateral geniculate where some 5-HT and NA terminals can be seen, it was found that 5-HT has a depressor effect, whereas NA has mainly an excitatory effect (PHILLIS et al., 1967).

5. Effect of Monoamines at the Cell Level

Little is known about the effect of transmitters upon the nerve cell itself (after the initiation of the depolarization or the hyperpolarization of the postsynaptic cells). Are macromolecular mechanisms also involved? A current theory is that NA would bind to specific receptors that activate adenyl cyclase to produce 3'5'-AMP which, in turn, is the molecular effector in the target cell (see RALL and GILMAN, 1970). In some cases, cyclic AMP iontophoretically injected may mimic the inhibitory action of NA upon the Purkinje cells (SIGGINS et al., 1971).

6. Possible Interactions between Monoamines and Acetylcholine

There is no *direct* evidence of synaptic contact between presynaptic terminals containing 5-HT or NA or ACH with other MA or ACH-containing perikarya or axons. However, this is a likely possibility if one considers the accumulation of CA terminals around the raphe system, and the response of raphe cells to 5-HT, NA and ACH. The possible physiological interactions between these neurotransmitters are still hypothetical. 5-HT may act upon acetylcholinesterase and thus facilitate the action of ACH (see APRISON, 1962). In vitro, 5-HT is an effective inhibitor of the enzymatic oxidation of NA and DA (VAN DER WENDE and JOHNSON, 1970a, b). ACH may liberate NA in postganglionic adrenergic fibers in the peripheral nervous system (see BURN and RAND, 1965; FERRY, 1966). But no *direct* evidence has been brought for such a mechanism in the central nervous system. On the other hand, NA may alter the activity of choline acetylase (SINGER and GERSHON, 1971). Considering so much indirect evidence of possible interactions between the MA systems, a question should be asked at the end of this chapter: Is the Dale Law applicable to the central nervous system? This law states (DALE, 1935) that a given neuron may only release its own transmitter: i.e. a 5-HT-containing neuron releases only 5-HT at the presynaptic terminal. However, indirect evidence indicates that this may not always be the case. After injection of a *large* dose of DOPA, associated with peripheral inhibition of DOPA decarboxylase, 5-HT perikarya become green as if they contain catecholamine (CONSTANTIDINIS et al., 1968) and there is much evidence that either 5-HT or NA may enter the "wrong" terminals (see FUXE and UNGERSTEDT, 1967, 1968; SHASKAN and SNYDER, 1970). On the other hand, exogenous 5-HTP may be decarboxylated in NA-containing neurons which contain the "non-specific" DOPA-5-HTP decarboxylase (see below, p. 201). Thus, some 5-HT may theoretically be released by "catecholaminergic" nerve endings. This possibility is obviously not welcomed in neuropharmacological experiments, as we shall see later.

C. Monoaminergic Neurons: Some Possible Strategies and Tactics of an Approach to their Role in the Sleep-Waking Cycle

Three main strategies are possible in studying the role of MA in sleep-waking. 1. To alter, as selectively as possible, the activity of a given MA system and to study the effect upon sleep-waking. 2. To alter sleep or waking and to study the possible alteration of the activity of MA system. 3. To study the possible parallel alteration of sleep and MA system during their natural circadian or ultradian variation.

Whatever might be the strategy involved, it must be pointed out that quantitative measurements of sleep or waking are necessary. This necessitates the use of polygraphic recordings during *chronic experiments*, which are *prerequisites* to the study of physiological sleep.

1. The Frontiers of the Sleep-Waking Cycle

The time has now arrived to delimit precisely the polygraphic frontiers of the sleep-waking cycle. Indeed, as "the swallow does not make a summer", one cortical slow wave does not make slow-wave sleep. Fig. 7 depicts the main polygraphic aspect of the sleep-waking cycle in the cat. Besides the alteration in the cortical activity, from arousal to drowsiness with some slowing of the frequency of the cortical waves, to stage I (with spindles) and stage II (high-voltage 2–3/sec slow waves), some subcortical patterns enable us to assess that the slowing of cortical activity belongs to physiological sleep. High-voltage sharp waves appear periodically in the pontine reticular formation, the lateral geniculate and occipital cortex (PGO). This PGO activity appears either during SWS (stage II) (isolated PGO), or immediately before and during PS. PS occurs spontaneously every 25 minutes during sleep, and there is some relationship between the amount of stage II and the amount of PS (URSIN, 1970). Thus, we shall use the term of physiological sleep (after any direct or indirect alteration of MA) only if we observe the same periodical and harmonious succession of the states of sleep which are observed under control conditions. *Hypersomnia* would be defined as a state in which *both* SWS and PS increase significantly above the control level, since the quantity of sleep is remarkably constant in laboratory cats (DELORME et al., 1964; STERMAN et al., 1965). Since many drugs or lesions may increase cortical synchronization only (but may suppress isolated PGO or PS), we shall call this state either increase of SWS (if isolated PGO still occurs) or increase of *cortical synchronization* (if the pattern of cortical activity is different from that of normal sleep). If this state is accompanied by an obvious behavioral sedation, associated with miosis and the posture of sleep, it will be called *sedation* or *decrease of waking*.

Waking is defined at the polygraphic level by a cortical fast low-voltage activity accompanied by an increase in muscle tone. In order to cover most of the spectrum of waking behavior, behavioral waking will be defined either as spontaneous (the animal standing up and having spontaneous motor activity) or as a response to environmental stimulation (orientating reaction). Any animal which lacks both spontaneous and reactive motor activity will be considered as comatose (we shall not consider in this review the more elaborate

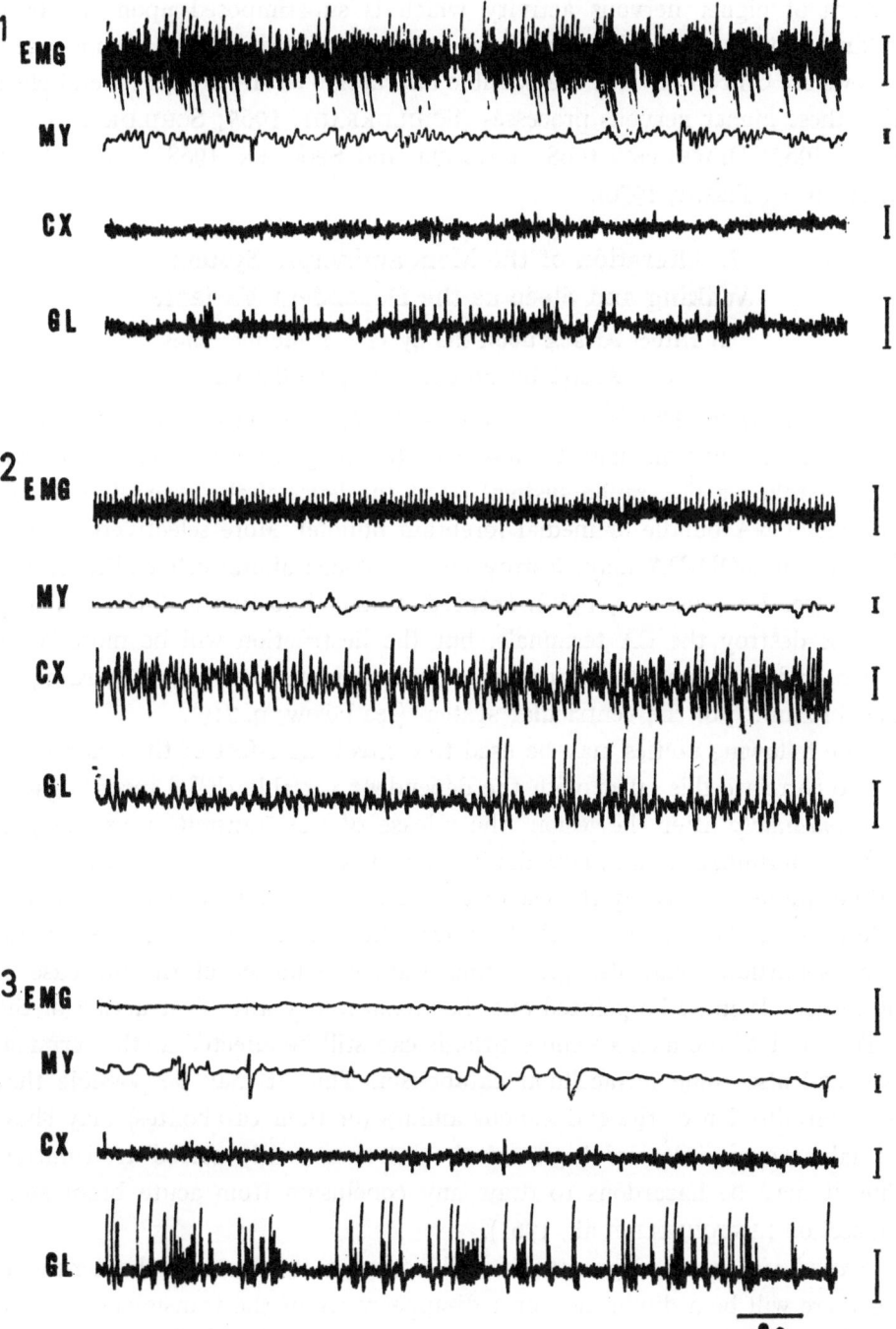

Fig. 7. *The frontiers of the sleep-waking cycle.* *1* Waking: low-voltage fast activity on the frontoparietal cortex (*CX*), increase of the electromyographic activity of the neck (*EMG*), rapid eye movement (*MY*) and eye movement potentials in the lateral geniculate (*GL*). *2* Slow-wave sleep: slowing of the cortical activity (stage II) and appearance of isolated high-voltage PGO in the lateral geniculate (slow sleep with phasic activity). *3* Paradoxical sleep: low-voltage fast cortical activity with total disappearance of the EMG of the neck. PGO activity, isolated or in bursts in the lateral geniculate, accompanies rapid eye movements. Calibration: 6 seconds—50 microvolts

pattern of higher nervous activity which is superimposed upon the basic waking state, i.e. learning, memory, emotion, affective state. Some recent reviews are concerned with the possible interaction of MA or ACH metabolism with these higher nervous processes (SCHILDKRAUT, 1965; SCHILDKRAUT and KETY, 1967; DEWHURST, 1968; MANDELL and SPOONER, 1968; DAVIS, 1970; KETY, 1970; TISSOT, 1970).

2. Alteration of the Monoaminergic System: Waking and Sleep as the Dependent Variable

i) Direct Attack upon the System of Monoamines and Acetylcholine-Containing Perikarya

a) Destruction. This is one of the classical approaches of neurophysiology. MA-containing neurons may be destroyed (by coagulation) either at the level of the perikarya (i.e. raphe system) or at the level of the ascending bundle (i.e. dorsal NA bundle or medial forebrain bundle). More selectively, in situ injection of 6-OH-DA may destroy either CA-containing cell bodies or terminals (see UNGERSTEDT, 1969). Intraventricular injection of 6-OH-DA may likewise destroy the CA terminals, but the destruction will be more widespread and may possibly affect the CA-containing terminals according to their proximity to the ventricular system (see below, p. 219).

The following tactics may be used to control the effect of the destruction and to correlate this effect with the dependent variable (EEG recordings).

Immediately after the lesion, the release of the transmitter at the presynaptic terminals is strongly diminished or suppressed. Theoretically this can be studied in vivo by the use of the push-pull cannula or the superfusion technique at the cortical level. In vitro, the determination of the "membranous outflow" can also give some indirect evidence of the decrease of the release. It must be pointed out that *immediately* after destruction of the perikarya or of the axons some synthesis can still be effected at the terminal level and also some intracellular catabolism. Thus it may be possible that the determination of the endogenous amines (or their catabolites) may show normal or even *increased* levels of the intraneuronally stored transmitter. Thus it may be hazardous to draw any conclusion from acute brain stem transection (ANTONELLI et al., 1961).

Secondarily, some degeneration will occur at the level of the terminals and there will be a diminution or a disappearance of the transmitter 8 to 10 days after the lesion (DAHLSTRÖM and FUXE, 1964). This decrease can be studied *in vivo* by push-pull or superfusion techniques measuring either the release, or the local synthesis of labelled precursor. *In vitro*, the diminution of synthesis can also be studied by incubating slices containing terminals with labelled precursors (see Fig. 8). Finally, post-mortem semi-quantitative fluorescence techniques or biochemical techniques may indicate the diminution of endogenous amine in the terminals. All these parameters can be tentatively correlated either with the state of the animal during the "in vivo" determination, or with the mean percentage of sleep-waking during survival (for the in-vitro or post-mortem determination). Thus, the lesion technique has many advantages but also some important drawbacks since, besides known MA neurons, other unknown systems may be destroyed.

b) Stimulation. The stimulation can be theoretically effected either to some known presynaptic afferent neurons to the MA perikarya, or directly (electrically or with a known presynaptic transmitter) upon the perikarya or the ascending MA bundle. The effect obtained at the terminal level may be assessed quantitatively in vivo by either the push-pull or superfusion technique. If the stimulation has continued long enough, it may even be possible to obtain some effect in vitro when studying the biosynthesis, provided that some induction of enzymatic activity has been obtained. Finally, post-mortem biochemical examination may show alteration of the endogenous level of the transmitter or its catabolites. The histofluorescence technique is not really useful in this condition since decrease of fluorescence at the terminals after stimulation of perikarya or ascending bundle can be obtained only if some inhibitor of synthesis has been simultaneously injected (ARBUTHNOTT et al., 1970, 1971). Since any drug injected at the periphery may itself act upon the waking-sleep mechanism, its effect could cancel out the physiological effect of stimulation. The advantages and disadvantages of stimulation techniques are well known for the central nervous system. If an electrical current is used, what frequency is the most "physiological"? How does one control the spread of current? Are electrically induced spindling or recruiting responses true physiological sleep? How long should the stimulation last in order to obtain a *steady state* of cortical activity compatible with parallel or subsequent biochemical evaluation?

If neurotransmitters are used to mimic physiological presynaptic transmitters (at the perikarya or at the receptor level), many questions are difficult to answer: How should it be injected—by iontophoresis, micro-injection, or micro-implant? In what dose? Is it certain that the transmitter in excess does not act upon other, non-"ceptive" neurons?

ii) Indirect Approach: Neuropharmacological Approach

The pharmacological approach lacks the anatomical parameter which is of paramount importance. It is nevertheless widely used because it is easier: "a little of anything should do something to everything". This is certainly true for the sleep-waking cycle. Any drug injected at the periphery will act not only upon the brain, but also upon other MA systems situated at the periphery. Intraventricular injection has so many nonspecific effects that its usefulness is somewhat questionable. Finally, the differences in pharmacological responses between species have made a purely neuropharmacological approach to sleep-waking difficult but not totally fruitless.

The main tactics concerning the pharmacological approach of MA system can be summarized as follows:

— *Inhibition of Synthesis*. This can be obtained either by decreasing the availability of the essential precursor amino acid (free diet or competition with other amino acids) or by inhibiting the rate-limiting enzymes responsible for the intraneuronal synthesis. This will theoretically lead to the secondary disappearance of the transmitter from the synaptic cleft.

— *Increase of the Transmitter at the Synapses*. Theoretically this can be obtained in many different ways.

1. Increase of the availability of the precursor amino acid, provided that the first step of the synthesis is not rate-limiting.

2. Injection of the direct precursor (if it can cross the blood-brain barrier e.g., 5-HTP or DOPA).

3. Injection of the transmitter itself, either into the blood (only in some special conditions, since in most cases MA does not cross the blood-brain barrier) or directly into the brain.

4. Inhibition of catabolism of the transmitter.

5. Inhibition of the reuptake of the transmitter. Each of these steps has its own drawback. Thus, the inhibition of MAO, while increasing 5-HT at the terminal level, will at the same time decrease the synthesis of 5-HT through a probable feedback inhibition (MACON et al., 1971). Other drugs may act through still unknown or very complicated mechanisms. This is the case with reserpine or the false transmitters (KOPIN et al., 1965, 1969). Their mechanism of action is not simple, since KOPIN lists at least six different mechanisms through which a false transmitter may alter the effectiveness of a nerve impulse. Finally, drugs acting upon the receptor itself may be used provided they are "specific".

We shall meet the problem of the multiple mechanisms of action of any drug when considering drug effects upon sleep-waking. To add further complexity, it should also be remembered that peripheral or central "side effects" of a drug might well counteract or conceal some other central effects. Hyperthermia (as obtained with MAO inhibitors) or diarrhea (after reserpine) can certainly interfere with the physiological mechanisms of sleep.

Finally, even if it is proved that some drug acts *selectively* upon a specific MA system, it remains to be shown whether its action upon all the perikarya and terminals might not conceal some effect which would happen if only a small group of neurons were affected. Even if sleep does seem to affect the electrical activity of the brain in totality, *this does not mean that the totality of a given MA system is involved in its triggering*. The organization of these systems is not yet sufficiently well known to refute the likely hypothesis that their complexity (ascending neurons and terminals) is the basis of many functions. 5-HT neurons have been implicated (besides sleep) in thermoregulation (see FELDBERG et al., 1966; MYERS, 1970; SHEARD and AGHAJANIAN, 1967; WEISS and AGHAJANIAN, 1971), sexual activity (FERGUSON et al., 1970; GESSA et al., 1971; SEGAL and WHALEN, 1970), feeding (MYERS and YAKSH, 1968, 1969), sensitivity to pain (TENEN, 1967) and endocrine function (see WURTMAN, 1971). Thus, any drug which would alter the liberation of 5-HT in all the terminals could well act only indirectly upon sleep (by increasing sexual drive or by lowering the threshold for painful stimuli).

3. Alteration of the Sleep-Waking Cycle: The Activity of Monoamines-Containing Neurons as the Dependent Variable

Several tactics are possible: the electrical activity of a known group of MA neurons can be recorded during the sleep-waking cycle. This approach would certainly be fruitful, provided that one could recognize with certitude that a MA-containing neuron is being recorded.

The central regulation of MA neurons can be studied *in vitro* (by killing the animal during various states of sleep or waking). In such conditions the determination of the endogenous level of amine is of lesser importance than the estimation of the biosynthesis, or the "release" of a particular amine on cerebral slices incubated with labelled precursor. *In vivo*, thanks to the use of the push-pull cannula and superfusion technique, the alteration of the release or of the turnover of MA may be "theoretically" related to the sleep-waking cycle. At least, four techniques have been proposed for estimating MA synthesis or turnover (see WURTMAN, 1971). In the first approach (LIN et al., 1969), labelled precursor amino acids are injected systematically and the rate at which ^3H MA accumulates is monitored as an index of MA synthesis. This technique, however, reflects both the synthesis and the turnover and may lead to underestimation of both rates. The second method utilizes the injection of ^3H NA or ^3H 5-HT into the ventricular system. The rate at which brain ^3H MA levels decline is taken as an index of MA turnover. However, ^3H NA or ^3H 5-HT may not be taken up by the specific terminals, so this technique is not absolutely "foolproof". In the third technique, an inhibitor of the first step of the synthesis may be injected, i.e. α.methyl-p-tyrosine (α.MPT) which blocks CA biosynthesis. The rate at which brain CA falls is monitored and it is assumed that this decrease is proportional to the amount of CA which would have been synthesized without α.MPT. In fact, since turnover and synthesis may be physiologically coupled, the administration of α.MPT may disturb the steady-state relationship that controls brain NA levels, and the subsequent decline in brain NA may reflect the excess of NA turnover over NA synthesis. Finally, in the fourth technique, MAO inhibitors are given and it is assumed that the rate at which brain MA increases is proportional to what MA turnover would be in untreated animal. But this method also induces a change in the steady state and does not take into account the fact that alteration in MA turnover may influence the MA synthesis. Moreover, when concerned with CA, the compartment of CA turn-over which is blocked by MAO inhibitor may be the compartment that is *not* related to the physiological activity of CA neurons. Finally, it is quite difficult to obtain a steady state of sleep during which turnover could be measured (even when using the rebound of PS following its previous selective suppression). For all these reasons, the results of the determination of brain MA turnover during the sleep-waking cycle must be always accepted with some caution.

4. Parallel Alteration of Monoamines and Sleep-Waking: The Circadian Approach

This very indirect approach does not provide any data concerning short-term regulation of the sleep-waking cycle. Nevertheless, this approach may give interesting data concerning the possible circadian alteration of regulation of the MA or ACH systems (uptake of amino acids, biosynthesis or release). These long-term regulatory mechanisms may be very important for an under-standing of the circadian variation of short-term regulatory mechanisms which might be superimposed.

5. Summary

Many of the techniques which have been described here have not yet been applied to the study of sleep-waking in chronic conditions because of obvious technical difficulties. Thus, most of the data which will be exhibited below come from rather indirect approaches (lesion or neuropharmacology). There will be converging indirect data rather than a unique direct proof. The ideal experiment remains to be done. In fact, such an experiment is almost impossible at the present time. Even if one were able to prove that the reversible suppression of release of 5-HT, NA or ACH at a given pre-synaptic site is followed regularly, and in a dose-dependent effect, by a predict-able effect upon sleep or waking, there would always be the problem of some still unknown transmitter, located in the same place, which could be the *specific* transmitter acting upon the receptor. Thus, an entirely skeptical or nihilistic approach is possible. Waking, sleep and dreaming are such com-plicated processes, involving so many neurons, subject to so many factors, that is would be naive to incriminate only one or two transmitters in their mechanisms. In this review I have deliberately taken the opposite view. This "optimistic" approach derives from the fact that it is now possible to induce total insomnia, lasting several weeks in chronic cats following very small lesions which coincide with the topography of the 5-HT perikarya or to obtain *any* predictable amount of physiological sleep (from less than 5% to more than 60%) for a long time (week) by interfering "only" with the biosynthesis of 5-HT.

D. 5-Hydroxytryptamine and the Sleep-Waking Cycle
1. Pharmacological Alteration of 5-HT Metabolism
i) Action of 5-HT and 5-Hydroxyacetaldehyde

In chicks, whose blood-brain barrier is permeable, (HANIG et al., 1970) the intravenous injection of 5-HT is able to induce "sleep-like" behavior and cerebral synchronization (SPOONER and WINTERS, 1965, 1966; SPOONER et al., 1968). The same effect is obtained with the deaminated metabolites of 5-HT, 5-hydroxyacetaldehyde (SABELLI and GIARDINA, 1970; SABELLI et al., 1969), 5-hydroxytryptophol (FELDSTEIN et al., 1970) and with melatonin (BARCHAS, 1968; HISHIKAWA et al., 1969). KOELLA and his co-workers (KOELLA et al., 1960, 1965; KOELLA and CZICMAN, 1966) have made numerous attempts to by-pass the blood-brain barrier, which is impermeable to 5-HT in mature mammals: by local application of 5-HT at the level of the area postrema, which is close to some 5-HT perikarya (FUXE and OWMAN, 1965), or in the ventricles, they were able to induce sedation and cortical synchronization or augmentation of the recruiting response in the cat. More recently it has been shown that the local injection of 5-HT (ROTH et al., 1970) into the arterial circulation, which supplies (ROTH and YAMAMOTO, 1968) the area postrema and the adjacent medulla, increases cortical synchronization, whereas the injection of NA induces cortical desynchronization. Finally, YAMAGUCHI et al. (1963) have obtained true physiological sleep by microinjection of 5-HT in the preoptic area and the N. centrum medialis thalamus, while MARCZYNSKI et al. (1964) have induced sleep in the cat by injection of melatonin in the hypothalamus and LEDEBUR and TISSOT (1966) have induced cortical synchronization in the rabbit by local injection of 5-hydroxytryptophan in the medulla.

These experiments are difficult to interpret, since it is hard to determine the concentration of exogenous 5-HT in the brain. 5-HT may mimic the effect of endogenous 5-HT at the receptor site [if the amount is sufficiently small), but it may also act upon other (non-receptor) cells through a pharmacological effect and, finally, it may enter other MA neurons] (AGHAJANIAN et al., 1966; FUXE and UNGERSTEDT, 1967, 1968; FUXE et al., 1968; SHASKAN and SNYDER, 1970).

ii) Action of Precursor of 5-HT

Transport of precursor amino acids to sites of biosynthetic enzymes.

Before its hydroxylation by tryptophan hydroxylase, tryptophan must be actively transported by catalytic mechanisms or permeases from the circulation to the intracellular site of the enzyme. Since permeases are non-specific, it is possible that some amino acids in excess may compete with tryptophan. Thus high levels of phenylalanine (SCHAIN et al., 1965) or leucine (RAMANA-MURTHY and SRIKANTIA, 1970) may inhibit tryptophan transfer across the membrane (and thus decrease brain 5-HT). At the present time, however, there has been no report concerning the effect of high concentrations of leucine or phenylalanine upon sleep.

—*Tryptophan*. JEQUIER et al. (1967; see also CONSOLO et al., 1965) have shown that tryptophan hydroxylase is the rate-limiting step in the biosynthesis of 5-HT. Furthermore, the high K_m[1] observed for tryptophan suggests that the enzyme may not be fully saturated with substrate so that the overall rate of 5-HT synthesis may be partially dependent upon availability of the substrate (MOIR and ECCLESTON, 1968). This may explain why loading doses of tryptophan increase the level of brain 5-HT in rats. In the rat, excess tryptophan induces a small decrease in total sleep duration accompanied by a significant increase in the frequency of PS, whereas a tryptophan-free diet (which also decreases sleeping time) reduces the number of PS episodes (HART-MANN, 1968b). Since these observations were made without any biochemical control it is difficult to interpret the findings. In man, high doses of L-trypto-phan (5–10 g) decrease the latency for the first PS episodes (EVANS and OSWALD, 1966; OSWALD et al., 1966), or significantly increase SWS in normal subjects and in insomniac patients, whereas PS is decreased (HARTMANN et al., 1971; WYATT et al., 1970).

—*5-HTP*. This direct precursor of 5-HT readily crosses the blood-brain barrier (UDENFRIEND et al., 1957) and its depressor effect upon behavior is well known (see WADA and McGEER, 1966). DL-5-HTP has been used in most pharmacological experiments (only L-5-HTP is the active precursor and nothing is known about the possible neurophysiological action of the D-isomer). The effect of DL-5-HTP depends upon the dose injected. In the cat, at low doses (1–5 mg/kg/I.V. or I.P.) there is a slight tendency toward more synchronization and some increase (through not significant) of PS during the 6 hours following the injection (DELORME, 1966).

Large doses (30–50 mg/kg) definitely induce in the normal cat, rabbit (MONNIER and TISSOT, 1958) and monkey (MACCHITELLI et al., 1966) a state of cortical synchronization which lasts continuously for 4–6 hours. The behavior of the animal resembles sleep, with fissurated miosis. However, the nictitating membrane is retracted. The polygraphic pattern also suggests that the cortical synchronization is different from the electrical pattern of physiological sleep. There is a tendency toward synchrony in the lateral geniculate nucleus where such activity is never seen during physiological sleep. Moreover, PGO activity

---

[1] K_m is the Michaelis constant.

and PS are totally suppressed for 5–6 hours, after which PS occurs with some rebound (DELORME, 1966). At higher doses, 5-HTP induces first a state of sedation followed by excitation (COSTA et al., 1960) and even sham rage (GREEN and SAWYER, 1964). This reversal of 5-HTP-induced sedation appears to be related to the production of bufotonine (MANDELL and SPOONER, 1969). Since 5-HTP may be decarboxylated, not only in the serotoninergic terminals but in brain capillaries and also presumably in catecholamine terminals (where it might act as a false transmitter) (see APRISON et al., 1968a), the effect of high doses of 5-HTP is difficult to interpret. It may be that exogenous 5-HTP has a synchronizing pharmacological action of its own, whereas its possible decarboxylation in catecholaminergic neurons might be responsible for the alteration of the mechanism of PS (which may depend upon noradrenergic mechanisms).

iii) Inhibition of the Biosynthesis of 5-HT

The most potent inhibitor of tryptophan hydroxylase now available is p-chlorophenylalanine (PCPA). This drug selectively decreases the level of 5-HT in the rat's brain without significantly altering the level of catecholamines (KOE and WEISSMAN, 1966; GAL et al., 1970). The relatively slow rate of depletion suggests that an active metabolite of PCPA, possibly p-chlorophenylpyruvic acid, might be responsible for the decrease in cerebral 5-HT. There is also experimental evidence that PCPA inhibits the synthesis of brain 5-HT at the level of tryptophan hydroxylase, in the cat, whereas the synthesis of 5-HT from 5-HTP is normal (Fig. 8).

The action of PCPA upon the sleep-waking state has been extensively studied in the cat (DELORME et al., 1966; MOURET et al., 1967; KOELLA et al., 1968; PUJOL et al., 1969, 1971; COHEN et al., 1970), rat (MOURET et al., 1968; TORDA, 1967; SHEARD, 1969; BRODY, 1970; FIBIGER and CAMPBELL, 1971), rabbit (FLORIO et al., 1968), monkey (WEITZMAN et al., 1968) and man (WYATT et al., 1969; SATTERLEE et al., 1970; SJOERDSMA et al., 1970). After a single injection of 400 mg/kg of PCPA in the cat, no apparent variation is observed in behavior or in polygraphic recordings during the first 18–24 hours. This demonstrates that the drug itself has no direct pharmacological action upon the brain. Following this period, an abrupt reduction in both states of sleep occurs and, after about 30–40 hours, almost total insomnia appears, as shown by a permanent and quiet waking behavior, mild mydriasis and an almost permanent low-voltage fast cortical activity. The recovery of sleep begins after the 40th hour and is accompanied by the appearance of permanent PGO waves in the lateral geniculate body and occipital cortex. Very discrete episodes of PS may appear, either following short episodes of SWS or even directly following waking. SWS episodes of longer duration gradually reappear at

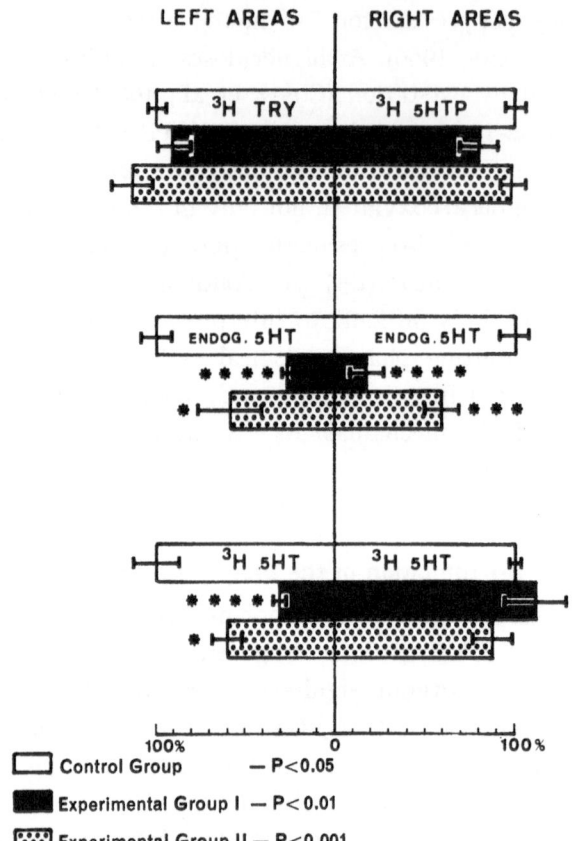

Fig. 8. *Decrease of endogenous 5-HT and alteration of the synthesis in vitro of (³H) 5-HT from (³H) Trypto-phan (left areas) and (³H) 5-HTP (right areas) in the cortex of the cat after administration of p-chlorophenyl-alanine (experimental group 1), or destruction of the raphe system (experimental group II).* Results are expressed in percentage of the mean value obtained for the control group ± SEM. In both experimental groups the synthesis of (³H) 5-HT from (³H) TRY is decreased whereas it is not significantly altered when (³H) 5-HTP is utilized as precursor (Pujol et al., 1971)

shorter intervals. Qualitatively and quantitatively normal patterns of sleep are resumed after about 200 hours (Fig. 9).

Under the influence of PCPA, a significant correlation has been found to exist in the rat between the decrease of SWS and the decrease of cerebral 5-HT (Fig. 10) whereas there was no significant alteration of catecholamine levels (Mouret et al., 1968). Koella et al. (1968) have observed the same correlation in the cat brain after careful neurophysiological and biochemical experiments.

Since PCPA inhibits only the first step in the synthesis of 5-HT at the level of tryptophan hydroxylase and since 5-HT may still be synthesized from 5-HTP, it is possible to by-pass the blocking action of PCPA and thus to re-establish a higher level of 5-HT by injecting 5-HTP. [In PCPA-pre-treated cats in which the level of 5-HT is decreased by 85%, the injection of 5 mg/kg of 5-HTP is followed by an increase of 5-HT to 60% of the normal

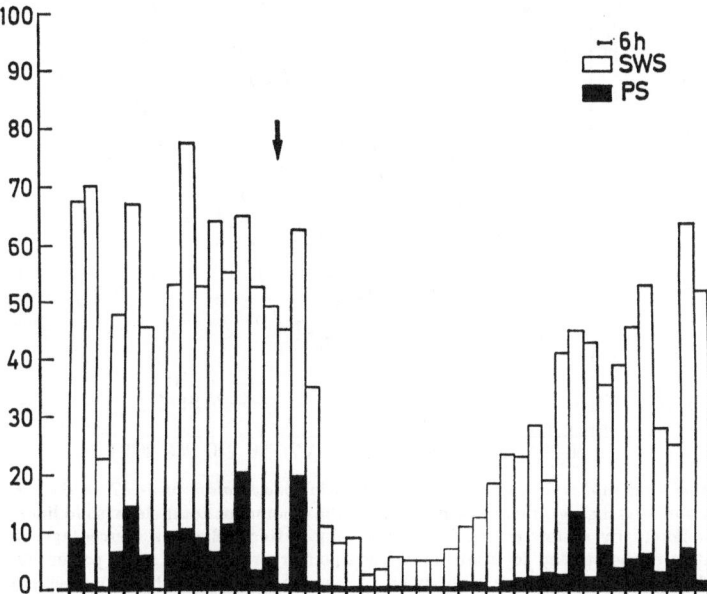

Fig. 9. *Effect of p-chlorophenylalanine upon the sleep-waking cycle in the cat.* Ordinates: percentage of slow-wave sleep (white rectangle) or paradoxical sleep (in black) every 6 hours (continuous recording). Abscissae: time (period of 6 hours). On the left: control recording. The arrow signals the intraperitoneal injection of 400 mg/kg of p-chlorophenylalanine. Insomnia appears after about 18 hours and lasts for about 60 hours, after which there is a progressive return to normal sleep. (PUJOL et al., 1971)

level one hour after the injection (HOYLAND et al., 1970).] With this procedure, it is possible to manipulate at will the state of sleep of the animal: thus, a single injection (I.V. or I.P.) of a very small dose of 5-HTP (2–5 mg/kg) given when the insomnia has reached its maximum (30 hours following the administration of 400 mg/kg of PCPA) *is able to restore a quantitatively and qualitatively normal pattern of both states of sleep for 6–8 hours* (Fig. 11) (MOURET et al., 1967; KOELLA et al., 1968; JOUVET, 1968; PUJOL et al., 1971). If a larger dose of 5-HTP is injected (30–50 mg/kg), only cerebral synchronization accompanied by sedation reappears during the first hours and PS is delayed for 4–6 hours. This suggests that, after the inhibition of the tryptophan hydroxylase, in the absence of endogenous substrate the 5-HT-containing neurons are able to synthesize 5-HT rapidly from *small* amounts of exogenous 5-HTP, whereas it appears possible that, with larger quantities, exogenous 5-HTP may interfere at some other site with the normal process of PS.

Other experiments have shown that a cat receiving balanced daily doses of 5-HTP with the dose of PCPA which would induce insomnia in a control cat may present normal or even hypersomnia during at least one week (MOURET et al., 1967). *These experiments show that sleep mechanism can be manipulated by interfering "only" with the synthesis of 5-HT.*

The insomnia due to the short-term administration of high doses of PCPA and its reversibility to normal sleep with very low doses of 5 HT has been

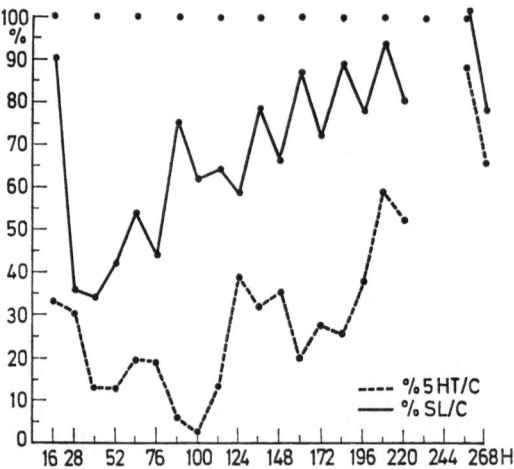

Fig. 10. *Effects of p-chlorophenylalanine on sleep and brain concentration of 5-HT in the rat.* After intra-peritoneal injection (500 mg/kg), there is a decrease in the amount of total sleep (solid line) and in 5-HT concentration followed by a slow return to normal values after 268 hours. Percentage of sleep (mean of 12 rats) relative to the amount for control rat; each point represents the mean percentage per 12-hour period (black dots at the top indicate night hours, 7 p.m. to 7 a.m.). Dashed line: percentage of 5-HT in the telediencephalon (relative to the concentration for a control rat). For the purpose of this analysis, two animals were killed every 12 hours. The relative level of catecholamines (NA or DA) did not vary significantly in the treated rats. (MOURET et al., 1968)

confirmed in several laboratories (KOELLA et al., 1968; COHEN et al., 1970; HOYLAND et al., 1970). Some conflicting results have, however, been obtained: in the rat, contrary to the findings of MOURET et al. (1968), RECHTSCHAFFEN et al. (1969) found only a 46% decrease in total sleep whereas brain 5-HT fell to a very low level. In the cat, daily administration of PCPA for one week leads initially to insomnia, but SWS and PS slowly reappear despite a very low level of brain 5-HT (COHEN et al., 1970). In this condition PCPA has been shown to decrease the turnover of cerebral NA (STOLK et al., 1969). Moreover, the return to cortical synchronization has been observed not only with 5-HTP but also with chlorpromazine, although in this case PS did not return to the normal level (COHEN et al., 1969). In man, PCPA given to patients with carcinoid tumor significantly reduces the duration of PS but not the duration of other stages of sleep. 5-HTP given to PCPA-pretreated patients restores PS to normal levels whereas tryptophan increases slow-wave sleep and decreases PS (WYATT et al., 1969, 1970). This finding is interpreted as an indication that tryptophan could act through the formation of kynurenine (another metabolite of tryptophan). But since urinary excretion of 5-HIAA is increased in these patients after tryptophan administration, it is evident that the inhibition of tryptophan hydroxylase was not total and that trypto-phan could also have been metabolized to 5-HT.

Since the suppressor effect of PCPA upon sleep and its reversibility to normal sleep by a low dose of 5-HTP is one of the *crucial* arguments favoring

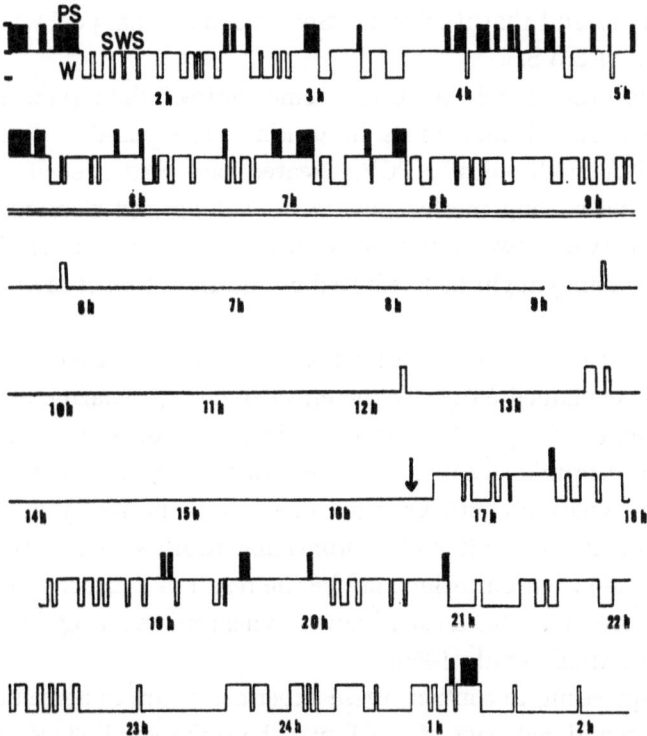

Fig. 11. *Effect of secondary injection of 5-HTP during insomnia produced by p-chlorophenylalanine in the cat.* Top two lines: normal pattern of sleep in a cat during control recordings. Bottom five lines: total insomnia 96 hours after the last of the three daily injections of p-chlorophenylalanine (400 mg/kg), then, following the injection (see arrow) of 5-HTP (5 mg/kg), reappearance of the normal pattern of sleep. Insomnia returns 8 hours after the injection. (Base line of each record: waking. White rectangles: slow wave sleep. Black rectangles: paradoxical sleep). (JOUVET, 1969)

the role of 5-HT in the mechanism of sleep, these conflicting data must be discussed in some detail.

1. The fact that in man PCPA decreases only PS does not invalidate the role of 5-HTP in SWS. It is likely that the doses used in man are rather low in comparison with the cat (in which the effect of low doses has not yet been studied). In man, low doses of PCPA might reduce the turnover of cerebral 5-HT with the result of decreasing PS only, since PS appears to depend for its priming upon the turnover of 5-HT (vide infra). On the other hand, it is difficult to compare the reaction of patients with carcinonoid tumor (who have very high levels of blood 5-HT) with that of normal subjects.

2. The return of a subnormal level of SWS after chronic PCPA treatment in cats with very low levels of brain 5-HT might also be explained through two different mechanisms:

a) The data of PUJOL (1970) and PUJOL et al. (1969, 1971) have shown that in the cat the "in-vivo" inhibition of tryptophan hydroxylase by PCPA is not total (only 80%). Thus, there is still the possibility that a small pool of 5-HT remains and that it may represent the functional pool which is critical for the return of SWS.

b) On the other hand, there are some indirect data (vide infra) which favor the role of catecholamines in waking. The possible decrease in the turnover of NA in chronically PCPA-treated cats (STOLK et al., 1969) might thus impair the waking mechanism. Since waking and sleep may represent the balance between two antagonistic systems, the return to sleep (or the reduction of waking) might be facilitated by the alteration of catecholaminergic mechanisms.

c) Finally, the return to cortical synchronization after chlorpromazine might also be explained either by the effect of chlorpromazine upon catecholamines or even by the possible effect of this drug upon 5-HT (BARTLET, 1965) (inhibition of reuptake leading to an increased concentration of the transmitter in the synaptic cleft). On the other hand, the fact that PS does not return to normal levels after chlorpromazine (COHEN et al., 1969) suggests that the cortical synchronization may be the result of a direct pharmacological action of chlorpromazine, a well-known synchronizing drug (BRADLEY and KEY, 1958; BRADLEY et al., 1966).

Though appearing to conflict, these experiments are in agreement in showing that the functional pool of 5-HT may be quite small since minute doses of 5-HTP are able to restore both quantitatively and qualitatively normal sleep for 4–6 hours after total insomnia in PCPA-pretreated cats. Thus the neuropharmacological model provided by PCPA and 5-HTP is, at the present time, the best model for testing the 5-HT hypothesis of sleep and it is likely that PCPA or even more specific inhibitors of 5-HT biosynthesis will provide us in the near future with a better structural and dynamic picture of the functional pool of 5-HT which is related to sleep.

3. p.chloromethamphetamine was the first drug known to deplete selectively brain 5-HT and its metabolite 5-hydroxyindolacetic acid without altering catecholamine levels (PLETSCHER et al., 1965, 1966). Its mechanism of action is still unknown and it is still uncertain whether it acts only upon the inhibition of 5-HT synthesis (FULLER et al., 1965; FULLER and HINES, 1970; SANDERS-BUSH and SULSER, 1970). This drug (20 mg/kg) induces in the cat a very marked arousal which lasts for 16–20 hours (DELORME et al., 1966b; KOELLA, 1969) and is followed by a gradual recovery of both SWS and PS (and by an almost permanent discharge of PGO spikes even during waking). However, a subsequent injection of high or low doses of 5-HTP (50–5 mg/kg) is not able to reverse the insomnia but is highly toxic and lethal in the cat (DELORME et al., 1966; see JOUVET, 1969b). This is in total contradiction to the effect of 5-HTP after PCPA and suggests that p-chloromethamphetamine may act through a mechanism different from the inhibition of tryptophan hydroxylase.

iv) Action upon the Storage and Release of 5-HT

Reserpine, almost 20 years ago, opened the gate to the so-called "mono-amines game" (MANDELL and SPOONER, 1968). Its sedative action led workers to consider 5-HT as a possible transmitter of the trophotropic system (BRODIE et al., 1957; BRODIE and SHORE, 1957). Among the numerous hypothesis which have been proposed concerning its mechanism of action, the most likely is that reserpine reduces 5-HT storage, presumably by acting on the storage particles, and releases 5-HT with increased catabolism by MAO. Unfortunately reserpine acts similarly upon catecholamine metabolism (see ACHESON, 1968). Thus, its action upon sleep and waking is very difficult to interpret.

In the cat, reserpine (0.5 mg/kg) suppresses SWS (in its EEG aspect) for 6–8 hours and PS for one day: 60 to 90 minutes after its injection, continuous PGO begins to appear even during waking. We shall discuss this most peculiar effect when considering PGO activity. 5-HTP (30–50 mg/kg), which replenishes the 5-HT store when injected after reserpine (CORRODI et al., 1967), is able to restore EEG synchronization almost immediately, whereas DOPA (30–50 mg/kg) hastens the reappearance of behavioral waking and PS. This finding led to the hypothesis that SWS and PS were associated with 5-HT and CA metabolisms (MATSUMOTO and JOUVET, 1964).

In rats and rabbits, reserpine suppresses PS only at a much higher dose (2 mg/kg) (GOTTESMAN, 1966; TABUSHI and HIMWICH, 1969). Contradictory results have been obtained in man, either decrease (HOFFMAN and DOMINO, 1969; COULTER et al., 1971) or increase of PS (HARTMANN, 1966) being reported.

v) Action upon the Catabolism of 5-HT

MAO inhibitors (MAOI) decrease the deaminated metabolites of both 5-HT and catecholamines. At the same time they increase 5-HT (and catecholamines) at the receptor level. This fact is probably responsible for the decrease in turnover of central amines (as measured by biochemical techniques) (MACON et al., 1970). MAOI also decrease the firing of postulated 5-HT nerve cells in the raphe system (AGHAJANIAN et al., 1970b). Their mechanism of action is thus very complex. These drugs have (in cats, rats, monkeys and man) the most dramatic inhibitory effect upon PS and the PGO activity (vide infra). In the cat, whereas harmaline has an amphetamine-like effect, nialamide and pargyline increase cortical synchronization (JOUVET et al., 1965; DELORME, 1966). In the rat, nialamide has an activating effect upon the EEG and suppresses both SWS and PS (MOURET et al., 1968b).

2. Direct Approach to 5-HT-Containing Neurons

i) Destruction of 5-HT-Containing Perikarya of the Raphe System

The destruction of the 5-HT-containing neurons of the raphe system was performed stereotaxically in chronically implanted cats (RENAULT, 1967;

Jouvet, 1969b). Following the operation, the animals were continuously recorded for 10–13 days (this being the critical period for the voiding of the serotonergic terminals). On the 13th day, the cats were sacrificed, always at the same time of day in order to avoid any circadian variation of brain MA. This method yields the following information: a valid quantification of the sleep states (obtained by the mean percentages of SWS and PS for 10–13 days recording), a measurement of the volume of the lesion by topographical analysis (represented by the percentage of the total raphe system destroyed), and an analysis of the monoamine levels in the brain (expressed as a percentage of the 5-HT, 5-HIAA, NA and DA found in normal cats sacrificed under the same conditions).

Following a subtotal (80–90%) coagulation of the raphe system, a state of permanent behavioral and EEG arousal is observed during the first 3–4 days. In the period that follows, the percentage of SWS does not exceed 10% of the nycthemeron. In these preparations, PS is never observed. However, continuous discharges of PGO spikes, recorded from the lateral geniculate or occipital cortex appear immediately following the destruction of the raphe, at a rate of 30–40/min. The pattern of discharge is similar to the one which follows injection of reserpine in the normal cat and has been called the "reserpinic syndrome". The rate of discharge of the PGO spikes declines on the 3rd day to 10/min. The injection of a low dose (5 mg/kg) of 5-HTP does not alter the permanent arousal which follows raphe lesion. Larger doses (30–50 mg/kg) induce a state of cortical synchronization accompanied by a waking behavior (Renault, 1967; Pujol et al., 1971). Partial lesions of the raphe system result in less pronounced insomnia (due to a gradual recuperation after the first two days). In these preparations, PS is found to occur only if the daily percentage of SWS exceeds 15%. There is thus a significant correlation between the quantity of 5-HT-containing cell bodies destroyed and the amount of sleep. The biochemical analysis of these insomniac preparations also revealed a significant decrease in cerebral 5-HT and 5-HIAA with no variation in catecholamines (NA or DA) in the telencephalon and diencephalon. It was therefore demonstrated that a significant correlation exists between the amount of destruction of the raphe system, the intensity of the resulting insomnia, and the selective decrease of cerebral 5-HT (Fig. 12).

Metabolism of 5-HT in Raphe-Destroyed Cats. By incubating cortical or brain stem slices in labelled precursor of 5-HT, some additional information has been obtained concerning the metabolism of 5-HT at the level of the terminals after destruction of the 5-HT-containing perikarya (Pujol, 1970; Pujol et al., 1971). Eighteen hours after the lesion, at the time of maximum insomnia, there is no alteration in either endogenous 5-HT or 5-HIAA. However, the synthesis of labelled 5-HT from labelled tryptophan and the activity of tryptophan hydroxylase are decreased. Moreover, there is a very significant

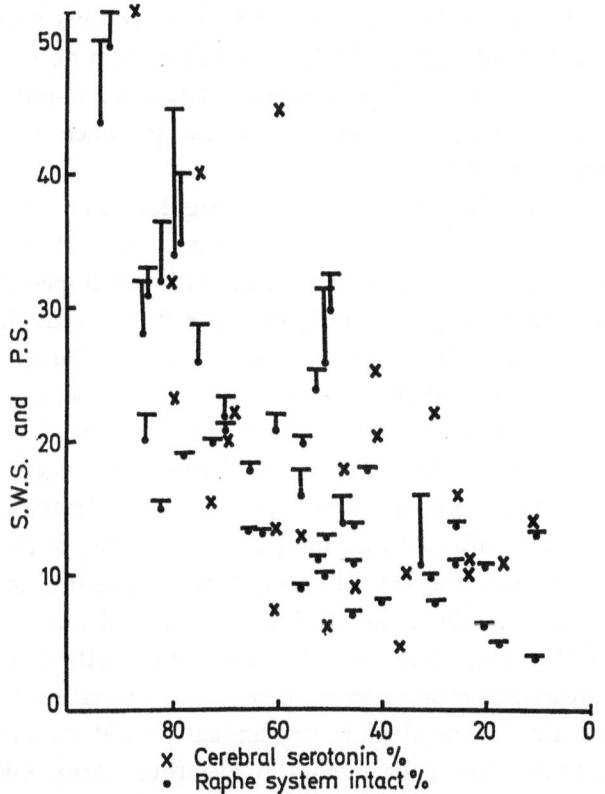

Fig. 12. *Correlation of the extent of the destruction of the raphe system and the resulting reduction of both states of sleep during the 10–13 days of survival.* Slow-wave sleep is represented by dots, paradoxical sleep by a vertical line and total sleep by a horizontal bar. Crosses: Correlation of the level of 5-HT in the brain rostral to the lesion (relative to the amount for control cat) and the resulting diminution of sleep. (Jouvet, 1969)

diminution of the spontaneous release "in vitro" of labelled 5-HT present in the incubation medium. This fact suggests that at the time of maximum insomnia there is a decline in the release of 5-HT at the terminal level after the destruction of 5-HT-containing perikarya. Ten days after the destruction of 5-HT-containing perikarya there is a very significant decrease in endogenous 5-HT and 5-HIAA at the level of the terminals which is correlated with the decrease in labelled 5-HT synthesized in vitro from labelled tryptophan. On the other hand, there is no significant alteration in labelled 5-HT synthesized in vitro from labelled 5-HTP. This can be explained by the fact that 5-HTP can be decarboxylated in other neurons, presumably catecholaminergic, or in the brain capillaries where the "non-specific" 5-HTP-DOPA decarboxylase is located.

This finding is in agreement with other experiments which show that only tryptophan is the physiological precursor of 5-HT in the brain (see Moir and Eccleston, 1968). Apparently exogenous 5-HTP at a very low dose can be decarboxylated preferentially in 5-HT-containing neurons only when endogenous 5-HTP is lacking (as after PCPA). It is thus possible that the high doses of 5-HTP (50 mg/kg) which are usually given do not really induce a physiological increase in 5-HT, but perhaps induce a non-physiological state due either to exogenous 5-HTP itself or to the presence of 5-HT in other monoamine-containing neurons.

The striking insomnia which follows the coagulation of the raphe has also been obtained in the rat where the destruction of the anterior raphe nuclei induces a state of agitation and cortical arousal which persists for several days. The increased waking is significantly correlated with the decrease in both 5-HT and 5-HIAA (Kostowski et al., 1968). In the cat, Michel and Roffwarg (1967) have obtained an almost total insomnia with a mid-sagittal splitting of the brain stem. They also observed permanent discharges of PGO in the lateral geniculate. Striking insomnia is also seen in "encéphale dédoublé" in which both the brain and the brain stem are split (Michel, 1967). Later Mancia et al. (1968) confirmed that mid-sagittal splitting of the pons induces insomnia in the monkey and cat (Mancia, 1969). According to Mancia, this lesion reduces sleep, not by lesioning the raphe nuclei but by interrupting, at the level of the pons, some crossing ascending pathways coming from synchronizing structures located somewhere in the caudal medulla. Although such an interpretation is possible, it is very unlikely (see Moruzzi—this review). A mid-sagittal brain stem split inevitably destroys many 5-HT-containing nerve cells in the raphe and decreases 5-HT in the telediencephalon. The insomnia which is observed after midpontine sagittal splitting is milder (20% of SWS and 3% of PS) than the insomnia observed after total destruction of the raphe system (10% of SWS, no PS). This can be explained by the fact that pontine splitting destroys only a part of the raphe cells.

On the other hand, it was shown previously by Cordeau and Mancia (1958, 1959) that homolateral transection of the brain stem at the pontine level may increase cortical activation on the ipsilateral cortex. This was interpreted as the demonstration of an ascending *ipsilateral* synchronizing influence coming from the caudal brain stem. It is difficult to reconcile these findings with the hypothesis that synchronizing fibers must cross the midline at the pontine level. On the other hand, the stimulation of the vago-aortic nerves which activate the synchronizing structures of the lower brain stem (nucleus of the solitary tract) is still followed by a phasic *ipsilateral* cortical synchronization after splitting of the brain stem. This demonstrates that the ascending synchronizing influences do not cross at the pontine level (Puizillout and Ternaux, 1971).

ii) Stimulation of the Raphe System

Contradictory results have been reported. In the rat, Aghajanian et al. (1967c) have demonstrated that the stimulation of the dorsal raphe causes dishabituation of an habituated skeletal motor response to loud auditory stimuli. They also observed a small decrease in 5-HT and a significant increase in 5-HIAA in the telencephalon. The hypothesis that 5-HT might serve some disinhibitory function was strengthened by the observation that prior treatment of the animal with PCPA abolished the effect of raphe stimulation, whereas normal response was restored by the administration of 5-HTP. However, CONNER et al. (1970) have found contrary results and reported that PCPA greatly impairs the process of habituation. Since AGHAJANIAN et al. (1967) did not record the EEG of their animals and since loud auditory stimuli would certainly have disturbed any possible sleep-inducing effect, it is difficult to interpret these findings. KOSTOWSKI et al. (1969), KOSTOWSKI and GIA-CALONE (1969), GUMULKA et al. (1971) have been able to induce behavioral and EEG sleep in the rat with low-frequency stimulation of the medial or dorsal raphe. The stimulation also induced a long-lasting increase of 5-HIAA in the forebrain without significant alteration of 5-HT. This was interpreted as an increase in central 5-HT turnover. Finally, stimulation of the midline structures of the brain stem in the rabbit led, on the contrary, to behavioral and EEG arousal (POLC and MONNIER, 1970). These conflicting data reflect our ignorance concerning the physiological triggering of the 5-HT-containing perikarya.

3. Effect of the Sleep-Waking Cycle upon 5-HT Neurons

i) Electrical Activity of the Raphe System during Sleep

The method and the technique developed by AGHAJANIAN et al. (1970b) to record and identify postulated 5-HT cell bodies in the raphe (by stopping their firing with LSD) has not yet been applied to chronic conditions during sleep. In chronic cat, by using micro-macroelectrodes and the integrating technique developed by ARDUINI and PINNEO (1962), BALZANO and JEANNEROD (1970) have recorded from the raphe nuclei some spindle activity during SWS and a tonic elevation of activity which preceded the appearance of PS by some minutes. There is no proof, however, that this raphe activity was due to 5-HT-containing nerve cells.

ii) Metabolism of 5-HT during Sleep

There has not yet been any direct determination of 5-HT release related to physiological sleep. SINHA et al. (1971) have sacrificed cats (by injecting potassium chloride into the heart through a chronic catheter) during waking, SWS or PS. They found some decrease of 5-HT in cats killed after 10 minutes of SWS and of NA in cats killed after 5 minutes of PS. They interpret these changes as an "increase in impulse traffic in serotoninergic fibers during SWS and in noradrenergic fibers during PS".

Since the sleep deprivation experiments deal mainly with selective PS deprivation, they will be considered in the appropriate chapter.

4. Circadian Alteration of 5-HT Metabolism

This chapter does not bring crucial evidence either for or against the role of 5-HT in sleep mechanisms since the changes in endogenous levels of 5-HT are difficult to interpret without a parallel determination of the turnover of the amine. Whereas in newborn rats which have not yet acquired a circadian rhythm of sleep there is no circadian alteration of brain 5-HT (Okada, 1971), in the adult rat, whose circadian rhythm of sleep is quite well established, there is an increase in endogenous cerebral 5-HT during the day (when rats sleep most of the time) and a decrease during the night (when rats are awake) (Dixit and Buckley, 1967; Scheving et al., 1968; Friedman and Walker, 1968; Quay, 1968) [see also Reis et al. (1969) for the circadian rhythm of 5-HT in the cat's brain]. Alterations in the feeding schedule (the food being available only during the day) have resulted in significant change in the circadian rhythm of sleep in rats, totally independent of light. After 3 weeks the amount of SWS is the same during day and night and the difference in endogenous 5-HT between day and night disappears (Bobillier and Mouret, 1971; Mouret and Bobillier, 1971). 5-HT has also been implicated in the process of hibernation in the golden-mantled ground squirrel (*Citellus lateralis*). There is a decrease in brain 5-HT when this animal enters hibernation, with a partial recovery after 5 days. This phasic decrease is interpreted as an increase in the utilization of 5-HT at the beginning of hibernation. On the other hand, PCPA and lesion of the raphe system, prevent hibernation in *Citellus lateralis* (Spafford, 1970). Finally, a decrease in 5-HT fluorescence in the hypothalamus of hibernating bats has also been observed by Constantinidis et al. (1970). All this indirect evidence favors the intervention of 5-HT in the still-unknown mechanisms of hibernation.

5. Summary

Although no direct evidence has yet been presented, there are many converging data which strongly favor the hypothesis that the 5-HT-containing neurons of the raphe system are involved in the sleep mechanism. The most convincing evidence may be summarized briefly as follows: 1. Inhibition of the synthesis of 5-HT (at the level of tryptophan hydroxylase) with p-chlorophenylalanine induces insomnia which is immediately reversed to normal sleep by the subsequent injection of small doses of 5-HTP, the immediate precursor of 5-HT, which by-passes the inhibition of the synthesis. 2. The destruction of the 5-HT-containing perikarya located in the raphe system induces insomnia, the intensity of which is correlated with the decrease in cerebral 5-HT.

E. Catecholamines and the Sleep-Waking Cycle
1. Pharmacological Alteration of CA Metabolism
i) Injection of the Amines themselves

Early pharmacological studies of the role of the catecholamines in central mechanisms were made after the systemic injection of the amines into adult animals (BONVALLET et al., 1954; see DELL, 1960). In moderate doses, noradrenaline and adrenaline produce, in the cat, a cortical activation paralleled by an increase in peripheral sympathetic tone (raised blood pressure). The activating effect of noradrenaline is abolished if the mesencephalic reticular formation is transected or lesioned. It was considered possible that the neurons of the reticular activating system might be adrenergic. Thus, at this time, catecholamines were related with waking (see DELL, 1960, 1963).

However, evidence was soon presented that neither adrenaline or NA injected directly into the blood stream in adult animals passes the blood-brain barrier in any significant amount except in the hypothalamus (WEILL-MAL-HERBE et al., 1959). Moreover, it was shown that the direct injection of NA into the carotid artery did not induce arousal (CAPON, 1959) (and did not increase blood pressure) whereas other experiments demonstrated that increased blood pressure per se could induce EEG arousal through postulated baroreceptors located in the brain stem reticular formation (BAUST and NIEM-CZYK, 1963, 1964; BAUST et al., 1963).

These experiments led to a total reversal of the situation and, for a time, the catecholamines were considered as "sleep neuromodulators". A good example in favor of the depressant or "soporific" action of catecholamines may be found in a recent paper by MANDELL and SPOONER (1968) who list 21 experiments (from 1914 to 1969 and from chicks to man) in which adrenaline or NA injected into the ventricles or directly into the brain produce "sedation", sleep, "deep anesthesia" and even "unconsciousness".

However, the results of intraventricular injections are very difficult to interpret. Very minute amounts of labelled catecholamines may be taken up by non-catecholaminergic neurons (LICHTENSTEIGER and LANGEMANN, 1966; FUXE and UNGERSTEDT, 1968; FUXE et al., 1968). Thus, it is likely that in some cases NA might have induced some response upon non-catecholamino-ceptive neurons. This might explain why the effect of NA depends upon the dose injected. Very small doses have a clear-cut behavior-activating effect (SEGAL and MANDELL, 1970). This is also true for the direct injection of NA into the brain stem of cats (CORDEAU, 1962; CORDEAU et al., 1963); larger doses may have a depressant effect (see ROTHBALLER, 1959; HERMAN, 1970).

Intravenous injections of adrenaline, NA, isoprenaline and dopamine in chickens not older than 28 days (whose blood-brain barrier is permeable to catecholamines) induce EEG and behavioral signs of sleep whereas NA, directly injected into the brain, may induce behavioral sedation without electrocortical synchronization (Key and Marley, 1962; Marley and Key, 1963; Spooner and Winters, 1966; Marley and Stephenson, 1970). In adult chickens, on the contrary, the same drugs elicit arousal (Dewhurst and Marley, 1965). However, most of the drugs injected at the periphery, both in young chicks and in older birds, caused hypertension. But since carbachol and 5-HT (which also triggered sleep) induced either hypotension or variable blood pressure responses (Spooner and Winters, 1966, 1967), it was concluded that the sleep-like state was due to direct central effects. Nevertheless, it is difficult to eliminate a possible reflex effect in response to a blood pressure increase. Indeed, increases of sino-aortic pressure via excitation of baro-receptors elicit a central shift toward synchronization of the EEG (Bonvallet et al., 1953). Even if these peripheral effects do not play a role, the results are difficult to interpret since it has not yet been demonstrated that intravenously injected catecholamines are selectively taken up by central NE terminals and immediately released into the synaptic cleft (see Marley, 1966).

ii) Postulated Increase of the Transmitter through the Peripheral Injection of the Precursor

—*Tyrosine*. Since tyrosine hydroxylase is a rate-limiting enzyme, it is not possible to increase brain levels of CA by increasing the blood tyrosine level (Sjoerdsma et al., 1965).

—*DOPA* is considered as the most likely precursor of DA and NA. DOPA is first converted to dopamine by DOPA decarboxylase, then DA is hydroxylated by dopamine β-hydroxylase to form NA.

—*L-DOPA* crosses the blood-brain barrier. It is, however, decarboxylated to DA to a considerable extent in the brain capillaries (Constantinidis et al., 1969, 1970). The administration of inhibitors of DOPA decarboxylase which act mainly in the periphery enables a larger quantity of DOPA to penetrate into the brain (Bartholini et al., 1967; Bartholini and Pletscher, 1968, 1969). There is evidence, in cat (Delorme, 1966; Jones, 1970) and rabbit (Monnier and Tissot, 1958; see Marley, 1966), that L-DOPA (30–50 mg/kg) induces a state of arousal which lasts continuously for 5–6 hours [however Kadzielawa and Widy-Tyszkiewicz (1970) have not consistently observed cortical activation with DOPA in the intact cat]. The activating effect of DOPA is much enhanced by the peripheral inhibition of DOPA decarboxylase (Bartholini et al., 1967; Gaillard et al., 1969). It is possible that L-DOPA acts by increasing DA or NA in the presynaptic CA terminals. According to Reis et al. (1970), there is indeed a high degree of correlation between the level of NA in the brain stem and the magnitude of excitement in cats receiving L-DOPA. This excitement would be due to the release of newly synthesized NA. However, many other mechanisms are possible. There is some evidence that a part of the exogenous L-DOPA may enter 5-HT containing terminals, undergo decarboxylation to DA and there displace the endogenous 5-HT from vesicular stores (BARTHOLINI et al., 1968; NG et al., 1970; COLBURN and KOPIN, 1971). Thus, L-DOPA could be a central false serotonergic transmitter and its arousing effect might be related not only to its role as a precursor of CA but also to a depleting action upon 5-HT stores. It must be also pointed out that, after high doses of L-DOPA associated with peripheral inhibition of DOPA decarboxylase, the 5-HT-containing perikarya of the raphe system become green in the histofluorescence method, as if they contained CA (CONSTANTINIDIS et al., 1968). Finally, it has been proposed that the behavioral excitation which follows the combined use of DOPA and inhibitor of DOPA decarboxylase may be due to the formation of some aberrant metabolite of DOPA such as DOPPA [3.4-dihydroxyphenyl-pyruvic acid (SCHECKEL et al., 1969)]. Thus, as previously with 5-HTP, we are confronted with the problem of the possible "non-specificity" of the 5-HTP-DOPA decarboxylase.

—*Dihydroxyphenylserine (DOPS)*. Both in vitro and in vivo (BLASCHKO et al., 1950) DOPS may be decarboxylated to NA. Contrary to DOPA, injections of DOPS increase both SWS and PS in the rat (HAVLÍČEK, 1967). Since it is still not certain whether DOPS is the physiological precursor of NA, this finding must be interpreted with caution.

iii) Inhibition of the Biosynthesis of CA

Inhibition of Tyrosine Hydroxylase. Tyrosine is hydroxylated by tyrosine hydroxylase, a rate-limiting enzyme, to DOPA (SJOERDSMA et al., 1965; SPECTOR et al., 1965; see ACHESON, 1966).

This enzyme is inhibited competitively by many analogs of tyrosine. At the present time, the inhibitor most widely used in behavioral experiments is α-methyl-p-tyrosine (αMPT), which induces a decrease in cerebral CA, as verified by biochemical and histofluorescent techniques (ANDÉN et al., 1966a). However, because of its nephrotoxicity (MOORE et al., 1967; HOOK and MOORE, 1969), this drug is not well tolerated in chronic experiments and the soluble methyl ester of α-methyl-p-tyrosine is less toxic and more suitable for chronic experiments. There is considerable evidence, from behavioral observations that an inhibition of tyrosine hydroxylase, which decreases both brain DA and NA, induces a state of "sedation" which can be counteracted by a subsequent injection of DOPA (WEISSMAN and KOE, 1965; RECH et al., 1966; FUXE and HANSON, 1967; HANSON, 1967a; SCHOENFELD and SEIDEN, 1969).

At the polygraphic level, the effect of αMPT has been studied in rats (MARANTZ and RECHTSCHAFFEN, 1967; TORDA, 1968; BRANCHEY and KISSIN, 1970), monkeys (WEITZMAN et al., 1969) and cats (ISKANDER and KAELBLING, 1970; KING and JEWETT, 1971). Although the effects upon PS were not the

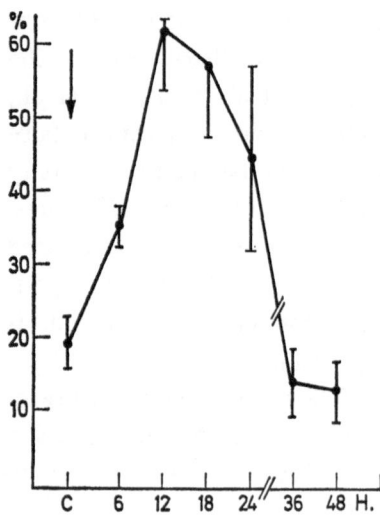

Fig. 13. *Decrease of waking in raphe-destroyed cats after α-methyl paratyrosine.* Ordinates: percentage of sleep (slow-wave sleep) during the experimental period (every 6 or 12 hours). Abscissae: time in hours. C Percentage of sleep (mean and SEM) in control conditions in 8 cats, whose raphe system was subtotally destroyed 2 to 6 days previously. The arrow signals the injection of 150–200 mg/kg of α-methyl-p-tyrosine methyl ester intraperitoneally. There is a very marked increase in cortical synchronization which lasts for about 24 hours after which there is a return to insomnia. (Jouvet, 1971)

same (see p. 235) the common effect observed in all species was decreased behavioral and EEG waking.

The suppressant effect of αMPT upon waking may be obtained even when arousal is dramatically increased through two different experimental methods: destruction of the raphe system or injection of amphetamines.

1. The injection of 200 mg/kg of αMPT was given to raphe-destroyed cats at a time when behavioral and EEG waking was almost permanent (2 to 6 days after the lesion); 4 to 6 hours after the injection, behavioral sedation appeared: the running movements stopped, miosis appeared and there was an almost continuous cortical synchronization which lasted for 24 hours, after which time there was a rapid return to behavioral and EEG insomnia (Fig. 13). This experiments provides some neuropharmacological evidence that the almost permanent arousal which follows the destruction of the 5-HT-containing neurons of the raphe system might be related to the increased turnover of central catecholaminergic neurons (Jouvet, 1971).

2. There is converging evidence that the amphetamines act upon CA metabolism (see the most recent reviews in Costa-Garattini, 1970 and in Stein and Wise, 1970). In brief, amphetamine effects a reduction of NA level and an increase of normetanephrine. This is interpreted as the result of competitive inhibition by this drug of the deamination of NA by MAO (so that more of the NA released from a still unknown pool is available for O-methylation by catechol-O-methyl transferases). The amphetamine excitation is probably mediated via a CA mechanism, since pretreatment of the animal with αMPT totally suppresses the excitation by amphetamine at the behavioral (Hanson, 1967b; Dominic and Moore, 1969; Stolk and Rech, 1970; Svensson, 1970) or at the polygraphic level (Jouvet, 1971) (Fig. 14).

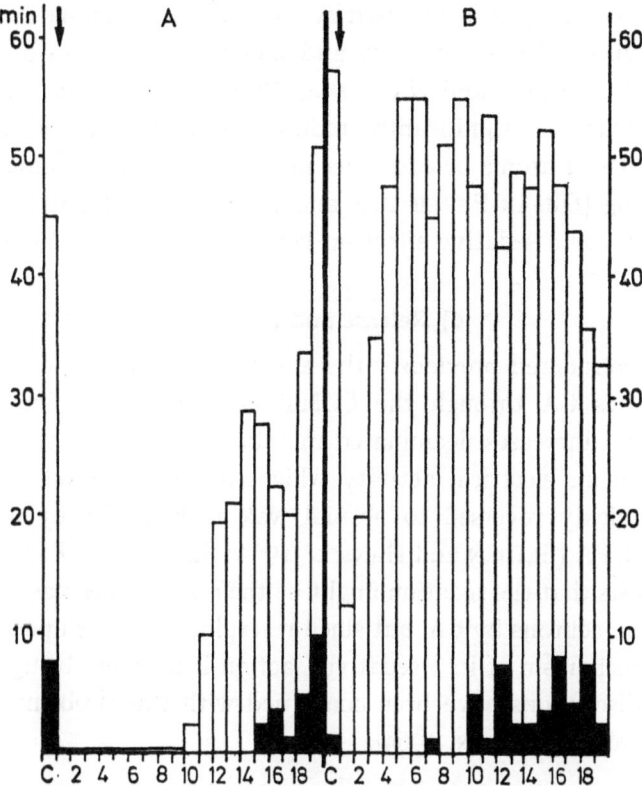

Fig. 14. *Suppression of the waking effect of DL amphetamine by α-methyl-p-tyrosine in cats.* Ordinates: time (in minutes) spent in slow-wave sleep (white) or paradoxical sleep (black) every hour. Abscissae: *C* Control recording during 6 hours: time in hours. *A* The arrow signals the injection of 5 mg/kg of DL amphetamine intramuscularly. There is long-lasting and permanent arousal during 10 hours followed by a return of sleep (mean of 5 experiments). *B* The cats were pretreated with α-methyl-p-tyrosine (200 mg/kg) 6 hours previously. The arrow signals the injection of 5 mg/kg of DL amphetamine. The waking effect of this drug is almost totally suppressed (mean of 3 experiments). (JOUVET, 1971)

Finally, it is probable that the CA-containing neurons which may be responsible for amphetaminic arousal are located in the ponto-mesencephalic reticular formation since its destruction or intercollicular transection suppresses the amphetamine-induced EEG arousal (HIEBEL et al., 1954; FUJI-MORI and HIMWICH, 1969). On the other hand, in the rabbit, the intravertebral injection of amphetamine is not followed by EEG arousal if the basilar artery is tied at the midpontine level (VAN METER and AYALA, 1961). Hence caudo-pontobulbar neurons cannot be responsible for the cortical arousal induced by amphetamine.

Although these experiments do not determine which catecholamine is involved (noradrenaline or dopamine), they strongly suggest that the catecholamine-containing neurons of the rostral pons or caudal mesencephalon are most likely to be involved in behavioral and/or EEG arousal.

Inhibition of Dopamine β-Hydroxylase. This enzyme catalyzes the final step of the biosynthesis of NE. This step is probably rate-limiting (THIERRY et al., 1971 b). The enzyme is inhibited by diethyldithiocarbamate or disulfiram. This drug reduces the NA content and increases the DA store (CARLSSON et al., 1966; GOLDSTEIN and NAKAJIMA, 1967; LIPPMAN and LLOYD, 1969). In the cat disulfiram significantly reduces both waking and PS, whereas striking episodes of atonia or catatonia may appear after chronic administration of the drug (DUSAN-PEYRETHON and FROMENT, 1968). In mice and rats, disulfiram also has a sedative effect on behavior (MOORE, 1969; ROLL, 1970).

iv) Release and Reuptake

We have considered previously the effect of amphetamine, a drug which possibly releases CA through the COMT pathway. Reserpine releases CA (and 5-HT) and increases deaminated metabolites. The sedation induced by reserpine (with an electrical activity which, however, differs from sleep in the cat) is quickly reversed to normal waking behavior by a subsequent injection of DOPA (SEIDEN and HANSON, 1964; SEIDEN and PETERSON, 1968).

This suggests that the sedation might be the result of decreased stimulation of postsynaptic neurons by CA, but since reserpine also acts upon 5-HT metabolism, the mechanism of its "sedative" action if far from being understood. Many lively discussions have been concerned with this problem (see BRODIE et al., 1960, 1966; CARLSSON, 1964).

v) Inhibition of the Reuptake

With regard to EEG studies of the sleep-waking cycle, only desmethylimipramine (DMI) and imipramine have been studied with polygraphic techniques. These drugs inhibit "only" the reuptake of NA and increase its concentration at the postsynaptic site (see GLOWINSKI et al., 1966). They probably also decrease the turnover of 5-HT neurons (see CORRODI and FUXE, 1969). Both imipramine and DMI strongly decrease EEG waking (and PS) (HISHIKAWA et al., 1965; KHAZAN and SULMAN, 1966; MANDELL et al., 1969; WALLACH et al., 1969a, b; KHAZAN and BROWN, 1970). This increase in cortical synchronization is accompanied by a marked mydriasis, due to the fact that these drugs also have a strong central and peripheral anticholinergic effect.

vi) Catabolism of CA

Inhibition of MAO. As shown previously, inhibitors of MAO (MAOI) act upon both 5-HT and CA metabolism and increase endogenous 5-HT and CA in the synaptic cleft. The degree of accumulation of endogenous transmitter depends of the species and on the structure of the MAOI. In the rabbit and rat, block of MAO is associated with an increase of CA, sympathomimetic effects are associated with the accumulation of NA, and an almost permanent arousal occurs after nialamide (COSTA et al., 1960; MOURET et al., 1968).

If DOPA is injected together with MAO, this induces an intense excitement with long-lasting cortical arousal in the cat (JONES, 1970) and in the rabbit (MONNIER and GRABER, 1963).

—*COMT*. There have been very few studies of the effects of inhibitors of COMT upon the sleep-waking cycle, probably because most of the COMT inhibitors are very toxic. Tropolone, however, is well tolerated by the cat. This drug is an effective inhibitor of COMT both in vitro and in vivo (GOLDSTEIN et al., 1964). The administration of tropolone alone (100 mg/kg) induces a long-lasting arousal for six to seven hours, subsequently followed by a rebound of both SWS and PS (JONES, 1970). The administration of DOPA (25 mg/kg) and tropolone (50 mg/kg) induces very dramatic behavior. In addition to waking, the cats present hallucinatory episodes between one and three hours after the injection. Continuous arousal is present for about eight hours. Continuous PGO appears during waking by the sixth hour (JONES, 1970, 1971).

vii) False CA Transmitter

α-methyl-DOPA was first believed to be an inhibitor of 5-HTP-DOPA decarboxylase (SOURKES, 1965) since it decreases brain 5-HT and NA. Now is assumed that α-methyl-DOPA is decarboxylated to form α-methyl-dopamine which is, in turn, hydroxylated to form α-methyl-NA. This substituted NA acts as a false transmitter and depletes NA by displacement (see CARLSSON, 1964; CARLSSON et al., 1968). This drug will be considered in detail when PS is discussed. In cats (100 mg/kg) it strongly decreased EEG and behavioral waking (and PS) (DUSAN-PEYRETHON et al., 1968).

viii) Pharmacological Destruction of CA Neurons

There is now good evidence that 6-hydroxy-dopamine (6-OHDA) may be "selectively" taken up by the CA nerve endings when injected intraventricularly. There is likely to be a gradient of uptake of 6-OHDA by CA terminals when the drug is injected into the ventricles. CA terminals close to the ventricle must be more affected than other terminals. Thus the ventricular approach may lack the anatomical dimension which is of paramount importance. Through a still unknown mechanism (see SANER and THOENEN, 1971), 6-OHDA destroys the CA terminals, as attested by the disappearance of fluorescence, the important but irregular decrease in endogenous CA (mostly NA), the decrease in synthesis of CA from labelled precursor and the degeneration of CA terminals verified by means of electron microscopy (LAVERTY et al., 1965; BURKARD et al., 1969; UNGERSTEDT, 1969; BLOOM, 1969; IVERSEN, 1970; IVERSEN and URETSKY, 1970; URETSKY and IVERSEN, 1970; BARTHOLINI et al., 1970a, b; BELL et al., 1970; BREESE and DENNIS TRAYLOR, 1970;

Laverty and Taylor, 1970; Uretsky et al., 1971). At the present time, however, there has not yet been any quantitative study of sleep-waking in rats with 6-OHDA injected into the ventricles. Their behavior was considered as just "normal" (Burkard et al., 1969). Moreover, subsequent injections of amphetamine in 6-OHDA-pretreated rats stimulate motor activity. However, in strongly DA-depleted rats the spontaneous motor activity was significantly reduced (Evetts et al., 1970). If polygraphic recordings confirm that normal sleep-waking could be quantitatively and qualitatively recorded in rats having a 90% decrease of brain NA, then two obvious conclusions could be drawn.

1. either CA has nothing to do with cortical or behavioral arousal mechanisms (in contradiction to the data obtained by inhibition of CA synthesis);

2. or a small functional pool of 10–20% of NA is sufficient to produce arousal since supersensitivity of the postsynaptic receptor, or an increased turnover of intact CA neurons might compensate for the destruction of other CA terminals.

In any case, the use of this drug, which also permits a *direct* attack upon CA neurons (p. 239 is a significant advance in our understanding of the function of the central catecholaminergic neurons.

2. Direct Approach to CA-Containing Neurons

i) Destruction of Dopamine-Containing Neurons of the Substantia Nigra
(Jones, 1969; Jones et al., 1969)

Stereotaxic coagulation of the DA-containing cell bodies (group A 9) (and also of the ventral NA bundle) produced a behavioral state of akinesia, an inability to initiate activity and unresponsiveness to external stimuli and a decrease in endogenous DA in the rostral brain (telencephalon, striatum), as shown by biochemical analysis 8–10 days after the lesion. In some severely lesioned cats, in which dopamine was greatly decreased (by more than 90%), behavior was almost permanently comatose. Despite such depressed behavior, *a quantitatively and qualitatively normal EEG record of alternating slow-wave and arousal activity persisted during the behaviorally comatose state.* Long-lasting cortical activation by peripheral stimulation was common even in the total absence of behavioral arousal.

These results confirm earlier findings (see Andén et al., 1964a, 1966b, e; Bedard et al., 1969a, b; see Hornykiewicz, 1966) that the catecholamine-containing cells of the substantia nigra contain dopamine. In view of the numerous anatomical data which suggest the existence of a nigro-striatal ascending system, it appears possible that dopamine could play a role in behavioral alertness and motor coordination, presumably by way of this nigro-striatal system. On the other hand, the dopamine-containing neurons do not play a significant role in either cortical synchronization or desynchronization since an almost total disappearance of dopamine from the telediencephalon does not induce a significant change in the spectrum of the EEG recording during the sleep-waking cycle. It should be pointed out that these data, obtained in cats, are in contradiction to the effect of the destruction of the substantia nigra in rats. In the rat, the destruction of the substantia nigra is followed by an increase of motility. On the other hand, this lesion greatly diminished the running activity induced by amphetamine (SIMPSON and IVERSEN, 1971).

ii) Destruction of NA-Containing Neurons

The coagulation of a group of CA-containing neurons of the mesencephalic reticular formation (dorsal NA bundle and group A 8) resulted in a large decrease of telediencephalic noradrenaline and dopamine accompanied by a significant decrease in the cortical low-voltage fast activity with *no impairment of behavioral arousal or motor function*. Sensory stimulation produced both behavioral orientation and cortical activation which were, however, contingent upon the presentation and duration of the stimulus. After mesencephalic lesions, the decrease in low-voltage fast activity was replaced by a deactivated EEG and to a lesser extent by an increase in SWS. There was also a slight diminution in PS (and a slight fall in telencephalic 5-HT, explained by the interruption of the central bundle of ascending serotonergic fibers to the telencephalon). Finally, in all these cats, there was a significant correlation between the decrease of noradrenaline in the rostral brain, and the *decrease in EEG desynchronization* (JONES et al., 1969).

After lesion of the dorsal NA bundle at the level of the isthmus, however, there was true *hypersomnia*, with an increase in both SWS and PS well above control. In these cats, an increase in telencephalic 5-HIAA and tryptophan paralleled the decrease in noradrenaline. There was no significant change in tele-diencephalic 5-HT or dopamine. These biochemical findings suggest that the destruction of a bundle of ascending NA axons might have increased the turnover of some 5-HT perikarya (PETITJEAN and JOUVET, 1970; PETITJEAN, 1970).

In summary, two mechanisms are implicated in these latter experiments concerning the maintenance of waking in the cat:

1. a dopaminergic mechanism of the nigro-striatal system which is responsible for the maintenance of behavioral arousal and alertness;

2. a noradrenergic system issuing from the anterior part of the locus coeruleus complex, ascending through the dorsal NA bundle and also probably situated in group A 8 of CA-containing neurons of the mesencephalic tegmentum, which appears to be responsible for the *tonic* cortical activation which accompanies waking. This system, however, does not play a role in the phasic cortical activation which accompanies external stimulation.

Since CA-containing neurons are much less concentrated than 5-HT-neurons, the coagulation technique is not selective enough and it is quite possible that other neurons, principally cholinergic, which might play a role in behavioral or EEG arousal may be destroyed by the lesion.

iii) Stimulation of CA Neurons

The Ventral NA System (group A 10 and ventral NA bundle). There is converging evidence that this system may be related, in the rat, with the so-called reward system of Olds (Olds, 1962; Olds et al., 1964; Routten-berg and Malsbury, 1969). Self-stimulation is obtained only from this area (Dresse, 1966) (and from the medial forebrain bundle) whereas stimulation of adjacent areas is not effective. The self-stimulation which is facilitated by central administration of NA (Wise and Stein, 1969) or suppressed by local injection of 6-hydroxy-dopamine (Stein and Wise, 1971) induces a decrease in NA or in the fluorescence (provided that the synthesis is blocked) in some rostral brain regions (septal area, hippocampus) (Arbuthnott et al., 1971) or release of NA, as seen with the push-pull cannula (Stein and Wise, 1969). However, a lesion of this area (together with the adjacent group A 9) does not interfere with the cortical sleep-waking cycle in the cat (Jones et al., 1969). So it is likely that this system is related to some more complex activity than the basic sleep-waking mechanism.

The Dorsal NA System. In the cat, the localization of group A 8 or of the dorsal NA bundle in the mesencephalic reticular formation makes it very likely that its stimulation would be followed by EEG arousal, since this localization corresponds to the physiologist's favorite site for stimulation-induced arousal (Moruzzi and Magoun, 1949). However, it is very difficult to assess whether the stimulation has excited only CA-containing neurons (and not adjacent reticular neurons). Theoretically, if these CA-containing neurons were implicated in the *tonic* EEG arousal which extends much beyond the stimulation of the mesencephalic tegmentum, then injection of 6-OHDA at the same site should greatly decrease the duration of arousal.

3. Effect of Waking upon CA Mechanisms

i) Electrical Activity of CA-Containing Neurons

Contrary to postulated 5-HT-containing perikarya, the activity of CA-containing cell bodies has not yet been recorded with extracellular micro-electrodes in the pons or mesencephalon in either acute or chronic experiments.

ii) Biochemical Activity of CA Neurons

Since, despite the ingenuity of experimenters, no instrumental methods have yet been found which suppress waking, most of our data come from a rather crude method of increasing waking, usually stress with electric foot shocks. If it can be safely postulated that a rat receiving electric foot shocks is awake, it is nevertheless difficult to draw a safe conclusion from the numerous data concerning CA metabolism. Stress may well trigger additional neuro-humoral mechanisms (i.e. the hypothalamo-hypophyseal complex) which may be less active during "physiological waking". In any case, these studies have shown that foot shock, exercise or stress lead to an increased turnover of NA in the brain (SCHECKEL and BOFF, 1964; BLISS and ZWANZIGER, 1966; GORDON et al., 1966; KETY et al., 1967; BLISS et al., 1968; WELCH and WELCH, 1968; THIERRY et al., 1968, 1970, 1971; MARK et al., 1969; SCHEKEL et al., 1969; STONE and DICARA, 1969) and that sham rage (a most excited state) is accompanied in the cat by increased utilization and release of NA (REIS et al., 1967; REIS and FUXE, 1969).

Finally, as in the peripheral nervous system, it is possible that CA mechanisms do not play a role under normal conditions but might be involved in an "emergency situation". It is fair to say that most of the experiments which have been done have simulated extraordinary emergency situations.

4. Circadian Periodicity of CA

The concentration of both endogenous NA and DA varies with time of day in different regions of the rat brain (SCHEVING et al., 1958; MANSHARDT and WURTMAN, 1968) and the cat brain (REIS et al., 1968; REIS and WURTMAN, 1968; REIS and GUTNICK, 1970). Using a refined technique (administration of labelled tyrosine) and calculating labelled CA in relation to the specific activity of the precursor, ZIGMOND and WURTMAN (1970) have found a significant diurnal fluctuation in CA. The size of the brain catecholamine pool, in which newly synthesized CA accumulates, is larger during the days (when rats tend to sleep more) than during night (when rats tend to be active).

5. Summary

There is general agreement, at least with regard to the cat, between the results of neuropharmacological and direct alterations of CA metabolism: the decrease in CA metabolism by inhibition of CA synthesis at the tyrosine hydroxylase level (with αMPT) or DA-β-hydroxylase level (disulfiram), or the destruction of NA-containing neurons of the mesencephalic tegmentum by coagulation induce a decrease in EEG waking (increase in cortical synchronization). DA does not seem to be involved in control of the cortical activity, since there is no significant alteration of the electrical aspect of the sleep-waking cycle after almost total disappearance of DA in the forebrain (by destruction of the DA-containing cell bodies of the substantia nigra). On the other hand, depletion of brain DA induces a state of behavioral unresponsiveness; this favors the view that DA might be involved in the control of extrapyramidal functions concerned with behavioral waking. An increase in CA levels is usually accompanied by behavioral or EEG waking, whereas increased levels of waking (stress, foot shock, amphetamine) are usually accompanied by increased turnover of CA. Thus, it is likely that both the dopamine nigro-striatal system and the noradrenergic neurons ascending in the dorsal NA bundle, and also the CA-containing group of the mesencephalic tegmentum, may be responsible for the maintenance of *tonic* behavioral and EEG arousal, whereas the ventral ascending NA system of neurons apparently is not concerned with the regulation of cortical activity but may be involved in the so-called "reward system" in the rat.

F. Acetylcholine and the Sleep-Waking Cycle

The fact that cholinergic transmission constitutes the first instance of proven chemical transmission demonstrated at the periphery has given pharmacologists an opportunity to study many drugs capable of acting as tools in studies of cholinergic transmission. By the same token, this long history of study of the cholinergic transmission has also led to the development of a healthy scepticism concerning the "over-exploitation of central cholinergic systems" (see RUSSELL, 1969; KRNJEVIC, 1969; WEISS and HELLER, 1969). As stated by KARCZMAR (1969), "there are many opportunities for sinning in the field of cholinergic transmission". It is evident that the same opportunities exist for monoaminergic transmission.

1. Pharmacological Alteration of ACH Metabolism

i) Increase of Availability of ACH to Cholinoceptive Neurons

The direct injection of ACH into the carotid artery induces a state of cortical activation (BREMER and CHATONNET, 1959; MONNIER and ROMANOWSKI, 1962). The same effect is obtained after intravenous injection (YAMAMOTO and DOMINO, 1967). On the other hand, physostigmine, nicotine and prostigmine significantly increase cortical and behavioral arousal in the intact animal (DOMINO and YAMAMOTO, 1965; YAMAMOTO and DOMINO, 1965, 1967; DOMINO et al., 1968; CUCULIC et al., 1968; BOKUMS and ELLIOTT, 1968; KAWAMURA and DOMINO, 1969). Since cholinoceptive neurons are present in the brain stem, diencephalon and cortex (see CURTIS, 1963; KRNJEVIC et al., 1964, 1966b), the mechanisms of the EEG activation induced by cholinergic agonists are still uncertain. The existence of possible muscarinic and nicotinic cholinergic receptors makes the picture even more complex (see DOMINO et al., 1968). RINALDI and HIMWICH (1955), BRADLEY and ELKES (1957) have suggested that the mesencephalic reticular activating system is cholinergic. Since the cortical arousal induced by nicotine is suppressed by lesions of the mesencephalic reticular formation, it is likely that nicotinic cholinergic agonists act upon the brain stem (KAWAMURA and DOMINO, 1969). But, on the other hand, physostigmine may activate the "chronic isolated hemisphere preparation" (VILLABLANCA, 1966c). This suggests a direct action at the cortical level. Finally, it has been suggested that it is the inhibition of pseudocholinesterase (which is concentrated in glial cells) and not the inhibition of true cholinesterase which might be involved in the mediation of cortical arousal even in the "cerveau isolé" preparation (DESMEDT and LA GRUTTA, 1957). Thus, besides their possible role as neurotransmitters, brain choline esters might act as a "local hormone" influencing the internal milieu of neurons in the cerebral cortex.

ii) Decrease in the Availability of ACH

– *Inhibition of Synthesis.* This can be achieved by hemicholinium 3 (HC₃) (GARDINER, 1961; SLATER, 1968). The intraventricular injection of this drug reduces EEG waking and produces a clear-cut dissociation between behavioral arousal and presence of cortical slow waves (HAZRA, 1970; DOMINO and STAWISKI, 1971).

—*Anticholinergic Drugs.*

—*Atropine and Related Compounds* (scopolamine-hyoscine). These drugs increase cortical ACH release very significantly (see CELESIA and JASPER, 1966; BARTOLINI and PEPEU, 1967). Several hypotheses have been put forward to explain this finding:

1. atropine increases outflow of ACH by occupying the cerebral postsynaptic receptor site and blocks some binding mechanism, thus preventing the reuptake of released ACH (SZERB, 1964; CELESIA and JASPER, 1966);

2. atropine acts by suppressing a possible negative cholinoceptive feedback so that the corticopetal cholinergic neurons fire at a higher frequency (McINTOSH, 1963).

The well-known dissociation produced by atropine and related compounds between cortical activity and waking behavior is a most puzzling fact (WIKLER, 1952). This dissociation between cortical slowing and waking has opened a great deal of lively discussion during the course of various symposia and meetings (see LONGO, 1966). This dissociation is, however, not constant, since fully atropinized cats (2 mg/kg) may be behaviorally asleep when recuperating from earlier sleep deprivation (VIMONT-VICARY, 1966; see JOUVET, 1968). On the other hand, the dissociation between cortical slowing and waking behavior may be less dramatic if one considers that in cat (ROUGEUL et al., 1969) and monkey (RICCI and ZAMPARO, 1965) there is a total disruption of conditioned behavior when the cortical activity is synchronized. On the other hand in the rat, scopolamine greatly impairs short-term memory (DEUTSCH and ROCKLIN, 1967; BERGER and STEIN, 1969). This might indicate that the subcortical mechanisms which are involved in the maintenance of waking are not impaired by atropine or scopolamine, whereas higher nervous processes (learning, memory) which take place at the neo- or paleocortex are "inhibited" by these drugs.

iii) Direct Injection of ACH into the Brain

Local injections of minute amounts of ACH or carbachol can elicit either sleep or waking (even sham rage), depending upon the loci of injection. From the extensive work of Cordeau (1962), Cordeau et al. (1963), Hernandez-Péon (1965 a, b), Hernandez-Péon et al. (1963, 1967), Hernandez-Péon and Chavez-Ibarra (1963), Morgane (1969) the following picture of two possibly antagonistic cholinoceptive systems can be schematized:

a) *A Sleep System* (in which a local injection of ACH may "trigger" SWS). This system has two components. *The descending component* which closely conforms to the limbic midbrain circuit described by Nauta (1958), i.e. it follows the trajectory of the medial forebrain bundle from the preoptic region through the lateral hypothalamus and into the limbic midbrain area (interpeduncular nucleus, ventral tegmental area of Tsai and the paramedial reticular nuclei of Bechterew and Gudden). An injection of ACH into this circuit if followed by sleep, whereas lesions or atropine injections made caudally to the injection of ACH suppress the sleep-inducing effect (Velutti and Hernandez-Péon, 1963). This system joins at the pontine level with an *ascending component* originating from the gray matter of the spinal cord (Hernandez-Péon et al., 1963, 1967). The localization of the cholinoceptive cells in the lower brain stem is uncertain, but according to Cordeau (1962) paramedial injection of ACH into the caudal medulla is regularly followed by cortical synchronization. According to Hernandez-Péon (1965), the cholinergic sleep system would operate as follows: the primary hypnogenic stimuli arise in the peripheral neurons of the ascending and descending segments as, for example, somatic sensory stimuli in the ascending component and conditioned stimuli (arising in the neocortex) in the descending component. These influences converge, then arise in the final pathway to cause a progressively spreading inhibition. As the inhibition begins to ascend, mesencephalic neurons become inhibited. Thus, the thalamic recruiting neurons, now disinhibited, organize the thalamo-cortical activity which is recorded as spindles and slow waves during SWS. The wave of inhibition continues to ascend and eventually reaches the thalamic recruiting nuclei. Inhibition of these structures would release the activity of the cortex which becomes rapid during PS. According to Hernandez-Péon, sleep is not only cholinergic but it is *unitary*. SWS and PS are not separate states but are simply different manifestations of the same basic processes. However, there is so much evidence (see p. 236) that SWS and PS may be selectively abolished by lesions at different sites of the brain stem, or can be altered independently by different drugs, that this unitary theory of sleep can no longer stand up.

b) *A Waking System* which corresponds to the ascending cholinergic reticular system described by SHUTE and LEWIS (1963, 1966, 1967; see p. 182). Local cholinergic stimulation of this system according to MORGANE (1969) produces arousal. The cholinergic (which should be called cholinoceptive) arousal system is topographically dissociated from the ascending sleep system in the midbrain, posterior to the interpeduncular nucleus. However, both sleep and waking cholinoceptive systems are topographically intermixed in the medial forebrain bundle and the lateral hypothalamic area, so that, in certain regions, alerting is produced by cholinergic stimulation while sleep may follow the same cholinergic stimulation just 1 mm away (see also BANDLER, 1969).

Evaluation of these results is difficult since the sleep-promoting effects in a freely behaving cat are very difficult to distinguish from the occurrence of spontaneous natural sleep. Besides, the cholinergic stimulation of many "hypnogenic loci" in the preoptic region and the hypothalamus of cats has failed to induce sleep in the experiments of MYERS (1964) or MACPHAIL and MILLER (1968). On the other hand, large doses of atropine do not suppress behavioral sleep. Last, but not least, the topography of neither the descending nor the ascending sleep cholinoceptive system would explain why total insomnia is obtained by a lesion of the raphe nuclei (a lesion which does not interfere with their postulated ascending or descending pathways). Only in the paramedial region of the caudal medulla, would the cholinoceptive sleep-inducing effect obtained by CORDEAU (1962) correspond to the caudal raphe (in a region where cholinesterase-containing neurons have been mapped out). In order to differentiate the possible cholinergic hypnogenic system from other non-cholinergic but possibly cholinoceptive systems; it would be interesting to see whether cholinergic stimulation of a given hypnogenic locus can still induce sleep in an insomniac cat after inhibition of 5-HT synthesis or destruction of the raphe system.

The waking effects obtained by cholinergic stimulation of the ascending cholinergic system of Shute and Lewis are probably much more significant, since long-lasting arousal is uncommon in the freely behaving cat. It remains to be seen whether the local injection of ACH or carbachol mimics the physiological release of ACH at the level of cholinoceptive receptors. The fact that many abnormal motor effects (catatonic-like posture, adynamia and even convulsions) can follow the injection of ACH in some parts of the "cholinergic ascending system" (ventral tegmental area of TSAI, lateral reticular formation of the midbrain) (HERNANDEZ-PÉON et al., 1963, 1967) is an indication that some unphysiological "side effects" may result from the chemical stimulation. In fact, the technique of mapping out any system in the brain with in situ injection is open to criticism. The insertion of a "few minute crystals" into a small area may well deliver a huge dose to the surrounding neurons. As pointed out by MANDELL and MANDELL (1965), a large amount of a polar compound like ACH may have physicochemical effects that are entirely different from the drug's usual effects. On the other hand ROUTTEN-BERG et al. (1968) have demonstrated clearly how non-specific the implantation technique can be. They studied the movement of carbachol, NA, and DA in rat brain. Their results showed not only localized spherical diffusion but also diffusion for relatively long distances along nerve tracts, even to the contralateral side. They state: "Any attempt to ascribe anatomical localization to behavioral effects resulting from chemical stimulation of the brain should take into account the widespread movement of chemical after their application to brain tissue."

2. Alteration of Sleep-Waking and ACH Metabolism

Two main techniques have been utilized to study brain ACH metabolism in relation to the "state of excitation": determination of the endogenous level of brain ACH by cortical biopsy (SZERB et al., 1970) or after killing the animal (bound ACH), or determination of the release of ACH (free ACH) through the use of the superfusion technique at the cortical level (McINTOSH and OBORIN, 1953). There is a general agreement that:

1. The cortical arousal provoked either by mediopontine transection (BAR-TOLINI and PEPEU, 1967) or by stimulation of the mesencephalic reticular formation (CELESIA and JASPER, 1966; COLLIER and MITCHELL, 1967; SZERB, 1967) and of the septal area (SZERB, 1967) by peripheral or central sensory stimulation (Fig. 15) (PHILLIS and CHONG, 1965; COLLIER and MITCHELL, 1966; TORU et al., 1966; APRISON et al., 1968b; NEAL et al., 1968; APRISON and HINGTEN, 1969) or by amphetamine (PEPEU and BARTOLINI, 1967) is accompanied by an increase in the release of ACH at the cortical level (and a decrease in "bound ACH" measured in the brain) (SZERB, 1967).

2. The cortical synchronization elicited by a precollicular transection (MAN-TEGAZZINI and PEPEU, 1964; PEPEU and MANTEGAZZINI, 1964) or even by physiological sleep (CELESIA and JASPER, 1966; BEANI et al., 1968) is accompanied by a decrease in the release of free ACH at the cortical level and an increase in bound ACH. Moreover, barbiturate anesthesia further reduces the release of ACH (KUROKAWA et al., 1963; CELESIA and JASPER, 1966).

These correlations, made in different laboratories, and under different conditions, are of great importance, since this is the first time that any significant alteration of the release of a neurotransmitter has been associated with the state of cortical activation and in some conditions even with the sleep-waking cycle. However, numerous difficulties exist when one considers the locus of origin of the ACH, which is released at the cortical level, and the possible physiological correlation between ACH release and cortical activation:

—The released ACH originates at least in part from the cortical tissue underlying the collection area, since acute undercutting of the cortex either

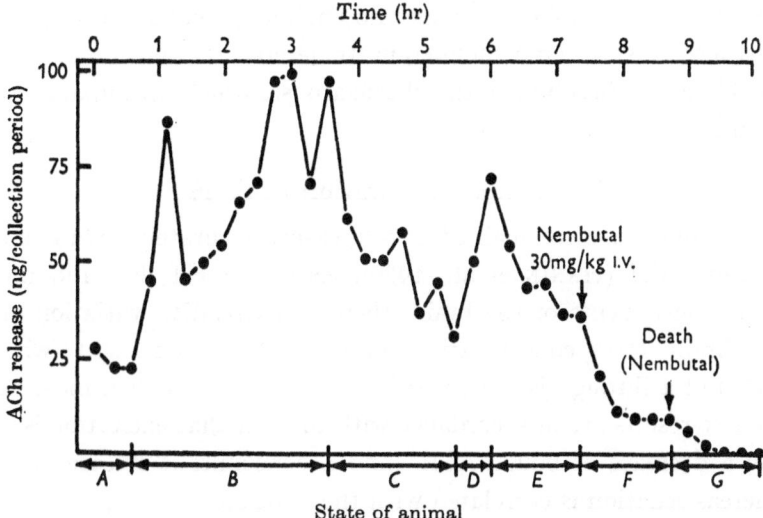

Fig. 15. *ACH release from the visual cortex of a free-moving rabbit during anesthesia and consciousness.* Perfusion system implanted in left visual cortex. Perfusion rate, 0.5 ml/15–18 min and samples assayed on recovery of 0.5 ml of perfusate. Allowance has been made for collecting-tube dead space of 0.5 ml. Legend on lower abscissa indicates state of animal as follows: *A* anesthetized after cannula implantation; *B* recovered from anesthesia and continuously active, exploring cage, eating and drinking; *C* quiet and moving only occasionally; *D* active; *E* quiet; *F* anesthetized after Nembutal (30 mg/kg i.v.) at arrow; *G* dead after lethal dose of Nembutal at arrow. (COLLIER and MITCHELL, 1967)

does not affect (SZERB, 1967; GREEN et al., 1970) or reduces (COLLIER and MITCHELL, 1967) the resting output of ACH, and since direct cortical stimulation increases ACH output (PHILLIS and CHONG, 1965). On the other hand, the increased release is concomitant with EEG activation, but the pathways involved in cortical arousal and increased ACH output may be different since the two phenomena do not always vary in parallel fashion when the reticular formation or other subcortical areas are stimulated (SZERB, 1967). Since hemisection of the midbrain tegmentum increases the ipsilateral ACH cortical content, this has been taken as proof that the cell bodies of the cholinergic neurons terminating in the cortex might lie above the level of the "cholinergic reticular formation" (SZERB, 1967). Indeed, had the section severed the perikarya from the ascending cholinergic fibers, a decrease in cortical ACH content should have been observed. However, these acute experiments provide no definite proof since a suppression of the release of ACH might increase endogenous ACH levels. Further work combining *chronic* lesions of the so-called "ascending cholinergic reticular system" and determination of the endogenous level of ACH or its release is necessary in order to evaluate the possibility of direct, or polysynaptic cholinergic synapses ascending to the cortex. The effectiveness of septal stimulation in increasing ACH release (SZERB, 1967) indicates that at least some of the corticopetal cholinergic fibers traverse this area. Finally, the fact that retinal, lateral or medial geniculate stimulation also increases the release of ACH in the primary receiving areas favors the view that, besides non-specific cholinergic corticopetal pathways, there must also exist specific cholinergic (thalamo-cortical) sensory pathways so that ACH may be involved in the phasic arousal which accompanies sensory stimulations.

3. Circadian Alteration of ACH

Recent techniques have enabled us to measure accurately circadian alterations of brain ACH (Hanin et al., 1970). As with 5-HT, NA and probably most of the constituents of the brain, there is a circadian variation of ACH in the rat brain. Peak concentrations occur at 2 hours of light (when rats are asleep) and a through is seen at 6 hours of darkness (when most rats are awake). These results are in accordance with the fact that excitation is accompanied by a decrease of bound ACH (and a concomitant release of cortical ACH) whereas sedation is correlated with the opposite.

4. Summary

There is considerable and concordant evidence that cholinergic mechanisms act at the cortical level to mediate cortical arousal or to modulate the catecholaminergic mechanisms implicated in arousal (see Fibiger et al., 1970). However, impairment of cortical arousal by "anticholinergic drugs" does not suppress *behavioral* arousal although it strongly impairs learning or memory. Thus, it is likely that some cholinergic intracortical mechanism plays a role in the higher nervous processes which take place during waking. Although this has not yet been proved by chronic lesion data, there are also some data which favor the view that ascending cholinergic subcortical mechanisms may participate in cortical arousal. On the other hand, the data favoring the intervention of cholinergic cortical or subcortical mechanisms in slow-wave sleep are not convincing, since neither atropine nor hemicholinium 3 suppresses behavioral sleep. However, cholinoceptive neurons might be involved in the triggering of SWS and of PS (see below).

IV. The Problems of Paradoxical Sleep

Although ontogenetically precocious in birds and mammals, since it exists in ovo (Klein, 1963) and in utero (Astic and Jouvet-Mounier, 1969), PS is phylogenetically recent since it is probably nonexistent in fishes and reptiles (Peyrethon and Dusan-Peyrethon, 1967, 1968). Thus, PS may represent the emergence of a new function during phylogenetic evolution. This could explain the complexity of its mechanism since, either at the neuropharmacological or at the neurophysiological level, PS appears to be the result of a chain of events including 5-HT "priming mechanisms" and both CA and ACH "executive mechanisms".

A. Evidence for the Participation of 5-HT Neurons in the "Priming" of Paradoxical Sleep
1. Neuropharmacological Evidences

—*Effect of P-Chlorophenylalanine (PCPA)*. The inhibition of the 5-HT synthesis with PCPA decreases both SWS and PS. PS usually disappears totally when the daily amount of SWS decreases below 15 % (KOELLA et al., 1968; JOUVET, 1969b; see DEMENT et al., 1970b). Thus, there is some correlation between the decrease of PS and the decrease of 5-HT biosynthesis after PCPA. During the recovery of both SWS and PS which follows the insomnia induced by PCPA in the rat, there is also a significant correlation between the increase in the daily *frequency* of PS episodes and the increase in the endogenous level of 5-HT in the brain stem (MOURET et al., 1968). Interestingly enough, the mean duration of PS episodes (which is longer than under controlled condition) is inversely correlated with 5-HT. This suggests that the *rate* of appearance of PS depends upon 5-HT priming mechanisms whereas the duration of PS (which depends upon the executive mechanism of PS) would depend upon other mechanisms.

—*Effect of MAO Inhibitors*. This group of drugs has the strongest suppressor effect upon PS in rats (MOURET et al., 1968b), cats (JOUVET et al., 1965; see JOUVET, 1967b), monkeys (REITE et al., 1969), and man (LE GASSICKE et al., 1965; CRAMER and KUHLO, 1967; WYATT et al., 1969, 1971; AKINDELE et al., 1970). There is a significant correlation between the decrease or disappearance of MAO in the pons (as measured with histochemical technique) and the suppression of PS in the cat and rat (DELORME, 1966; MOURET et al., 1968b). Short-lasting suppressions of PS by MAO inhibitors in cats (3–5 days) are not followed by any rebound whereas long-lasting suppressions (several weeks) are followed by a secondary rebound in man (AKINDELE et al., 1970). The suppression of PS is even observed when the need for PS is much, enhanced, as after instrumental deprivation in the cat (JOUVET et al., 1965). Although the suppressant effect of MAO inhibitors upon PS is the most prolonged observed with any drug, its interpretation is difficult since it may be obtained via different mechanisms: the drugs may be acting upon 5-HT metabolism to decrease the turnover of central 5-HT neurons (see AGHAJANIAN et al., 1970b; MACON et al., 1970) or to decrease the deaminated 5-HT catabolites (5-hydroxyacetaldehyde) which may play a role (see JOUVET, 1969b). On the other hand, MAO inhibitors may also act in the same way upon CA-containing neurons.

The result with both MAO inhibitors and PCPA suggests indirectly that the common denominator of the suppression of PS might be the decrease in 5-HT turnover. Two other experimental results indicate that an increase in 5-HT turnover may lead on the contrary to an increasing rate of appearance of PS.

1. When 5-HTP is injected at low doses (5–10 mg/kg) into an insomniac cat pretreated with PCPA, both SWS and PS return to normal levels or even increase as compared with control levels (Pujol et al., 1971). Larger doses of 5-HTP (30–50 mg/kg) lead, on the contrary, to a transitory suppression of PS which may be due to the decarboxylation of exogenous 5-HTP in non-serotoninergic neurons (see p. 201).

On the other hand, instrumental deprivation of PS is accompanied by an increase of the synthesis of ^{3}H-5-HT from ^{3}H-tryptophan and by an increase of incorporation of labelled tryptophan (whereas there is no increase of ^{3}H-5-HT from labelled 5-HTP). This indicates that PS deprivation (which will secondarily lead to an increase in the frequency of PS during the rebound) is probably accompanied by an increase of 5-HT turnover (since 5-HT endogenous levels are not altered). It should be pointed out that PS deprivation acts not only at the tryptophan hydroxylase level (increase of ^{3}H-5-HT from ^{3}H-tryptophan) but also at the level of the capture of tryptophan (Hery et al., 1970). Apparently this step is very important in the regulation of 5-HT synthesis (see Pujol, 1970; see also Tagliamonte et al., 1971).

Finally, indirect data suggest that there is also an increase in the turnover of 5-HT during the rebound of PS which follows its previous suppression (increase of cerebral 5-HIAA without alteration of endogenous cerebral 5-HT) (Weiss et al., 1968). As might have been expected, PCPA suppresses the rebound of PS when injected at the end of instrumental deprivation (see Dement et al., 1970b).

These neuropharmacological and biochemical experiments, which suggest that a decrease in 5-HT turnover decreases PS, whereas an increase in 5-HT turnover increases the frequency of PS, are admittedly indirect, but they are corroborated by similar neurophysiological evidence.

2. Lesion of the Raphe System

Total lesion of the raphe system totally suppresses PS (and strongly decreases SWS). However, there is some regional specificity of the 5-HT neurons as related to PS. The destruction of the anterior part of the raphe (N. raphe dorsalis or raphe centralis) induces a state of permanent arousal during the first two to three days but PS still appears (without preceding SWS) periodically for 5–10% of the time (Renault, 1967). This phenomenon is similar to the narcoleptic attacks described in man (Dement et al., 1966). Thus, the anterior raphe neurons are probably related more to SWS mechanisms than to PS. On the other hand, destruction of the caudal raphe (N. raphe pontis, magnus) is followed by an almost total disappearance of PS, whereas SWS is decreased to only 40% of control level. Thus, the caudal raphe neurons may represent the "priming" 5-HT neurons which project to

the dorso-lateral part of the pontine tegmentum where the "executive mechanisms" are located.

3. Lesion of the Isthmus

This lesion interrupts a contingent of the ascending dorsal NA bundle at
the junction between the pons and mesencephalon. It is followed during the
first 4–5 days by a significant increase of both SWS and PS. However, biochemical analysis of the forebrain and mesencephalon reveals a decrease of NA,
and a significant increase of both tryptophan and 5-HIAA without alteration
of 5-HT (PETITJEAN and JOUVET, 1970). This result strongly suggests an
increase in central 5-HT turnover. Since terminals from NA-containing neurons
of the anterior part of the locus coeruleus may be found in the anterior raphe
(LOIZOU, 1969), it is possible that the lesion of the isthmus may have suppressed an inhibitory control from NA ascending axons upon the raphe system
and that true hypersomnia (with increases of both SWS and PS) is the result
of the increase in 5-HT turnover.

4. Relationship between SWS and PS

Since SWS also depends on 5-HT mechanisms (see p. 208), it is not surprising to discover a very close relationship between the amount of SWS
and PS. This correlation has been demonstrated between stage II of SWS
and PS in normal cats (URSIN, 1970). It is well illustrated in Fig. 16 (see legend).
After raphe lesion the quantity of PS is significantly correlated with the amount
of SWS. In this case, PS occurs only if the daily amount of SWS exceeds
15–20% of the recording time. On the other hand, it is also evident that
a normal amount of SWS may be accompanied by a total suppression of PS
in the case of lesions of the dorso-lateral pontine tegmentum. This fact demonstrates that PS depends not only on "priming mechanisms" but also upon
other "executive mechanisms" which are located in the pontine reticular
formation.

B. Participation of Catecholamines in the Executive Mechanisms of Paradoxical Sleep

1. The Neuropharmacological Data

indirectly favoring the role of CA mechanisms in PS may be summarized as
follows:

a) DOPA (30–50 mg/kg), given to reserpine-pretreated cats, causes PS to
recur much earlier (1–3 hours) as compared with 24 hours without DOPA
(MATSUMOTO and JOUVET, 1964). Thus the "refilling" of some pool in the
CA terminals (since CA synthesis still occurs after treatment with reserpine)
(see GLOWINSKI et al., 1966) is a condition for the reappearance of PS.

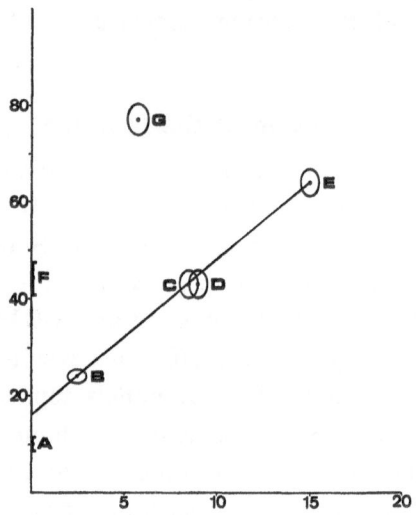

Fig. 16. *Relationships between paradoxical sleep and slow-wave sleep after different lesions of the monoamine systems.* Results are expressed in mean ± SEM. *A* Total or subtotal lesions of the raphe system (16 cats). *B* Partial lesions of the raphe system (13 cats). *C* Control lesions of the brain stem outside monoaminergic system in the ventral pons, caudally to the locus coeruleus or in the tectum (11 cats). *D* Sham-operated cats recorded under the same conditions as the experimental animals during 10–13 days (11 cats). *E* Hyper-somniac cats with increase of both SWS and PS after lesion of the isthmus destroying the dorsal NA bundle (9 cats), see Fig. 3. *F* Subtotal lesion of the locus coeruleus complex selectively suppressing para-doxical sleep (15 cats). *G* Lesion of the mesencephalic tegmentum destroying group A 8 of CA-containing neurons (decrease of waking but not true hypersomnia since PS is slightly decreased). In the group of cats *B, C, D, E,* the correlation between PS and SWS may be obtained with the following formula:

$$\% \ PS = \frac{\% SWS}{3.2} - 16.$$ This means that under a minimum amount of SWS which is equal to 16% of recording time, no PS can occur. Ordinates: percentage of cortical synchronization during the 10–13 days of the post operative survival. Abscissae: percentage of paradoxical sleep during the same duration

b) α-methyl-DOPA (200 mg/kg) which leads to the false transmitter α-methyl NA (and displaces NA from the stores) (see Carlsson, 1964; Carlsson et al., 1968; see Andén et al., 1969, 1970) suppresses PS in the cat for 16–20 hours (Dusan-Peyrethon et al., 1968). This drug also permits us to differentiate between the "priming mechanisms" (which are not altered) and the "executive mechanisms". Given every day, at a dose of 200 mg/kg during the 3 days of instrumental PS deprivation, this drug does not suppress the subsequent rebound. Thus the 5-HT mechanisms which are supposed to be involved in the priming of the rebound are not altered by the drug. However, if α-methyl-DOPA is given at the end of the deprivation procedure, it delays the reappearance of PS for 14–16 hours. Hence it must be postulated that the CA terminals are filled with the false transmitter and that, despite normal priming mechanisms (increased turnover of 5-HT), the effector CA neurons are unable to trigger a physiological response from the postsynaptic cells. However, due to the increased turnover of central CA neurons (as measured after the cisternal injection of labelled NA) which takes place during the rebound of PS (Pujol et al., 1968), the stored α-methyl-NA is metabolized more rapidly and PS suppression after α-methyl-DOPA is shorter than under normal conditions.

c) Disulfiram (which impairs NA synthesis) leads to a decrease in PS (and a decrease of PGO rate during PS). This phenomenon is paralleled by a decrease in waking (DUSAN-PEYRETHON and FROMENT, 1968).

d) α-receptor blocking drugs, which cross the blood-brain barrier of the rat, suppress PS (MATSUMOTO and WATANABE, 1967).

e) In partial contradiction to the preceding results are those obtained with α-methyl-p-tyrosine (αMPT). This drug, which inhibits the synthesis of CA at the level of tyrosine hydroxylase, decreases waking and suppresses PS in monkeys (WEITZMAN et al., 1969) and rabbits (FUJIMORI and HIMWICH, 1971). However, the suppression of PS is only short-lasting in cats (ISKANDER and KAELBLING, 1970; KING and JEWETT, 1971) and may be followed by a secondary rebound of PS. In the rat, αMPT does not decrease PS nor suppress the PS rebound which follows previous instrumental PS deprivation (MARANTZ and RECHTSCHAFFEN, 1967; MARANTZ et al., 1968). The discrepancies between the results obtained in different species are difficult to explain and the fact that PS is not suppressed by αMPT in the cat is a most puzzling fact when one considers the neuroanatomical data which favor the role of the NA-containing neurons of the dorsopontine tegmentum in the executive mechanisms of PS. Much more work is certainly needed in order to establish, on neuropharmacological grounds, that PS depends upon CA mechanisms. A possible explanation would be that αMPT does not deplete all CA stores uniformly and that some organs in which NA metabolism is not continuously excited (as in the seminal vesicles) are not depleted by αMPT (ANDÉN et al., 1966f). It thus remains possible that PS-executive pontine CA neurons, which are normally excited only 15 % of the time, might be much more resistant to the effect of αMPT than the ponto-mesencephalic NA neurons which are involved in waking mechanisms. Another possibility would be that the inhibition of CA synthesis with αMPT might increase 5-HT turnover (and thus the priming mechanisms of PS).

2. Neurophysiological Localization of the Executive Mechanisms of PS

The pontile origin of both ascending (fast cortical activity, PGO waves) and descending (inhibition of muscle tone, rapid eye movements) components of PS is demonstrated by the following experiments which will be briefly summarized since they have been the subject of recent reviews (JOUVET, 1962, 1965a, b, 1967a; ROSSI, 1963; ZANCHETTI, 1967).

i) Brain Stem Transections

The removal of all the brain rostral to the pons (including the hypothalamo-hypophysis) does not suppress the periodic appearance of PS in the chronic pontile cat (see Jouvet, 1962, 1965 a; Villablanca, 1966 b; Matsu-zaki et al., 1967; Matsuzaki, 1969). The PS episodes are characterized by a total loss of muscle tone and by bursts of pontine waves which accompany lateral eye movements (triggered from the VIth nerve). The duration of PS episodes (6 minutes) is the same as in normal cats, but the periodicity is longer (40 minutes). Thus, the nycthemeral percentage of PS is 8–10% of recording time, which is lower than the average percentage of a normal cat (15%), but similar to the percentage in cats sham-operated in the pontine or cerebellar region (10%) (Renault, 1967). The transection of the brain stem at the caudal third of the pons (caudo-pontine transection) suppresses the appearance of periodical atonia (Jouvet, 1962). Thus the anterior two-thirds of the pons is implicated in the executive mechanism of PS.

ii) Localized Lesions of the Pons

In early investigations, it was concluded that the dorso-lateral part of the pontine reticular formation was responsible for the executive mechanisms of PS, since its destruction totally suppressed PS without interfering significantly with SWS (Jouvet and Mounier, 1959; see Jouvet, 1962; Carli and Zanchetti, 1965; Rossi et al., 1961). The discovery of some groups of neurons containing MAO and catecholamines (see p. 179–181) in this part of the brain was a great step forward to a much more precise approach to the structures responsible for PS (Jouvet and Delorme, 1965).

iii) Direct Attack upon CA-Containing Neurons of the Dorso-Lateral Pontine Tegmentum

Coagulation. The NA-containing neurons of the dorso-lateral pontine tegmentum (group A 5, A 6, A 7) are concentrated in the nucleus locus coeruleus (see Russell, 1965, for the cytoarchitectonics of this nucleus), subcoeruleus and adjacent nuclei. From a series of more than 100 chronically implanted cats in which coagulations were made systematically in these groups and in control regions (Jouvet, 1965 a; Jouvet and Delorme, 1965; Roussel, 1967; Roussel et al., 1967; Buguet, 1969; see Jouvet, 1969 b; Buguet et al., 1970) the picture of a very intricate system of neurons, leading to descending and ascending axons (Fig. 17) can be briefly summarized as follows:

a) The bilateral lesion of the caudal part of N. locus coeruleus (situated in the plane P 3 and P 4) suppresses *only* the motor inhibition which takes place during PS. During the first days after the lesion some permanent PGO may occur (as after reserpine) and no PS. Subsequently, around the 8th to 10th day, "hallucinatory behavior" occurs periodically during sleep (Jouvet, 1965 c). The ascending components of PS are present (activated EEG, PGO activity, rapid eye movements, miosis, total relaxation of nictitating membranes). However, to a naive observer, the cat, which is standing up, looks

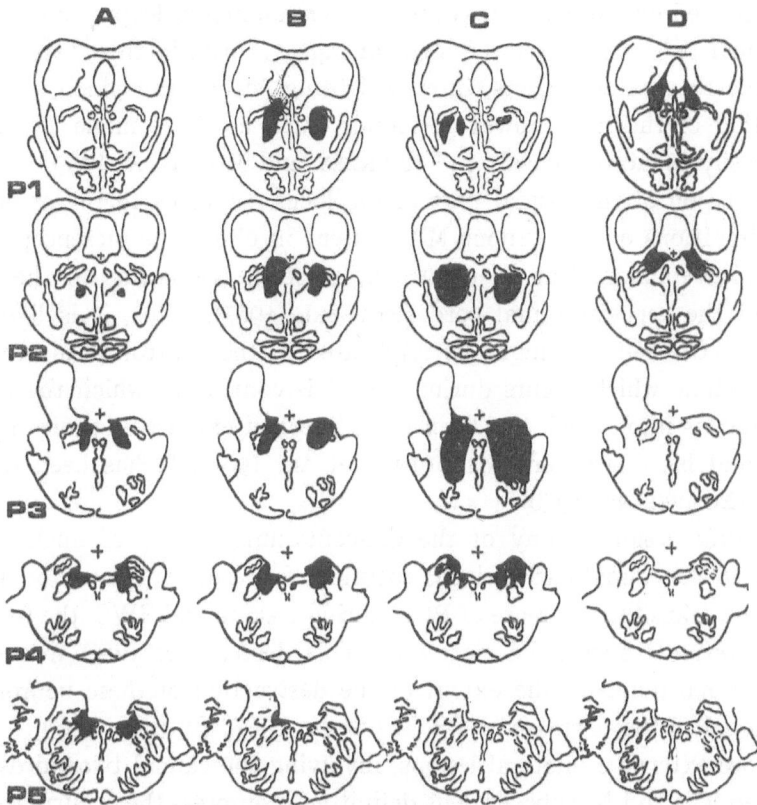

Fig. 17. *Effects of lesions of the dorso-lateral pontine tegmentum upon sleep.*—The lesions are represented in black upon frontal section according to the Horsley Clark coordinates (see Fig. 2 for the localization of NA-containing neurons of the pontile tegmentum). *A* Lesions of the caudal part of the N-locus coeruleus which suppress only muscular inhibition during paradoxical sleep. After such a lesion, the cat exhibits "hallucinatory" behavior during PS. *B* Subtotal lesions of the locus coeruleus which suppress totally paradoxical sleep. PGO still occur during slow wave sleep. *C* More extensive lesions which suppress definitively paradoxical sleep and PGO. *D* Lesions of the most rostral part of the locus coeruleus and of the dorsal NA bundle which increase both slow-wave sleep and PS (see Fig. 3)

awake since it may attack unknown enemies, play with an absent mouse, or display flight behavior. There are orienting movements of the head or eyes toward imaginary stimuli, although the animal does not respond to visual or auditory stimuli. These extraordinary episodes (which are a good argument that "dreaming" occurs during PS in the cat) last for 3–5 minutes and result in sudden awakening and return to SWS. This behavior is immediately suppressed by MAO inhibitors (nialamide or pargyline). The suppression of PS after similar lesions has also been observed by CARLI and ZANCHETTI (1965) who did not observe similar "hallucinatory" behavior. However HENLEY and MORRISON (1969) described the same dramatic episodes after similar lesions involving the dorso-lateral part of the pons including part of the locus coeruleus: "... Three cats demonstrated more elaborate behavior in that after the initial head and paw movements they often righted themselves, stood up and progressed forwards or backwards, either running or leaping convulsively. One animal twice demonstrated signs of rage ... cats in this state were unresponsive to moving objects, intense lighting, strong tail pinch and tactile stimulation of the face, however, sounds, such as those made by creaking knees, easily induced arousal ..." (HENLEY and MORRISON, 1969, l.c.).

After lesion of the caudal part of the locus coeruleus there are no significant alterations of endogenous MA content in either the mesencephalon or the telediencephalon. But a 30–40% decrease of NA occurs in the ventral pons and the cervical spinal cord (ROUSSEL, 1967). Thus, it is likely that descending CA neurons might be implicated in the control of the total loss of muscle tone which occurs during PS. This control, in which the reticulospinal tract is implicated, has been the subject of extensive studies by POMPEIANO and his co-workers (GIAQUINTO et al., 1964). It has been reviewed recently (POMPEIANO, 1970).

b) *Partial* lesion of any of the CA-containing groups of nuclei located rostrally to the caudal part of locus coeruleus decreases but does not suppress PS. After a transitory increase of PGO during waking and SWS, the frequency of PGO activity during PS decreases. This decrease of PGO frequency is roughly proportional to the extent of the destruction of these neurons, and proportional to the decrease of PS (ROUSSEL, 1967; BUGUET, 1969).

c) More extensive bilateral lesions, involving the caudal two-thirds of the locus coeruleus and N. subcoeruleus definitively suppress the occurrence of PS and of "pseudo-hallucinatory behavior". However, PGO activity is not suppressed and may still occur during SWS. This lesion induces a 30–40% decrease of NA in the telediencephalon without any alteration in the 5-HT or DA levels.

d) Bilateral *total* lesions of the entire group of nuclei containing CA neurons are followed by a decrease in waking and an immediate and permanent suppression of PGO activity (and of PS). This lesion causes a more significant decrease of NA in the mesencephalon and telediencephalon (BUGUET, 1969; BUGUET et al., 1970).

e) Lesion of the rostral third of N. locus coeruleus or of the ascending dorsal NA bundle at the level of the isthmus is accompanied by a temporary increase in SWS and PS, lasting for 4–5 days, without alteration of the PGO activity. The decrease of NA in the telediencephalon and mesencephalon is accompanied by a significant (30%) increase of telencephalic 5-HIAA and tryptophan (PETITJEAN and JOUVET, 1970).

f) Control lesions located either ventrally, medially, laterally or caudally to these groups of neurons (the entire group of vestibular nuclei) do not result in significant alterations of SWS, PS or of the amine content of the rostral brain.

From these data it may be concluded that most of the neurons located in the dorso-lateral part of the pons play a role in the execution of PS, and that there is a very complicated organization which is not yet understood. The most caudal part of the pontine CA neurons is related to descending mechanisms, whereas phasic PGO activity and fast EEG activity appear to depend most upon the neurons located in the caudal two-thirds of the "coeruleus complex". Only the group of CA neurons located in the anterior third of the coeruleus complex does not seem to participate in the PS mechanism, being apparently concerned with the control of waking.

The coagulation technique, however, is not selective enough to verify that CA neurons are the only ones responsible for the mechanisms of PS, since it is likely that "cholinergic" neurons are situated in the locus coeruleus complex (SHUTE and LEWIS, 1966).

iv) Effect of Microinjection of 6-Hydroxydopamine in the Dorso-Lateral Part of the Pontine Tegmentum

As stated above, this drug, when injected in the vicinity of CA-containing perikarya, is taken up by the nerve cells and a subsequent decrease of CA occurs in the perikarya and the terminals (UNGERSTEDT, 1969; see also UNGERSTEDT and ARBUTHNOTT, 1970). The injection of this drug into the dorso-lateral part of the pontile tegmentum has given the following results (BUGUET, 1969; BUGUET et al., 1970): during the first four days there was an increase of PGO (reserpinic syndrome) during waking and SWS, whereas the rate of PGO during PS decreased together with the amount of PS. On the 6th day, PS and PGO were totally suppressed. In the cats sacrificed on the 8th day there was a significant decrease of NA in the mesencephalon (but also a decrease of 5-HT which may indicate that some raphe cells might have taken up 6-OHDA and that, at least in the cat, this drug may not be as selective as in the rat) (see p. 219). In control cats to which microinjections of Ringer solution of the same pH, were given, there was also some temporary increase in PGO activity during waking and SWS, and a slight decrease in PGO frequency during PS. PS also decreased by 30% as compared with preinjection controls.

These experiments do not present the crucial proof that *only* CA-containing neurons are responsible for PS, since the mechanical lesions which follows any direct microinjection into the brain might have affected other neurons. There is, in fact, some indirect evidence that both the control of the inhibition of muscle tone and of the cortical activation are shared by the cholinergic mechanisms, as shown by the following pharmacological experiments.

C. Neuropharmacological Evidence for the Participation of Cholinergic Mechanisms in Paradoxical Sleep

The intervention of cholinergic mechanisms in PS was first suggested when it was demonstrated that PS could be selectively suppressed by the systematic injection of atropine, both in normal and pontile cats (Jouvet, 1962), or in rabbit (Khazan and Sawyer, 1964; Loizzo and Longo, 1968). The same effect is obtained by intraventricular injection of hemicholinium-3 at very low doses (Hazra, 1970; Domino and Stawiski, 1971). On the other hand, eserine, which induces a state of excitation with activated EEG in normal cats, may increase significantly (100%) the duration of PS in mesencephalic or pontile cats (Matsuzaki and Kasahara, 1966; Matsuzaki et al., 1967) or in rats (Khazan et al., 1967).

Moreover, direct injections of carbachol in the vicinity of the locus coeruleus (George et al., 1964) or in the 4th ventricle may considerably enhance the duration of PS (up to one hour) or induce only total atonia (Baxter, 1968, 1969). This suggests that the caudal part of the locus coeruleus responsible for the control of muscle tone is concerned with both ACH and NA mechanisms, or that the CA neurons are cholinoceptive. Thus, the sequence of events which primes and triggers PS appears to be the following: the 5-HT priming mechanism of the caudal raphe system would trigger a cholinergic link which in turn would activate the executive NA-neurons impinging upon the effector neurons. On the other hand the PGO activity does not seem to be under the direct control of cholinergic mechanisms, since some PGO spiking still appears under full atropinization (2 mg/kg) during the recovery of sleep after PS deprivation in the cat (Vimont-Vicary, 1966). Thus the cholinergic or cholinoceptive executive mechanism seems to concern mainly the tonic components of PS (either the cortical activation or the decrease of muscle tone). The fact that eserine can induce PS in a reserpine-pretreated cat is in accordance with this hypothesis (Karczmar et al., 1970). Finally, the intervention of possible cholinergic mechanisms is suggested by the finding that PS deprivation induces a 35% decrease in "bound" telencephalic ACH, but no change in the brain stem or diencephalon (Bowers et al., 1966; Tsuchiya et al., 1969). However, it is possible that the decrease in bound ACH was associated with the release of free ACH at a cortical level during the excitatory state which accompanied PS deprivation.

D. Short Chain Fatty Acids and Paradoxical Sleep

It may be surprising to meet the fatty acids in a review devoted principally to monoamines. In fact, these relatively simple compounds may well lead to the understanding of some mechanisms of PS by shedding some light upon the functioning of the membrane of the MA terminals.

1. Paradoxical Sleep-Inducing Effect of Short Chain Fatty Acids

The narcotic action of short chain fatty acids (from C 4 to C 10) was known even before the introduction of modern polygraphic recordings (Samson et al., 1956; White and Samson, 1956); later attention was focused upon

C 4 compounds: gamma hydroxybutyrate (GOH) and gamma butyrolactone (GBL) (BENDA and PERLES, 1960; see LABORIT, 1964). These drugs may induce both states of sleep in normal cats at a low dose (50 mg/kg) (JOUVET et al., 1961; MATSUZAKI et al., 1964; MATSUZAKI and TAKAGI, 1967a, b), but the effect is not constant (WINTERS and SPOONER, 1965; MARCUS et al., 1967). At a higher dose (100 mg/kg) they induce a state of anesthesia in both cats and rats which is quite different from normal sleep and whose polygraphic patterns has been compared with the effect of chloralose (MARCUS et al., 1967). The most interesting aspects of the C 4 compounds, and also of C 5 and C 6 compounds (valerate and caproate) (MATSUZAKI et al., 1964), is their PS-inducing effect in chronic *decorticate, 'mesencephalic* or *'pontine* cats (JOUVET et al., 1961; JOUVET, 1965 d; MATSUZAKI and TAKAGI, 1967b; TAKAGI and MATSUZAKI, 1968). Under appropriate conditions, these drugs are the *only ones known at the present time which always induce typical PS episodes.* In order to induce PS, the C 4, C 5 or C 6 compounds must be injected (50mg/kg at the earliest after a refractory period which is equal to half the duration of the PS period (i.e. 20 minutes in pontile cats whose periodicity is 40 minutes). If the injection is performed during the refractory period, there is no triggering of PS but only a short-lasting atonia of the neck without pontine PGO. The delay between the intravenous injection and the appearance of PS is shorter after C 6 than after C 5 or C 4, whereas C 3 compounds (propiolactone or propiobutyrate) have no effect (DELORME et al., 1966b). That C 4–C 6 compounds act at the level of the pons is well illustrated by the fact that they still induce PS in a pontile cat whereas they are ineffective after a medio-pontine transection (MATSUZAKI, 1969) or after a lesion of the dorso-lateral pontine tegmentum (JOUVET et al., 1961).

The C 4–C 6 compounds may counteract the inhibitory effect of atropine upon PS in pontile cats but are unable to induce PS during the time of the suppressor effect of MAO inhibitor (3–4 days after injection of 10 mg/kg of nialamide) (DELORME et al., 1966).

2. Mechanisms of Action

The numerous mechanisms of action which have been proposed to explain the effect of these drugs may be summarized as follows:

i) Direct action of the fatty acid anion upon nerve cells. The fatty acid salts would inhibit the metabolic activity of cerebral tissue as they do in muscle and yeast (SAMSON et al., 1965).

ii) Action upon cerebral metabolism by affecting the pentose shunt (see LABORIT, 1964) or the metabolism of GABA (MANDEL and GODIN, 1965; GODIN et al., 1968; ROTH and GIARMAN, 1969).

iii) GOH and GBL may also be the precursors of normal constituents of the brain since, despite some controversy, GOH is normally found in the brain (BESSMAN and FISHBEIN, 1963; GIARMAN and ROTH, 1964; ROTH and GIARMAN, 1970).

However, these two latter hypotheses are unlikely since C 6 compounds act much more rapidly than C 4 compounds.

iv) *Action upon Brain ACH and MA.* GOH increases the level of ACH in the region of the brain stem underlying the colliculi (i.e. the dorsal pontine tegmentum (GIARMAN and SCHMIDT, 1963) and increases brain DA either by stimulating its synthesis or blocking its release, whereas it has few significant actions upon endogenous levels of either NA or 5-HT (GESSA et al., 1968; ROTH and SUHR, 1970). The increase in DA is mainly observed in the striatum, both biochemically and with the histofluorescent technique (AGHAJANIAN and ROTH, 1970). Since it is possible that the locus coeruleus complex contains both ACH and dopamine (GERARDY et al., 1969; SHUTE and LEWIS, 1966) and since there is good evidence that these pontine neurons participate in the triggering of PS, the triggering effect of C 4–C 6 compounds might well be related to their regional action upon the pons. In the intact animal the effect upon striatal DA could well conceal the pontine effect. The intimate mechanism of action of C 4 compounds upon brain DA and ACH might also be due to their action upon the cell membrane.

v) *Action upon the Membrane.* Although the evidence is still indirect, the fatty acids may act at the membrane level, since it has been suggested by two different groups (DAHL, 1968; RIZZOLI et al., 1969; RIZZOLI and GAL-ZIGNA, 1970) that fatty acid anions may react with membrane lipids and interfere with the movement of critical inorganic ions. More specifically, it has been suggested that butyrate molecules may bind to the lecithin of synaptic structures, by a true absorption process. Butyrate is also capable of binding with 5-HT and DA by forming a molecular interaction complex when it passes through synaptic membranes. The formation of this complex may alter the equilibrium in the synaptic cleft and induce the secretion of additional transmitter, or it may interfere with ACH release (RIZZOLI and GALZIGNA, 1970) (such a hypothesis would not be in contradiction with the increase in DA synthesis).

The membrane hypothesis is a very interesting one, since it might also explain the absolute refractory period during which it is impossible to trigger, PS either pharmacologically or electrically (see JOUVET, 1962; MONTI, 1970) and when probably not enough new transmitter has been synthesized in the terminals. Once a critical level of transmitter has been reached, the fatty acid anions would be able to trigger the release of the transmitter before its physiological release. Whatever their intimate mechanism of action, the short chain fatty acids must certainly be taken into account when considering humoral factors which may facilitate the onset of sleep. Every one has experienced the sleep-inducing effect of a good meal with good wines (in fact, Bordeaux wines contain GBL!) (RIBEREAU-GAYON and SAPIS, 1965).

E. The Riddle of Ponto-Geniculo-Occipital Activity

In the cat, the central phasic electrical phenomena (PGO) which herald and accompany PS may be altered predictably by numerous drugs which act upon brain MA. However, the mechanisms of PGO activity still remain an unsolved and most challenging problem, yet this is probably one of the principal clues for the understanding of the function of PS. We shall first *summarize* briefly some of the phenomenological aspects of PGO activity before considering their anatomical organization and their possible relationship with brain MA.

1. Phenomenological Aspects

PGO activity was first discovered in the pontine reticular formation in the chronic mesencephalic cat during PS (JOUVET and MICHEL, 1959). Subsequently, monophasic waves were recorded from the lateral geniculate nucleus (MIKITEN et al., 1961; see BROOKS, 1967) and the occipital cortex (MOURET et al., 1963). Thence the name of "deep sleep waves" (BIZZI and BROOKS, 1963) or of "ponto-geniculo-occipital" activity which was coined to describe these phasic events (JEANNEROD et al., 1965 b). This abbreviation does not imply that the activity is propagated from the pons to the occipital cortex through the lateral geniculate, but it summarizes the three major localizations where high amplitude PGO can be recorded with macroelectrodes. PGO activity can also be recorded with macroelectrodes from the nuclei of the VI, VII and III nerves (COSTIN and HAFEMANN, 1970), superior colliculus and from the parieto-occipital cortex (with transcortical electrodes) [activation waves (CALVET et al., 1964)]. At the cellular level, it is evident from numerous studies that most of the neurons in the structures which have been investigated are under the influence of PGO activity during PS. In the lateral geniculate and occipital cortex most of the cells are activated during the first part of the wave (VALLEALA, 1967; JEANNEROD and PUTKONEN, 1970; McCARLEY and HOBSON, 1970). Even in regions where PGO activity is not recorded with macro-electrodes, more than 60% of the recorded neurons (in the thalamus) or associated area of the cortex (see complete references in BENOIT, 1971) are influenced (facilitated or inhibited) at the time of PGO activity. Hippocampal (BELUGOU et al., 1968) and callosal (BERLUCCHI, 1965) neurons are a very interesting exception since they are inactive during PS. PGO activity also influences the excitability of some relay nuclei and may alter the transmission of afferent stimuli in the brain. Paradoxically, PGO is accompanied by facilitation of the postsynaptic response in the lateral geniculate nuclei at a time when optic tract terminals are depolarized and when one would

16*

expect the presynaptic inhibition to block or to reduce synaptic transmission (Dagnino et al., 1965; Bizzi, 1966a, b; Sakakura and Iwama, 1966; Iwama et al., 1966; Baldissera et al., 1966; Carli et al., 1966; Sakakura, 1968; see Benoit, 1966, 1971). Similar observations have also been made at the level of the N. ventro postero lateralis thalami (Dagnino et al., 1966).

PGO activity is strongly correlated with the rapid eye movements of PS. This is well illustrated by the fact that every PGO discharge in the pontine reticular formation is followed by an EMG discharge in the external rectus (even in the absence of recorded eye movements with EOG techniques) (Michel et al., 1964). Geniculate or occipital PGO discharges do not depend upon the reticular or the muscular mechanism of the eye movements, since they persist for some days in the lateral geniculate nucleus and for many weeks in the occipital cortex after total removal of the contents of the orbits (Jeannerod et al., 1965 b; Brooks, 1967). Moreover geniculate PGO discharges precede or coincide with the onset of the eye movements (Kiyono and Jeannerod, 1967). These characteristics and the fact that cortical PGO are not influenced by darkness (Brooks, 1969) permit us to differentiate them from the so called "eye movement potentials" (EMP) which are observed during waking.

2. PGO and Eye Movement Potentials

When an animal is attentive and moves its eyes toward an external stimulus (orienting reaction), EMP can be recorded with macro-electrodes from the same region as PGO (Feldman and Cohen, 1968; Cohen and Feldman, 1968; Cohen et al., 1969; Brooks, 1969). However, numerous differences exist: geniculate EMP are of smaller amplitude (and distribution) than PGO (Brooks, 1969); the amplitude of cortical EMP is correlated with the velocity of the eye movement (Jeannerod and Sakai, 1970) and is decreased in the dark (Brooks, 1969); the coagulation of the optic nerve suppresses cortical EMP (Brooks, 1969); finally, geniculate EMP always follows eye movement (Cohen et al., 1969; Jeannerod and Sakai, 1970; Jeannerod and Putkonen, 1970) and have no or little action upon the transmission of evoked activity in the lateral geniculate (Kawamura and Marchiafava, 1968; Cohen and Feldman, 1969). EMP are considered as "corrolary discharges" (see Teuber, 1971) originating from the pontine oculomotor structures during the process of visual attention. Thus, they are observed only during cortical arousal in chronic condition.

In summary, although EMP and PGO may share some common mechanisms, two main characteristics enable us to differentiate between them:

a) geniculate EMP follow eye movements whereas geniculate PGO precede them (or coincide with them);

b) the amplitude of cortical EMP decreases in the dark whereas the amplitude of cortical PGO is unaffected.

3. Quantitative Aspect of PGO Activity

PGO never occur during waking under normal conditions. They may occur periodically during slow-wave sleep (slow sleep with phasic activity) (THOMAS and BENOIT, 1967). They are isolated, with a mean frequency of 10/min. PGO usually precede cortical activation of PS by 30–60 seconds. Their frequency increases to 60/min during PS. [Isolated PGO contribute 23 %, while double PGO contribute 24 % and bursts of PGO (more than five separated by less than 100 msec) contribute 53 % to the total number of PGO.] The average frequency of discharge is rather constant from one PS to the other and does not depend upon the duration of PS or the preceding or following interval (BUGUET, 1969). Under normal conditions, the total daily number of PGO is rather stable ($13\,000 \pm 1\,500$). This "daily quota" of PGO is apparently subject to long-term regulatory mechanisms. Indeed, when PS is suppressed for three days by instrumental methods (swimming-pool technique), few PGO occur during SWS (700/24 hours). During the rebound of PS which follows in the next two days, PGO occur during SWS and PS at a higher frequency and the "PGO debt" is almost totally repaid (75 to 90 %) taking into account the obligatory daily quota of 13000 PGO (DUSAN-PEYRETHON et al., 1967; see DEMENT et al., 1970a, b).

4. Anatomical Organization

That PGO depends upon a *pontine "pacemaker"* is proved by the following experiments:

i) Pontine PGO still persist during PS in the chronic pontile cat (see JOUVET, 1962). ii) Dorsal prepontine transection of the brain stem suppresses PGO in the lateral geniculate but not in the pons (HOBSON, 1965). iii) The stimulation of the pontine reticular formation during PS triggers lateral geniculate PGO (whereas the stimulation is not effective during waking or SWS). This fact was interpreted as a gating effect (BIZZI and BROOKS, 1962; BROOKS and BIZZI, 1963). The geniculate evoked PGO show "all-or-none-type" responses according to whether the intensity of the pontine stimulation is below or above "threshold" (MALCOLM et al., 1970). iv) The bilateral coagulation of the dorsolateral pontine tegmentum at the level of the "locus coeruleus complex" immediately and definitively suppresses the cortical and geniculate PGO (BUGUET, 1969; BUGUET et al., 1971). v) Finally, micro-injection of 6OHDA in this region subsequently suppresses PGO (BUGUET et al., 1971).

Thus, there is valid experimental evidence that the PGO waves are under the control of a group of neurons located in the dorso-lateral part of the pontine tegmentum. However, the *pattern* of PGO discharges is also under the influence of other structures. 1. Bilateral destruction of the medial and descending vestibular nuclei suppresses bursts of rapid eye movements and bursts of PGO in the lateral geniculate nucleus, whereas isolated PGO and eye movements still occur during PS (Morrison and Pompeiano, 1966). How the vestibular nuclei may influence geniculate activity is still obscure, since a total transection of the medial longitudinal fasciculus (the main ascending pathway from the vestibular nuclei) alters neither PGO activity nor bursts of rapid eye movements during PS, whereas the horizontal eye movements are suppressed during waking (Perenin and Jeannerod, 1971). 2. There is also evidence that a descending neo-cortical control upon PGO activity is acting upon the pontine "pacemaker" since the patterns of discharges of pontine PGO in a pontine or decorticate cat is totally different from that of a normal cat [bursts of 4–5 waves separated by silent intervals of 4–5 seconds (see Jouvet, 1962; Gadea-Ciria and Jouvet, 1971) (see Fig. 18)]. Finally, the organization of the ascending pathways mediating the PGO activity is still obscure. It is probable that at least two main pathways ascend from the pontine "pacemaker" since the almost total destruction of both lateral geniculate nuclei does not suppress cortical PGO (Hobson et al., 1969). But the topography of the ponto-geniculate and ponto-occipital pathways has not yet been mapped out. Curiously enough, macro-electrode recordings in the mesencephalic tegmentum between the pons and the lateral geniculate nucleus have consistently failed to record PGO activity during sleep (or under reserpine, see below).

5. PGO and Monoamines

Neuropharmacological Experiments

The first evidence that PGO could occur outside PS was obtained under mild pentobarbital anesthesia during which isolated PGO (or even bursts of PGO after PS deprivation) may occur periodically even in the total absence of other signs of PS (Jouvet and Delorme, 1965; Malcolm et al., 1970b). It was later shown that numerous drugs which act upon brain MA may totally dissociate the appearance of PGO from PS:

The Reserpinic Syndrome. Reserpine (0.5–1 mg/kg), after a latency of 60–90 minutes, triggers the continuous appearance of typical PGO at a frequency of 15–30/min (Delorme et al., 1965; Jeannerod, 1965; Brooks and Gershon, 1971). Those PGO_R[2] are accompanied by discharges in the external rectus, and small lateral eye movements. PGO_R thereafter appear continuously, during either waking or cortical synchronization (Fig. 19). When PS reoccurs (usually after 24 hours) the PGO frequency increases, but the end of the PS

---

2 We shall now call PGO occurring during PS, PGO_S, and PGO occurring after reserpine, PGO_R.

Fig. 18A and B. *Control of the pontine PGO activity by supra pontine structures.* A Pattern of the discharge of pontine PGO activity recorded in the VIth nucleus during paradoxical sleep in a normal cat. B Same cat: 8 days after total removal of the brain rostral to the pons, during an episode of paradoxical sleep. The pattern of the discharge is strikingly different. Calibration: 1 second. 50 microvolts. (GADEA-CIRIA and JOUVET, 1971)

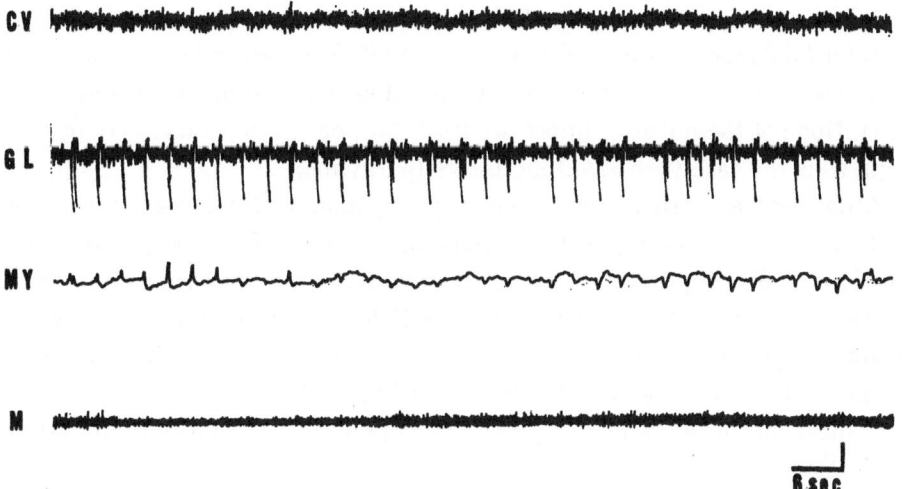

Fig. 19. *The reserpinic syndrome in the cat.* Regular discharges of PGO in the lateral geniculate (*GL*) accompanied by small lateral eye movements (*MY*), 6 hours after the intraperitoneal injection of 0.5 mg/kg of reserpine. The discharge of PGO is accompanied by a polygraphic pattern of waking with low voltage fast cortical activity (*CV*) and persistence of the EMG of the neck (*M*). Calibration: 6 sec. 50 microvolts. (DELORME, 1966)

episode is followed within minutes by a total suppression of the waves. The recording of PGO_R in acute experiments with curarized animals made it possible to study their distribution and origin in more detail than in chronic experiments (JEANNEROD, 1965; see JOUVET, 1967b). The cortical or geniculate PGO_R are not suppressed by retropontine transection or the total destruction of the vestibular or raphe nuclei. But they are suppressed by prepontine transection or by a large coagulation of the dorso-lateral pontile tegmentum. The lesion which suppresses PGO_R is similar to the lesion which suppresses PGO_S. Thus, it is likely that the "pacemaker" of PGO_R is the same as for PGO_S. Furthermore, numerous experiments have shown that PGO_R, like to PGO_S, induce excitability changes in the reticular formation and the visual

system (JEANNEROD et al., 1965a; BENOIT, 1967; JEANNEROD, 1967; JEAN-
NEROD and KIYONO, 1969). Nevertheless, it is possible that some EMP may
contribute to the discharge of phasic waves under reserpine. They can be
distinguished by their decrease in the dark (BROOKS and GERSHON, 1971)
and their relationship with eye movements as shown by the composite histo-
gram of the latency between phasic waves in the lateral geniculate and the
onset of eye movements (KIYONO and JEANNEROD, 1967). It is likely that
the phasic geniculate activity which is triggered by external stimuli or stimulation of central afferent pathways also belongs to the EMP (JEANNEROD and
KIYONO, 1969). The reserpinic syndrome has proved invaluable in the screening
of drugs which act upon the PGO_S activity in the cat. At the present time,
all the drugs which have been found to suppress immediately PGO_R also
suppress PGO_S (and, of course, PS).

PGO_R are suppressed by the following drugs: 5-HTP (20–30 mg/kg), MAO
inhibitor (nialamide, pargyline) (DELORME, 1966; BROOKS and GERSHON, 1969),
LSD, methysergide, α-methyl-DOPA and chlorimipramine (FROMENT et al.,
1971). None of these drugs suppresses EMP but, on the contrary, may increase
them, as is the case with MAO inhibitors (pargyline).

After reserpine, other drugs: p-chlorophenylalanine (DELORME et al., 1966;
see DEMENT et al., 1970), p-chloromethamphetamine (DELORME et al., 1966),
tropolone + DOPA (JONES, 1970, 1971) and finally 6-hydroxydopamine intraventricularly injected (LAGUZZI et al., 1971) have been found to trigger continuous "PGO_R-like" activity outside PS. A consideration of the known biochemical effects of drugs which trigger continuous PGO activity as compared
with those of drugs which suppress it supports the following hypothesis
(Table 1):

5-HT appears to play an inhibitory role in the triggering of PGO. Thus,
a decrease in 5-HT at the synaptic cleft would "open the gate" of the "gating
effect" (BROOKS and GERSHON, 1971). This might explain why PGO appears
continuously after reserpine, p-chlorophenylalanine, p-chloromethamphetamine
or lesion of the raphe system. On the other hand, an increase in 5-HT would
"close the gate": this would explain the disappearance of PGO after 5-HTP,
chlorimipramine, and in some cases after MAO inhibition. If it is accepted
that LSD and methysergide also act specifically at serotoninergic receptors
(ANDÉN et al., 1968; BOAKES et al., 1970a, b), such a hypothesis is still possible.
However, the fact that the decrease in 5-HT may be responsible for "opening
the gate", does not explain the origin of the PGO discharges. Two main
hypotheses are thus possible: 1. PGO result from tonic depolarization which
would permit the firing of the interneurons of the pontine reticular formation
when receiving random bombardment of afferent impulses (JEANNEROD and
KIYONO, 1969). 2. Some specific transmitter (tonically inhibited by 5-HT)
would be released at the presynaptic cleft.

Table 1. *Experimental alterations of the PGO activity*

	5-HT	5-HIAA	CA	CA (MAO)
I. Facilitation of the PGO activity				
Reserpine	—	+	—	+
P-chlorophenylalanine	—	—	=	.
P-chloromethamphetamine	—	—	.	.
Lesion raphe	—	—	=	.
60HDA (1)	—	—	+	+?
II. Suppression of PGO_R, PGO_S, and of PS				
MAO inhibitor	+	—	+	—
αMDOPA	.	.	—	—
5-HTP (2)	+	+	.	.
Total lesion locus coeruleus and sub-coeruleus	=	=	—	—
60HDA (3)	—	.	—	—
LSD	.	.	.	.
Methysergide	.	.	.	.

In this table, the drugs have been classified according to their action upon PGO activity in the cat.

The first group induces the appearance of PGO outside PS (i.e. PGO appear continuously either during waking or during cortical synchronization). The second group totally suppresses PS and PGO in normal conditions. It also suppresses PGO_R induced by a previous injection of reserpine.

The hypothetical concentration of transmitter or its metabolites at the synaptic cleft (according to current theory) is summarized by the (+): increase, (—): decrease, (.): unknown, (=): no change.

(1) A reserpinic syndrome follows the injection of 60HDA intraventricularly and lasts for 24–48 hours. It may be due to the release of CA from the synaptic vesicles.

(2) Suppression of PGO is obtained after mild or large doses (10—50 mg/kg).

(3) An important decrease of PGO during PS appears secondarily (after 48 hours) following 60HDA intraventricularly.

Although many interpretations are possible, and many unknown transmitters may be involved, it seems possible that 5-HT may exert an inhibitory role upon PGO whereas CA (or possibly their desaminated metabolites) may have a facilitatory action (see Fig. 20).

Hence, each PGO would be the result of the release of a quantum of the transmitter. Four main facts favor this hypothesis:

a) The "all-or-none" aspect of PGO also suggests some type of quantal release (MALCOLM et al., 1970b).

b) "Specific" destruction of CA-containing neurons by in-situ injections of 60HDA in the latero-dorsal pontine tegmentum suppress PGO secondarily and totally. Intraventricular injections of 60HDA in the cat result first in a transitory reserpinic syndrome, followed by a very drastic reduction of both PS and of the frequency of PGO during PS (LAGUZZI et al., 1971). Thus, the CA-containing neurons of the pontine tegmentum appear to be likely candidates for the release of this transmitter.

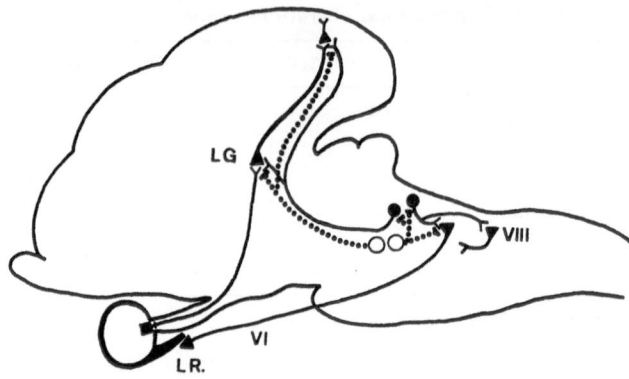

Fig. 20. *Hypothetical mechanism of the PGO activity*. CA-containing neurons (black dots) located in the locus coeruleus complex might be responsible for the quantal release of some deaminated metabolites of CA which trigger a postsynaptic response in the oculomotor nuclei (*VI*) or in the vestibular nuclei (*VIII*) leading to isolated or bursts of rapid eye movements during PS. Similar events occur in the lateral geniculate (*LG*) or in the occipital cortex. During waking and slow-wave sleep serotonin-containing neurons (white dots) exert a tonic inhibition upon the CA neurons, either at the level of the perikarya or the terminals. *LR* lateral rectus. It is also possible that PGO activity might be transmitted from the interneurons of the pons to lateral geniculate and cortex through nonaminergic pathways

c) After a prolonged reserpinic syndrome (obtained with reserpine, PCPA or DOPA associated with tropolone) the frequency of PGO during the recovery of PS is always decreased as if the newly synthesized transmitter had been depleted.

d) Finally, the long-term regulatory mechanisms which maintain the daily quota of PGO constant might be explained more easily if the PGO result from quantal release of a transmitter rather than from some unspecific gating effect permitting tonic depolarization of pontine neurons.

If this hypothesis is correct, deaminated metabolites of CA would be the most likely candidates for the transmitter involved in the triggering of PGO (see Dusan-Peyrethon, 1968; Jones, 1970) (as shown in Table 1). However, this hypothesis does not explain why αMPT does not suppress either PGO_R or PGO_S. The negative results obtained with αMPT have been discussed previously. These negative results are, however, compensated by the fact that α-methyl-DOPA, disulfiram, and 6OHDA suppress or decrease PGO. Needless to say, much work remains to be done concerning the mechanisms of PGO activity. We have tentatively put forward a working hypothesis in Fig. 20 synthesizing most of the results obtained from neurophysiological and neuropharmacological experiments.

V. A Monoaminergic Theory of the Sleep-Waking Cycle: A Synthesis

In this final synthesis, I shall first briefly summarize a theory according to which 5-HT neurons are involved in the sleep mechanism, whereas CA-containing neurons are responsible for both the "executive mechanisms" of PS

and the maintenance of behavioral and EEG waking (Fig. 21). I shall subsequently test this theory against the neurophysiological data which have
been reviewed by MORUZZI (this journal).

1. 5-HT and Sleep

According to this theory, SWS is initiated by the release of 5-HT at some
central serotoninergic synapses. Thus the sleep system would be composed
of some of the 5-HT-containing perikarya located in the raphe system. 5-HT
neurons located mostly in the caudal raphe (N. raphe pontis, N. raphe magnus,
N. raphe pallidus, N. raphe obscurus) would be responsible for the priming
of PS and would send terminals to the caudal two-thirds of the complex
of the locus coeruleus responsible for the "executive mechanism" of PS.
On the other hand, 5-HT perikarya located mostly in the anterior part of
the raphe system (nucleus centralis superior, N. raphe dorsalis) would be
responsible for SWS. Serotoninergic terminals exist at different levels of the
brain: in the spinal cord, the N. of Edinger-Westphal, the substantia nigra,
the mesencephalic reticular formation, the anterior hypothalamus, the preoptic area and the gyrus cingulate, etc. It is likely that the release and possible
binding of 5-HT upon the postsynaptic receptors might be responsible for
both behavioral (miosis, decrease of muscle tone) and EEG aspects of SWS
(cortical synchronization). It is also possible that 5-HT terminals might interfere directly with the presynaptic CA mechanism responsible for the tonic
maintenance of waking, either at the level of the CA-containing perikarya
of the anterior part of the coeruleus complex, or at the level of group A 8
of the mesencephalic tegmentum. The ascending and descending pathways
of the 5-HT system are not yet mapped out with accuracy. Some neurons
may follow the medial forebrain bundle in close proximity to ascending NA
and DA neurons but it is likely that a more dorsal 5-HT bundle exists in the
midbrain tegmentum, since lesions in this area decrease telencephalic 5-HT.
On the other hand, it is possible that some descending pathways to the spinal
cord may contribute to the modulation of the postural tone during SWS.

*Summing up, according to this theory, caudal raphe neurons play a role
in priming PS mechanisms whereas rostral raphe neurons are involved in behavioral and EEG aspects of SWS.*

2. Executive Mechanisms of Paradoxical Sleep

Most of these mechanisms are triggered from the caudal two-thirds of
the N. locus coeruleus complex (N. locus coeruleus, subcoeruleus, and possibly
N. parabrachialis medialis). The caudal third of N. locus coeruleus is responsible for the CA mechanisms involved in the control of the total inhibition
of muscle tone. Such a control may be direct, since heavy projections from

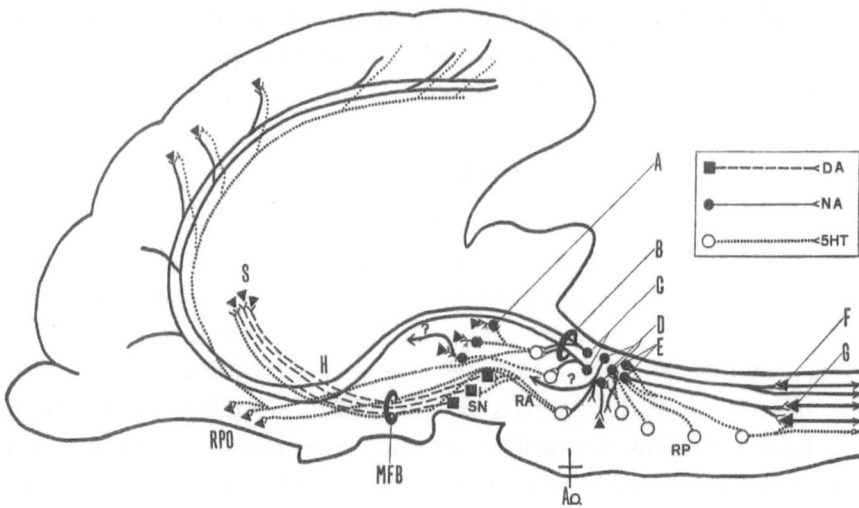

Fig. 21. *Organization of the MA systems responsible for the sleep-waking cycle.* On a sagittal map of the brain of the cat (Ao is the reference Horsley Clark zero), are highly schematized some possible organizations of the MA systems. The posterior or caudal raphe system (*RP*) is responsible for the priming of PS and sends terminals to the locus coeruleus complex (*CDE*). The caudal part of the locus coeruleus (*E*) is responsible for the CA mechanisms which are responsible for the total inhibition of muscle tone, either by acting upon n. nervi accessorii (*F*) which innervates the neck muscle, or by acting upon the origin of the reticulo-spinal tract (*G*). The medial part of the locus coeruleus and subcoeruleus (*D*) is responsible for the innervation of the reticular formation of the pons and for the triggering of PGO activity and rapid eye movements during PS (see Fig. 20). From the anterior part of the locus coeruleus (*C*) starts the ascending NA bundle (*B*). This bundle is responsible for an important NA innervation of the tele-diencephalon since its destruction is followed by a 50% decrease of NA. It is also possible that collaterals from this bundle may act upon the anterior raphe system (*RA*) since there is an increase of tryptophan and 5-HIAA after lesion in B (see Fig. 3). Thus the anterior part of the locus coeruleus is concerned with the control of waking (and not with PS). The anterior or rostral raphe (*RA*) is responsible for the maintenance of slow-wave sleep, either by acting upon the waking system [substantia nigra (*SN*) or group A 6 and A 8 of the Ponto-mesencephalic tegmentum (*A*) (*C*)], or by acting upon effector mechanisms located in the preoptic area (*RPO*). Some 5-HT neurons ascend through the medial forebrain bundle (*MFB*) but others certainly ascend more dorsally in the cat. EEG arousal is tonically prolonged by the CA mechanisms located in group A 6 and A 8 (*A*). (*C*). These groups contribute to the CA innervation of the adjacent mesencephalic reticular formation. (Note that a lesion destroying group A 8 may also destroy the dorsal NA bundle in the mesencephalic tegmentum.) Behavioral waking is under the influence of the dopaminergic nigro-striatal system which ascends from the substantia nigra (*SN*) to the striatum (*S*) through the medial forebrain bundle. The lesion of this system decreases DA in the striatum and strongly impairs behavioral alertness, whereas it does not interfere with the cortical activity. Lesions of the lateral and medial hypothalamus (*H*) may interfere with both ascending NA and DA systems and thus alter EEG and behavioral waking but this does not imply that the hypothalamus is a waking center. The ventral NA pathway originating from the medulla is not represented in this schema. (See text for further explanations)

the N. locus coeruleus to the nuc. N. accessori (responsible for the innervation of neck muscles) have been demonstrated in the rat (OLSON and FUXE, 1971). This control may also be exerted via the bulbar reticular formation and the reticulo-spinal tract (see POMPEIANO, 1970).

The medial third of the "coeruleus complex" would correspond to the "pontine pacemaker" of PGO activity and is responsible for both the phasic and the tonic ascending components of PS. From this area, CA-containing axons might be responsible either for triggering isolated eye movements (terminals located in the oculomotor region of the pons and mesencephalon) or for the bursts of eye movements (terminals impinging upon medial and descending vestibular nuclei; see MORRISON and POMPEIANO, 1966). The pathways and the mechanisms responsible for cortical activation during PS are not yet known. It is unlikely that NA axons ascending in the dorsal NA bundle participate in the cortical activation of PS, since after their destruction, which decreases cortical arousal, there is still cortical activation and geniculate PGO during PS (PETITJEAN et al., 1970).

Possible 5-HT—5-HT Interactions in the Raphe System. It is possible that short 5-HT axons might effect some interaction between caudal and rostral raphe systems. Thus, the lesion of the anterior raphe is followed for 2–3 days by continuous waking interrupted by narcoleptic attacks probably triggered by priming mechanisms of the caudal raphe; this is followed subsequently by a decrease in PS (RENAULT, 1967). It is thus possible that any large lesion bearing upon either 5-HT perikarya or terminals might secondarily affect the activity of other 5-HT perikarya via some transynaptic mechanism (see p. 265).

Possible Interaction between 5-HT Priming Mechanisms and the Executive Mechanism of PS. A two-way interaction exists between the raphe system and the locus coeruleus complex, since pathways and terminals originating from the raphe system have been demonstrated going in the direction of and terminating in close vicinity to the locus coeruleus, and since medial fibers projecting from the locus coeruleus terminate in the raphe system (OLSON and FUXE, 1971; LOIZOU, 1970). Although no experimental proof exists, the following biochemical mechanism could explain some of the unsolved problems concerning cortical activation during PS. If it is conceivable that 5-HT priming mechanisms might trigger the locus coeruleus complex, one may also speculate that this complex, in turn, might affect 5-HT neurons. Since the PGO activity which heralds and accompanies PS appears to be associated with a decrease of 5-HT in the terminals (see above p. 248), it is possible that the terminals from the locus coeruleus may act upon 5-HT perikarya in order to suppress the release of 5-HT or start the active process of reuptake. Then the fast cortical activity which is observed during PS would be due to the combined effect of possible ascending CA or ACH mechanisms and the phasic decrease of 5-HT. The fact that chlorimipramine, which "selectively" suppresses the reuptake of 5-HT, also suppresses PS is in agreement with the hypothesis. If this hypothesis is true, then the increased firing of postulated 5-HT neurons from the raphe system, as demonstrated with macro-micro electrode techniques by BALZANO and JEANNEROD (1970), would be concomitant with some phasic reuptake mechanism of 5-HT at the terminal level, triggered from the cell bodies. *Alternation between SWS and PS would thus appear to be the alternation between the release of 5-HT in the synaptic cleft, its possible binding to the receptor, and its active reuptake.* It is quite conceivable that the increase of 5-HT turnover might accelerate this process and increase the frequency of PS episodes observed during the rebound which follows PS deprivation. Only the determination of extraneuronal and intra-neuronal catabolism of 5-HT, together with the measurement of the uptake of exogenous labelled 5-HT during SWS and PS in chronic experiments might verify this hypothesis.

3. CA and Waking

At least three different systems of CA-containing neurons are involved in the maintenance of tonic cortical or behavioral arousal:

i) The NA Pontile Waking System

These neurons are situated in the anterior part of the locus coeruleus. They send axons to the dorsal NA bundle which ascends near the grey matter, and collaterals to the anterior part of the raphe system (Loizou, 1969; Maeda, 1970, Olson and Fuxe, 1971). They ascend further in the mesencephalon to contribute to the innervation of the telencephalon. The destruction of these cells, or of the dorsal NA bundle at the level of the isthmus, is followed by a significant increase of both SWS and PS, whereas biochemical analysis reveals a decrease of NA in the tele-diencephalon and an increase of both tryptophan and 5-HIAA (Petitjean et al., 1970). It is thus possible that some collaterals which issue from the anterior N. locus coeruleus may exert a control upon the rostral raphe system during waking. The suppression of this control by a lesion would cause an increase in activity of the raphe system, leading to a temporary increase of both SWS and PS. It must be pointed out that, after such localized lesions, both behavioral and EEG arousal are normal but their spontaneous occurrence is much less frequent.

ii) The CA Mesencephalic Waking System

This system corresponds to group A 8, which sends numerous fibers with varicosities to the adjacent mesencephalic reticular formation. The destruction of this system decreases cortical arousal. The time originally spent in tonic arousal is occupied by a deactivated EEG. Nevertheless, phasic cortical arousal and waking behavior are still possible. But they are mostly contingent upon the presentation of external stimuli, and the cortical arousal does not long outlast external stimulation. Lesions of this group are accompanied by a decrease of both DA and NA and also a small decrease of 5-HT (which may indicate that ascending 5-HT fibers ascend through this area). There is no increase in PS. Thus, the result of the destruction of the mesencephalic CA neurons is a decrease of EEG waking but not a true hypersomnia. These mesencephalic CA neurons appear to be implicated in the maintenance of the tonic EEG activation mediated by the mesencephalic reticular formation rather than in a possible control of the raphe system. Finally, as shown by numerous pharmacological experiments and determination of the release of ACH, it is likely that cortical cholinergic mechanisms constitute the final common link mediating cortical arousal.

Destruction of the Substantia Nigra

in the cat strongly impairs the possibility of waking behavior (orienting reaction toward external stimuli, spontaneous motility). Depending upon the size of the lesion in the substantia nigra and on the intensity of the decrease of striatal DA, the behavior of the cat may vary from catatonia with long-lasting freezing immobility to almost comatose behavior. On the other hand, there is no significant alteration of the EEG, and long-lasting cortical arousal is possible after sensory stimulation, even in the absence of behavioral waking. According to the projection of this system established by ANDÉN et al. (1964), the DA fibers ascend in the lateral part of the medial forebrain bundle and the posterior lateral hypothalamus. Thus, it is likely that any lesion impinging upon this system might also induce comatose behavior.

iii) Possible Interaction between Sleep and Waking Systems

There exist many anatomical possibilities for such an interaction, since CA terminals are found in close proximity to the raphe system, and vice versa. On the one hand, the release of 5-HT at the terminals impinging directly upon CA-containing perikarya or upon CA terminals may be responsible for the onset of sleep. On the other hand, the NA terminals from the anterior part of the locus coeruleus may tonically inhibit the anterior raphe system during waking, since their destruction is followed by transient hypersomnia.

The problem of long-term regulatory mechanisms is still open. It is well known that, after total deprivation of sleep, the duration of the recovery SWS is not proportional to the lack of sleep and that after 300 hours of sleep deprivation, in man, one "good night's sleep" (14–16 hours) is sufficient to pay the debt of SWS. However, the rebound of PS may continue for days or weeks (GULEVICH et al., 1966).

The same is true after "selective" deprivation of PS (DEMENT, 1960; VIMONT et al., 1966). Since there are many indirect experimental data which favor the increase of 5-HT turnover in the augmentation of PS (HERY et al., 1970), it is possible that some enzymatic induction may occur as a consequence of enforced waking, either at the level of the entry of tryptophan into the cell or at the tryptophan hydroxylase level; these are the principal regulatory steps of 5-HT synthesis. The increase of enzymatic activity would favor an increased turnover of 5-HT during sleep (and thus increase the priming mechanisms of PS). On the other hand, a possible increase of CA turnover during the rebound of PS (PUJOL et al., 1968) might also be responsible for the increased frequency of PGO activity.

iv) The Problem of the Onset of Sleep

The waking system is almost indefatigable, since a subtotal insomnia of at least 2 weeks duration may be obtained after raphe lesion (Renault, 1967). Thus the onset of sleep has to be triggered by the active triggering of the 5-HT sleep system (and not only by a possible circadian dampening of the turnover of the CA waking neurons). It must be recognized that, besides the possible triggering of 5-HT neurons by the waking system, almost nothing is known concerning the mechanisms which activate the 5-HT system. It is possible that 5-HT-5-HT synapses may exist from N. paragigantocellularis to the medial raphe, whereas raphe cells (but not necessarily 5-HT peri-karya) may apparently also respond to iontophoretic administration of NA and ACH (Couch, 1970). Thus the 5-HT perikarya could be either serotoni-ceptive, catecholaminoceptive or cholinoceptive. In this case, the "doors of SWS" might also be opened by other transmitters or neural systems whose anatomical organization is still obscure. Is the raphe system connected to the synchronizing lower brain stem structures? (see Moruzzi, this journal). The fact that stimulation of the vago-aortic nerves is still able to induce phasic cortical synchronization (and miosis) after destruction of the caudal raphe system or inhibition of 5-HT synthesis with PCPA (Puizillout and Ternaux, 1971) is an indication that true neural mechanisms are able to trigger EEG signs of sleep even in the possible absence of 5-HT in the terminals. It remains to be shown, however, whether true physiological sleep, outlasting the duration of the stimulus, can be obtained in these conditions. In any case, some signi-ficant advances in the knowledge of the process of the onset of sleep will probably come in the near future from the study of the interactions between the raphe system, and the nucleus of the solitary tract or its afferents. Finally, some relationships are also possible between the raphe system and the area postrema, which plays a role in the synchronizing mechanism. According to Koella (1969), 5-HT could act via the area postrema as an agent which facilitates or increases "gain" in the "transfer function" located in the solitary tract nucleus. This would enhance the inhibitory feedback from this nucleus and depress the arousal level and possibly enhance the output from the "thalamic hypnogenic area". Such a hypothesis is difficult to reconcile with the fact that the destruction of the thalamus (Naquet et al., 1965; Angeleri et al., 1969) does not interfere significantly with either state of sleep.

Besides true neural mechanisms, it is also quite likely that the triggering of the 5-HT sleep system could be facilitated (at the perikarya or at the terminal level) by true humoral influences. Among them, the level of blood tryptophan might play a role, since tryptophan hydroxylase is not a true rate-limiting enzyme and it is possible that a study of the intimate mechanism regulating tryptophan uptake by the synaptosomes might elucidate some forms of pathological insomnia or hypersomnia. It must be admitted that

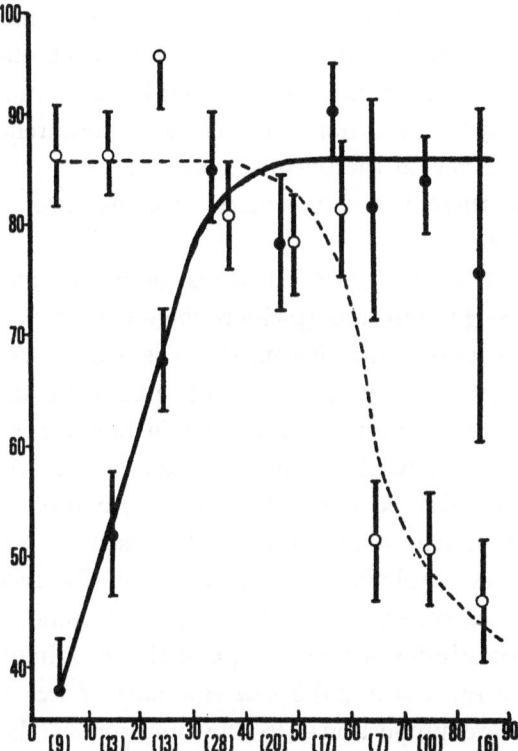

Fig. 22. Ordinates: percentage of 5-HT (black circle) and NA (white circle) in the telediencephalon relative to control cat. (Mean and SEM.) Abscissae: percentage of cortical synchronization during the 10–13 days of survival after the lesion. The numbers in brackets refer to the number of cats in each group. Further explanation in the text. The absolute values (100%) for 50 control cats are respectively 622 ng/g ± 22 for 5-HT and 399 ng/g ± 10 for NA. (JOUVET, 1971)

considerable work is needed to clarify the mechanism by which 5-HT neurons are activated.

4. Summary

Two antagonistic systems are strongly implicated in the mechanisms of cortical synchronization and desynchronization (JOUVET, 1971). This is illustrated in Fig. 22, which summarizes all the biochemical data obtained from 133 cats (which were all operated, recorded, sacrificed and biochemically analyzed by the same technique). They were subjected to lesions of the midline raphe, or of the bulbar, pontine, mesencephalic tegmentum or of the substantia nigra. Control lesions were also made outside MA-containing neurons, and twelve sham-operated animals served as controls (groups of animals are classified according to the percentage of EEG synchronization during the 10–13 days of postoperative survival and the mean value of the percentage of 5-HT and noradrenaline in the telencephalon and diencephalon is given for each group). It is clear that 5-HT and noradrenaline fit two opposite curves: cats with insomnia (less than 30% of cortical synchronization) have a lowered tele-diencephalic 5-HT and normal telencephalic noradrenaline content. Cats with normal or subnormal levels of synchronization (30–60%) have normal or subnormal values for both 5-HT and noradrenaline, whereas cats with increased synchronization have decreased noradrenaline with normal or subnormal 5-HT levels in the telediencephalon.

i) According to these two antagonistic systems, two different kinds of insomnia are possible:

a) Insomnia after lesion of 5-HT neurons (perikarya, axons, terminals) or after inhibition of 5-HT synthesis (p-chlorophenylalanine). The common biochemical mechanism underlying this insomnia is a decrease in central 5-HT turnover. In most cases, this insomnia is accompanied by an increase in PGO activity in the cat (the subjective equivalent in humans is not yet known). Such insomnia is *never* followed by any secondary rebound of SWS nor of PS.

b) Insomnia may also be induced by the activation of the ascending noradrenergic ponto-mesencephalic system. This insomnia is induced by direct stimulation of the mesencephalic reticular formation (Frederikson and Hobson, 1970), by nociceptive stimuli, or by drugs (amphetamine) and is either short-lasting, or necessitates increasing quantities of stimuli or drugs. It is *always followed by a rebound of SWS and especially of PS*, the intensity and duration of which are proportional to the duration of the insomnia. This suggests a possible long-term regulation of biosynthesis and turnover of 5-HT in the serotoninergic neurons by the increased activity of noradrenergic neurons.

Finally, both forms of insomnia have a common characteristic. *They can be reversed immediately to sedation by inhibition of the synthesis of catecholamines by α-methyl-para-tyrosine which decreases the turnover of central catecholaminergic neurons.*

ii) Two different forms of reduction of waking are also possible:

a) True *hypersomnia* with increases of both SWS and PS. Such hypersomnia (above control level) has been obtained with small doses of exogenous 5-HTP but only if endogenous 5-HTP has been utilized after the inhibition by PCPA of tryptophan hydroxylase. Transitory hypersomnia with more than 80% total sleep (and 20–25% PS) may also be observed after limited lesions of the dorsal NA bundle at the ponto-mesencephalic junction. The fact that this insomnia is accompanied by an increase of telediencephalic 5-HIAA and tryptophan suggests that there is increased turnover of central serotoninergic neurons and that some noradrenergic neurons might control the synthesis of 5-HT in the anterior part of the raphe system. Theoretically, this form of hypersomnia should be suppressed by PCPA.

b) *An increase of cortical synchronization* (but not true hypersomnia) is observed after lesion of the mesencephalic catecholaminergic neurons (group A8 and dorsal NA bundle). After such a lesion, the time originally spent in waking (50%) is occupied by an increase of neither SWS nor PS, but by a "deactivated EEG".

The reduction of waking which is observed after inhibition of the synthesis of catecholamines by α-methyl-para-tyrosine, inhibition of DA-β-hydroxylase or by the use of α-methyl-DOPA cannot be considered true hypersomnia, since the increased synchronization is not accompanied by a parallel increase in PS (but usually by a decrease or total suppression). From these data some rather logical teleological deductions can be made. Waking mechanisms are much more powerful than sleep-inducing mechanisms. Thus, total insomnia can be immediately reversed to sedation by inhibition of the synthesis of catecholamines (αMPT), or by lesioning the waking system (RENAULT, 1967). On the other hand, it is very difficult to counteract the reduced waking due to lesion or to inhibition of the synthesis of CA neurons by destroying 5-HT neurons. This predominance of the waking system is well illustrated by the fact that it is easier and faster to wake a sleeping cat than to induce sleep in an excited cat. One wonders that would have happened to evolution if these systems had worked the other way!

5. The "Monoamine Theory" and the Effects of Brain Lesions upon the Sleep-Waking Cycle

The monoamine theory can be briefly summarized again as follows: two antagonistic histochemical sleep and waking systems, whose perikarya are located in the brain stem but whose terminals are disseminated throughout the brain, are responsible for the alternation of the sleep-waking cycle. 5-HT neurons are responsible for SWS and for priming PS, whereas ponto-mesencephalic CA-containing neurons are responsible for both the executive mechanisms of PS and the *maintenance* of tonic behavioral and EEG arousal. *Thus, each level of integration of the central nervous system, from the pontile cat to the normal animal, is submitted to the presynaptic action of these neurotransmitters. The behavioral and EEG aspects of the sleep-waking cycle thus depend upon the organization of the postsynaptic efferent neurons:* i.e. sleep and waking behavior will be more primitive in the pontile cat than in the mesencephalic or decorticate cat. *But, provided that active terminals from both systems are present in a truncated brain stem preparation, there should be, at least biochemically, some sleep-waking cycle.* This theory also states explicitly that PS depends upon some priming 5-HT mechanism, and that SWS and PS must be in some way correlated. This theory emphasizes the crucial importance of the brain stem for the modulation of the sleep-waking cycle. Since other structures, mainly concentrated in the hypothalamus and the basal forebrain area, may also be involved in the regulation of sleep-waking (see MORUZZI, this journal). I shall review successively the sleep-waking cycle after total brain stem transections,

17*

and the experimental insomnia or coma induced by localized encephalic lesions (Fig. 23).

i) The Sleep-Waking Cycle after Brain Stem Transections

a) Caudo-Pontine Transection (Jouvet, 1962). Only two chronic preparations survived such a transection for 6–7 days. Caudally, they did not exhibit any periodic disappearance of muscle tone (this can be explained by the fact that the locus coeruleus complex was situated rostrally to the transection). Rostrally, these cats exhibited well recognizable SWS and waking behavioral and EEG patterns and a "stade d'interprétation difficile" with low-voltage fast activity, miosis and some spontaneous eye movement. It is possible that this stage might have been cerebral PS but PGO activity, which could identify this state in the lateral geniculate nucleus, was not investigated at this time.

b) Medio-Pontine Preparation (Batini et al., 1959; see Moruzzi, this journal). The increased waking of this preparation is a well established fact. It can be explained by the fact that about two-thirds of the raphe system is situated caudal to the level of the transection.

In fact, selective lesions of the raphe system caudal to the level of the medio-pontine transection induce during the first three days about the same amount of insomnia as the medio-pontine transection (Renault, 1967). Since the medio-pontine transection destroys at least the caudal half of the locus coeruleus complex, it is unlikely that, caudally, any periodical decrease of muscle tone related to the PS "executive mechanisms" may appear. Besides its effect on sleep mechanisms, the medio-pontine preparation also serves to delimit the most caudal transection compatible with tonic arousal or increased waking. Indeed, in such a case, the transection should pass caudal to the most caudal "waking" neurons. This is exactly the case if one considers that the rostral third of the locus coeruleus complex (situated in the Horsley-Clark plane P2P1) is situated just in front of the transection. On the other hand, the medio-pontine transection destroys the ventral NA bundle originating from groups A1–A4 of NA-containing neurons in the medulla. This suggests, that this ascending component of NA neurons does not participate in the maintenance of tonic EEG arousal but is probably concerned with "reward mechanism" superimposed upon waking behavior.

c) Rostro-Pontine Transection: Pontile Cat. It is easily understandable that PS occurs periodically in these preparations since the essential priming and executive structures are located below the transection. "SWS" exists at the biochemical level but cannot be recognized since no postsynaptic effector cells are present to identify it on EEG and behavioral criteria: indeed, the slowing of the electrical activity of the brain stem, which depends upon descending cortical influences (see Jouvet, 1962 and Moruzzi, this review), is suppressed by the transection. On the other hand, the N. of Edinger-

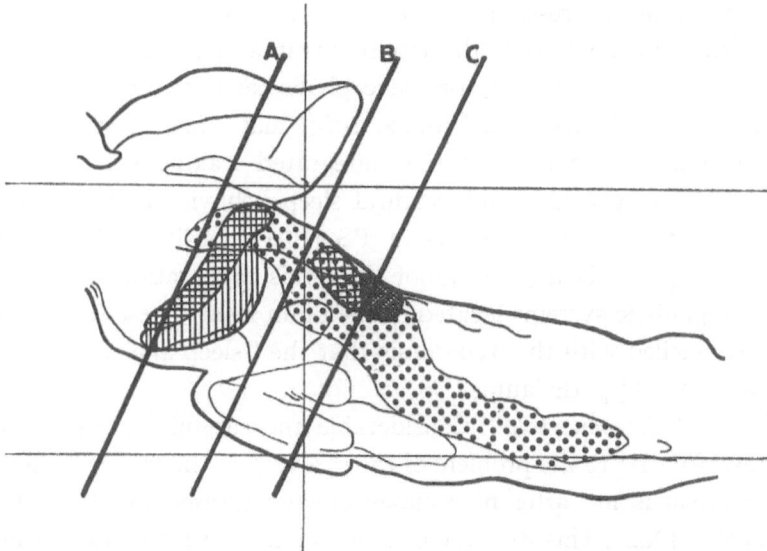

Fig. 23. Upon a sagittal section of the brain stem (the vertical line represents the frontal plane zero, the horizontal lines represent the horizontal planes HC 0 and HC –10), are projected schematically the serotonin-containing neurons of the raphe system (in dots) which appear to be responsible for sleep. The noradrenaline-containing neurons are represented by crossed lines. The neurons which are responsible for cortical desynchronization (anterior part of the locus coeruleus) in the pontine tegmentum and group A 8 of the mesencephalic tegmentum are represented by lighter crosses than the caudal part of the locus coeruleus whose destruction suppresses muscular inhibition during PS. In vertical hatching: projection of the dopamine-containing neurons of the substantia nigra.—Line A represents the intercollicular transection of the brain (cerveau isolé) which totally suppresses cortical activation.—Line B represents the most caudal brain stem transection (rostro-pontine transection) which increases cortical synchronization. This section is just in front of the group of noradrenaline-containing neurons of the pons. The rostro-pontine transection is also the most caudal transection which still permits the periodical appearance of PS (in the pontile cat).—Line C represents the medio-pontine transection which increases cortical desynchronization. This transection is just caudal to the activating neurons and suppresses most of the serotoninergic influences from the caudal raphe system. The lesion of the raphe system caudal to transection C induces an insomnia similar to the medio-pontine transection. The lines marking the transection are ideal lines. Usually, a transection would destroy the brain to a width of 1–1.5 mm. (Jouvet, 1971)

Westphal, which is either removed or situated rostrally to the transection, cannot receive direct or indirect serotoninergic information so no miosis can occur. In fact, it is difficult to recognize apart from PS any well-defined state in a chronic pontile cat, since it remains in a state of total immobility and does not exhibit either sleeping or waking posture but only some reflex head or leg movements which follow nociceptive stimulation. Most activating presynaptic NA systems of the pons and effector mesencephalic reticular formation are, in fact, destroyed by the lesion.

d) High-Mesencephalic Cat. Since a chronic mesencephalic cat has some neural structures which can be influenced by both the 5-HT sleep systems and the ponto-mesencephalic CA waking and PS executive systems, it is not surprising that it presents the ocular behavior of waking, SWS and PS (see Villablanca, 1966a). On the other hand, some modulation of its behavior can occur thanks to some intact extrapyramidal structures (mesencephalic reticular formation, rubro-spinal tract, colliculi) which may be subjected both to the phasic influences from the environment and to some CA and 5-HT-innervation. The fact that a high-mesencephalic cat (Bard and Macht, 1958) or rat (see picture 4, p. 639 in Woods, 1964) may present, some days after the transection, clear-cut and rather integrated waking behavior followed successively by true ocular and postural sleep behavior and almost all the central and peripheral components of PS (PGO pontine spikes, rapid eye movements, etc...) is a good proof than this preparation is subjected to the two antagonistic systems located in the brain stem. These findings cannot be easily reconciled with the hypothesis that the "sleep and waking centers" are located in the hypothalamus.

e) Chronic "Cerveau Isolé". Considerable and careful discussion has been devoted by Moruzzi to the problem of the recovery of the fast cortical activity in the "cerveau isolé" after high-mesencephalic transection (Batsel, 1960; Villablanca, 1962). This discussion is one of the cornerstones of the hypothesis that the hypothalamus is essential to waking. However, the secondary recuperation of fast EEG activity by the "cerveau isolé" does not contradict the hypothesis that cortical arousal depends on ascending ponto-mesencephalic CA influences, for the following reasons:

As rightly pointed out by Moruzzi, the main problem is to explain the recuperation of a tonic fast cortical activity, since the cortical synchronization of the "cerveau isolé" can be explained by the disconnection of the cortex from activating structures (mesencephalic reticular formation) which are normally subjected both to the phasic (sensory stimulations) and tonic (ponto-mesencephalic CA neurons) influences. It is most unlikely, as also emphasized by Moruzzi, that the low-voltage cortical activity of the "cerveau isolé" belongs to some cerebral PS autonomous phenomenon. The experimental evidence that PS depends upon pontine structures is so obvious that the search for PS activity in the "cerveau isolé" is a desperate venture.

Taking into account that the recurrence of fast cortical activity is a secondary phenomenon, several mechanisms may explain it (without necessarily considering the hypothalamus as a waking structure).

i) As shown by Soderberg (1962), Rothballer (1956), Sharpless and Rothballer (1961) and Monnier and Fallert (1967a, b) (see p. 174), there is good evidence that some blood-mediated influences can activate the cortex. Thus, the stimulation of the mesencephalic reticular formation in a donor rabbit may activate the cortical activity of a recipient rabbit in crossed-blood circulation experiments. In chronic "cerveau isolé" preparations, sensory stimulations may activate the ascending NA ponto-mesencephalic systems, and either CA or their metabolites may be taken up by the blood and activate effectors situated rostrally to the transection which might be in a state of postsynaptic supersensitivity (Trendelenburg, 1966). The fact that cortical activation may follow exposure to cold is a good argument in favor of this hypothesis (BATSEL, 1960). The more rapid recovery of the cortical arousal and of the ocular signs of wakefulness in the low "cerveau isolé" cat (see VILLABLANCA, 1962) can be readily explained by the fact that most of the CA-containing cell bodies and terminals of the group A8 (which extend rostrally to the frontal plane A5) are situated in front of the midbrain transection together with the rostral mesencephalic reticular formation. In such a case, it is evident that olfactory stimulation may activate the forebrain phasically, and that CA mechanisms may contribute to the tonic prolongation of the cortical and ocular arousal.

ii) Total transection of a MA bundle is followed by an immediate increase of fluorescence of the severed axons distally to the transection, then a strong diminution of fluorescence follows (degeneration) up to the 12th–15th day. However, recent evidence suggests that some sprouting of central CA neurons exists in the brain (KATZMAN et al., 1971). The sprouting starts after the 7th day and attains its maximum on the 15th–17th day. Thus it is likely that, after a high-mesencephalic transection which interrupts MA ascending axons, *some sprouting of these neurons may occur and new connections be established*. The fact that DOPA may desynchronize the electrical activity of the "cerveau isolé" is a strong argument in favor of the existence of a catecholaminergic mechanism which would activate CA receptors situated in front of the transection (MANTEGAZZINI and GLASSER, 1960; KADZIELAWA and WIDY-TYSZKIEWICZ, 1970b).

iii) Last but not least, it is likely that degeneration occurs in the "chronic" "cerveau isolé" (due to the interruption of afferent pathways and the almost inevitable damage to the caudal thalamus by lesion of the venous circulation). Hence, it is difficult to compare the electrical activity of postsynaptic nerve cells undergoing a pathological process with the cortical arousal of a normal cat.

For all these reasons, the capricious recovery of some low-voltage fast activity in the "cerveau isolé" cannot be taken as proof for the view that activating neural structures are located in the hypothalamus. On the contrary, the readily observable alternation of the three states, waking, ocular behavior of SWS, and PS, in the chronic mesencephalic cat is a much more powerful proof that the essential mechanisms for these three states exist in the total absence of tele-diencephalon.

Finally, the crucial proof that the "tonic arousal" of the high or low "cerveau isolé" depends upon CA mechanisms can easily be tested by the following experiment: if the MA theory is valid, then injection of α-methyl-para-tyrosine, which inhibits CA synthesis, should suppress the appearance of low-voltage fast activity in the chronic "cerveau isolé".

f) The Decorticate Cat. A chronic decorticate cat may sleep behaviorally without any slowing of activity in the diencephalon or the brain stem (see

JOUVET, 1962; MORUZZI, this journal). This indicates that the postsynaptic receptors which are implicated in both cortical and subcortical synchronization require the neocortex (and even more the orbital cortex) (VELASCO and LINDSLEY, 1965). Since 5-HT terminals are located in this region (and in the preoptic area), it is possible that release of 5-HT in the presynaptic cleft of these structures may be necessary for the initiation of the nervous mechanisms which are responsible for cortical synchronization. This hypothesis would unite the "ascending" and "descending" theories concerning SWS which have been the subject of so many discussions (see MORUZZI, this journal).

ii) Experimental Insomnia or the "Search for a Sleeping Center"

The characteristics of the insomnia which follows a subtotal raphe lesion (see p. 208) may be summarized as follows: the insomnia is immediate and *total* for at least three to four days, SWS does not exceed 10–15 % of the nycthemeron, and PS is absent. The insomnia is relatively well tolerated since cats survived for two weeks with less than 10 % of total sleep.

According to classical neurophysiology, three main structures are implicated in the mechanisms of SWS (see MORUZZI, this journal). The lower brain stem synchronizing structures, the thalamus, and above all the anterior hypothalamus or preoptic region.

a) Lower Brain Stem Synchronizing Center. Although it has been shown, in acute experiments, that cooling (BERLUCCHI et al., 1965), lesion or disconnection of the region of N. of the solitary tract may increase spontaneous arousal or increase the duration of phasic cortical arousal induced by reticular stimulation (BONVALLET and BLOCH, 1961; BONVALLET and DELL, 1965; BONVALLET and ALLEN, 1963), there are no data concerning insomnia due to chronic lesion of these structures. On the contrary, BONVALLET has stated explicitly that cats with lesions of the lower brain stem synchronizing structures "*ne semblent pas plus éveillés que les animaux témoins*" (BONVALLET and DELL, 1965, p. 149). Thus, there are no experimental data favoring the intervention of these structures in the *maintenance* of sleep.

b) Thalamus. Although, on the basis of results obtained with stimulation techniques (see the discussion in MORUZZI, this journal) the thalamus has often been implicated in sleep mechanisms (see HESS, 1944; AKERT et al., 1952; KOELLA, 1967), the total destruction of the thalamus, which suppress cortical spindles, neither suppresses cortical slow waves during SWS, nor significantly alters the sleep-waking cycle (NAQUET et al., 1965; ANGELERI et al., 1969).

c) The Anterior Hypothalamic Region. i) In the experiment of NAUTA (1946), which were done before the introduction of modern polygraphic techniques, after a transverse section situated on the rostral half of the hypothalamus, the operated rats were awake for the entire survival time averaging three days "after which the exhausted animal fell into a state of coma which soon ended in death" (NAUTA l.c., p. 303).

These experiments are difficult to interpret since no quantitative assessment was made of the sleep-waking cycle. On the other hand, the fact that such insomnia was followed after three days by death favors some unspecific mechanism. Had a true hypothalamic sleeping center been destroyed, one would have expected these rats to have lived awake much longer (like the insomniac rats with rostral raphe lesions described by KOSTOWSKI et al., 1968).

ii) The lesions made by McGINTY and STERMAN (1968) on the basal forebrain area in cats induce only a secondary insomnia, but since these cats were recorded only one day a week, it is difficult to gain a dynamic picture of the development of their insomnia. On the third day, total sleep amounted to 43 % of the recording time. Significant insomnia was seen only after one week and two weeks (with amounts of SWS as low as 20 % and 11 % of the nycthemeron and suppression of PS). No permanent recordings throughout the survival period were made in the two cats which died on the 10th day; they were recorded only on the 7th day, so one does not know whether the insomnia developed immediately after the lesion or only secondarily on the 7th day. The long delay before the appearance of insomnia argues for the existence of a transynaptic mechanism but not for the existence of a preoptic sleeping center, since one would have expected immediate insomnia in such a case. On the other hand, some unspecific effect may be induced by the lesion, such as some alteration of the temperature regulation which might interfere with sleep mechanisms (SQUIRES and JACOBSON, 1968). According to the MA hypothesis, this secondary insomnia might be explained in the following way.

The preoptic region receives an important contingent of 5-HT nerve fibers ascending through the medial forebrain bundle. It is likely that the destruction of these 5-HT fibers and terminals *secondarily* induces lower turnover of the 5-HT system (with a resulting decrease of SWS below 20 % and thus a suppression of the priming mechanism of PS). The fact that the locus coeruleus receives some afferent connections from the hypothalamus does not, however, exclude possible descending influences from the hypothalamus upon the "executive mechanisms" of PS (MIZUNO and NAKAMURA, 1970).

It is certain that the basal forebrain area (and the orbital cortex) play an important role in the induction of cortical synchronization and secondarily in that of sleep (WYRWICKA et al., 1962; CLEMENTE et al., 1963; STERMAN and CLEMENTE, 1962a, b). MORUZZI has rightly pointed out that the effectiveness of high-frequency stimulation of this area by far surpasses the "coefficient of credibility" of the stimulation of any other central "hypnogenic area". Since the pretreatment of cats with p-chlorophenylalanine (which depletes 5-HT terminals) suppresses the sleep-inducing effect of the basal forebrain

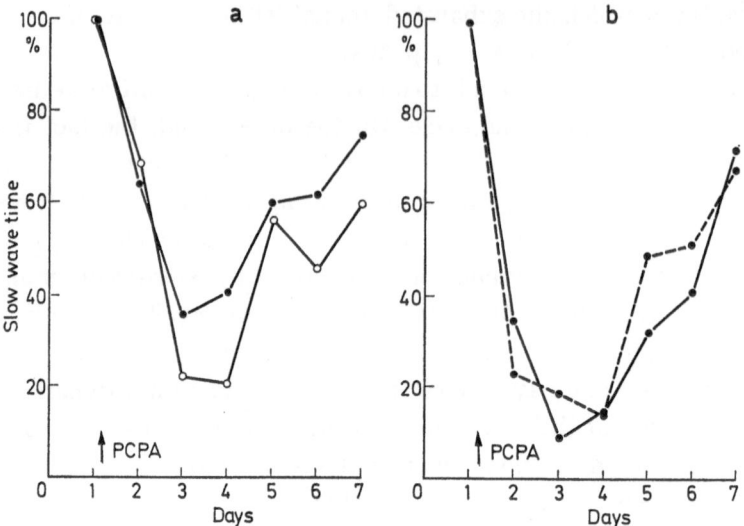

Fig. 24. Chronological alteration of slow wave activity after p-chlorophenylalanine (200 mg/kg IP) (PCPA). a Mean slow wave activity of two animals each with (open circles) or without (closed circles) basal forebrain stimulation. b Sustained slow wave activity after basal forebrain stimulation in cats No 456 (solid line) and No 460 (broken line). (WADA and TERAO, 1970)

area (WADA and TERAO, 1970) (Fig. 24) this constitutes strong indirect evidence that some 5-HT terminals are involved in the presynaptic mechanism which mediates the cortical synchronization. This would also explain why local injection of 5-HT in this area may induce sleep (YAMAGUCHI et al., 1963). Thus, the basal forebrain area would appear to be mostly a strategic post-synaptic ensemble of neurons where the 5-HT terminals act to trigger or "modulate" the neural mechanisms synchronizing the cortical activity. In any case, *a comparison between the immediate long-lasting and total insomnia following the destruction of the raphe system and the secondary hyposomnia following lesion of the basal forebrain area does not justify the term hypothalamic "sleep center" for this latter structure.*

iii) "Experimental Coma or the Search for a Waking Center"

According to the MA theory, long-lasting cortical arousal requires the integrity of NA ascending neurons from the pontile or mesencephalic group, whereas local cholinergic cortical mechanisms probably enter into play at the cortical level. *Phasic* arousal is mediated (as rightly emphasized by MORUZZI) by the mesencephalic reticular formation. Waking behavior is under the control of the dopaminergic nigro-striatal system ascending in the medial forebrain bundle and the lateral hypothalamus. The fact that a comatose cat with a hypothalamic lesion (FELDMAN and WALLER, 1962) or destruction of the substantia nigra (JONES et al., 1968) exhibits long-lasting arousal after sensory stimulation suggests that behavioral and cortical arousal mechanisms may be separate. Since I reviewed the problem of cortical arousal when considering the chronic "cerveau isolé", I shall now concentrate briefly on the problem of comatose behavior.

a) *Midbrain and Hypothalamic Lesions*. The common denominator of the lesions which induce a state of coma, unresponsiveness, catatonic state, or lethargic syndrome is that these lesions must be ventrally situated and destroy either the ascending nigro-striatal pathway or the descending extrapyramidal fibers or both. This is particularly evident when we consider the ventral extension of the lesion of the midbrain reticular formation of the comatose cat operated by LINDSLEY et al. (1950), in which the substantia nigra was totally destroyed. This is also the case for the comatose cat reported by FELDMAN and WALLER (1962) in which the lesion, situated in the posterior hypothalamus, destroyed the medial forebrain bundle. The common lesion which induced "cataleptic immobility" and a strong decrease of mobility in the cats described by SWEET and HOBSON (1968) also destroyed the lateral hypothalamus and the rostral part of the ascending nigro-striatal tract. As rightly pointed out by SWEET and HOBSON (1968), the usefulness of a hypo-thalamic "waking center" is somewhat doubtful, since lesions of the lateral hypothalamus also destroy descending motor extrapyramidal fibers. This also applies to the lesions made by RANSON (1939) or NAQUET et al. (1963). The fact that a transient behavioral and EEG lethargy lasting for only four days may follow extensive lesions of the dorsal posterior hypothalamus (McGINTY, 1969) cannot be taken as sufficient proof that the waking center is situated in the hypothalamus. The search for a center responsible for behavioral waking by means of localized brain stem lesions is seriously hampered by the fact that most of the lesions may act upon descending extrapyramidal pathways and thus induce lethargic behavior, even in the presence of waking EEG. Only extensive lesions of both the medial and lateral hypothalamus increase cortical synchronization and decrease behavioral waking (SWEET and HOBSON, 1968). This can readily be explained by the fact that most ascending activating influences from the reticular formation are impaired by the lesion, together with other ascending CA fibers and descending extrapyramidal pathways. A similar comatose behavior lasting in the cat for at least twelve days may be obtained by the total destruction of the substantia nigra, sparing the pes pedunculi, and decreasing by more than 90 % the DA level of the striatum, whereas the same amount of cortical synchronization (80 %) may be obtained by destruction of ascending ponto-mesencephalic CA-containing neurons (JONES et al., 1968; JONES, 1969). The posterior lateral hypothalamus in which both DA and NA ascending axons are concentrated appears to be a strategic locus where both systems may be destroyed (together with extra-pyramidal descending pathways). But this does not justify the term of hypo-thalamic waking center.

Summing up: in the light of the persistence of sleep-waking behavior in the chronic mesencephalic cat or rat, neither the secondary hyposomnia which follows basal forebrain lesions (as compared with the immediate, total and long-lasting insomnia which follows raphe destruction), nor the transient lethargic syndrome obtained after posterior hypothalamic lesion (as compared with the permanent comatose behavior following destruction of the substantia nigra) justifies the hypothesis that the structures which appear to be directly and critically responsible for both sleep and wakefulness are localized in the diencephalon. On the contrary, if it is possible that some of the effects which take place at the terminals of 5-HT and CA ascending systems mimic many "trophotropic" and "ergotrophic" effects which are believed to take place in the diencephalon, the existence of the cell bodies of these two antagonistic systems in the lower brain stem makes this part of the brain of paramount importance for the control of many events believed to be of hypothalamic origin.

Even if many effector neurons, responsible for the most elaborate behavioral and electroencephalographic aspects of sleep and waking, are located in the tele-diencephalon, it is likely that most of the mechanisms which regulate and adapt the activity at the terminal level are located at the place where the enzymes are synthesized, i.e., in the cell bodies of the 5-HT-containing neurons of the raphe system and in the catecholamine-containing cell bodies of the ponto-mesencephalic tegmentum.

VI. A Concluding Essay on the Function of Sleep

It has become almost a tradition among sleep physiologists to conclude any review of the mechanisms of the sleep-waking cycle with some remarks concerning the function(s) of sleep. This field is so fascinating, and we know so little about it, that the more speculative the hypothesis, the deeper the ignorance.

A. Slow-Wave Sleep

The MA theory does not shed very much light upon the function of SWS and it does not help very much to know that, among the serotoninoceptive neurons of the brain stem, diencephalon and cortex, 30% are facilitated, 30% are inhibited and 30% respond both ways (as in most iontophoretic studies). It would certainly help to know *which* system of cells is sensitive to 5-HT and which, if any, "second messenger effect" takes place in the perikarya of these cells, but there is no technique available at the present time to solve this problem. The neocortex is certainly a target for the function of SWS. Indeed "quiet sleep" (but not slow-wave sleep) is short-lasting in mesencephalic or decorticate preparations (see Jouvet, 1962; Villablanca, 1966b) and in newborn, immature mammals whose cortical mantle is still undeveloped (see Jouvet-Mounier et al., 1970). Besides its action upon cortical cells, the 5-HT sleep system may accomplish some restorative function directly upon the brain stem or diencephalic neurons and indirectly upon the body [muscular rest, release of somatotrophic hormone (IRIE, 1968; SASSIN et al., 1969a, b)]. But the main problem of the function of SWS may reside in its action upon the cortical cells. There is indeed plenty of evidence that, the higher the differentiation of the cortical mantle, the more complex the patterns of slow-wave sleep. There is some indirect evidence suggesting that the action of SWS upon cortical cells may be two-fold: a biochemical consequence of waking, and a priming mechanism for the occurrence of PS.

1. *SWS as a Consequence of Waking*. Although there have not yet been many experiments in humans and animals, it is surprising that so little SWS may follow long periods of enforced waking: under normal conditions, in man, 1 000 min of waking are followed by about 500 min of sleep (400 min of SWS, since PS constitutes 20% of total sleep time). Long-term enforced waking (during "wakathon") (BRAUCHI and WEST, 1959; WILLIAMS et al., 1959; LUBY et al., 1960; WEST et al., 1962; GULEVICH et al., 1966) lasting up to 10000 min (i.e. seven days) is followed by a recuperative SWS lasting only 15–16 hours (i.e. 1 000 min) after which only the duration of PS is prolonged in the following nights. Thus the duration of "recuperative SWS" is increased. two-fold when the preceding waking is increased ten-fold. The duration of SWS appears to be a function of the log of the duration of waking (SWS = $K \log W$) (whereas the duration of PS is directly proportional to the duration of waking; see below). This logarithmic relation suggests that some *non sequential* enzymatic processes may "repairs" at the level of the perikarya, the effects of waking possibly mediated by cyclic AMP. It is likely that these processes are evoked in the cell perikarya themselves and that they do not affect, at least in any sequential way, the synaptic network of the interneurons.

2. Another function of SWS might involve the priming of the cortical cells, or of some synapses, for the necessary appearance of PS episodes which appear to involve a *sequential process*.

B. Paradoxical Sleep

Three main aspects of the physiology of PS may serve to delimit its function:

1. its neurophysiological organization, 2. its evolutionary aspect, 3. the problem of rebound.

1. Schematically, PS is a process during which an increase of cerebral activity (brain storming) occurs, as demonstrated by the increase of unit firing, controlled by the pontine "pacemaker" (PGO activity). In this sense PS appears to be an arousal of the brain and, without the powerful process of inhibition of muscle tone triggered from the caudal part of the locus coeruleus,

PS would appear as a true arousal, as demonstrated by the striking "hallu-cinatory" behavior during PS of the cats subjected to destruction of the caudal locus coeruleus.

2. From the evolutionary aspect, PS is not present in either fishes or reptiles and seems to appear only with birds. Thus, in some way, PS appears to be the result of some emerging function correlated with the increasingly complex organization of the brain. On the other hand, PS constitutes almost all the "sleeping" time of immature mammals, almost 100 % of sleep in the newborn rat and 90 % in the newborn kitten, whereas in mature newborn mammals, it constitutes only 20 % to 7 % of sleep (lamb, guinea pig) (VALATX et al., 1964; see JOUVET-MOUNIER, 1968; JOUVET-MOUNIER et al., 1970). The hypothesis that PS accompanies maturation has been verified in utero in the guinea pig, since acute or chronic recording of foetal guinea pigs in utero has shown significantly more PS (up to 50 % of recording time) at a time when the cerebral maturity is equivalent to that of a newborn kitten (ASTIC and JOUVET-MOUNIER, 1969). Two conclusions may be drawn from these data: 1. PS may be genotypically programmed in utero; 2. PS may initially serve some function in the maturation of the higher nervous structures (tel-encephalon), according to the hypothesis of ROFFWARG et al. (1966). Thus, at least during maturation, one may speculate that PS is a "genotypic arousal" during which both indolamine and catecholamine-containing neurons would serve to stimulate and organize the delicate synaptic circuitry which is needed for some instinctive behavior which has to be ready immediately after birth. It is likely that "PS brainstorming" under the lead of the pontine pacemaker may affect mostly Class II neurons, as hypothetized by JACOBSON (1970). These neurons (mostly interneurons) remain unspecific until late in ontogeny. Their connections may be modified or maintained by function and experience. They are the opposite of Class I neurons, which are highly specialized early in development, after which their connectivity is strictly constrained and unmodifiable. The observation that both MA systems exist in utero (at least, as visualized by MAO staining, as early as nine days in the chick) (see MASAI et al., 1965) is in accordance with the hypothesis that both systems may be functional in utero and that CA-containing terminals, by releasing CA or its metabolites in some synapses of the maturing nervous system, might serve to stimulate Class II neurons according to some "genotypic coding". The fact that PS may serve some role during ontogeny does not, however, explain why it still persists in adult life.

The process of the rebound of PS may shed some light upon this problem. 1. PS rebound apparently requires the presence of the neocortex, since instru-mental deprivation of PS in the pontile cat (by electric shock) is not followed by a subsequent rebound either in duration or in the frequency of PGO pontine spikes which would "make up" for the debt (see JOUVET, 1965 a).

2. Long-term enforced waking or PS deprivation is followed by a subsequent rebound both of duration of PS and frequency of PGO (DEMENT, 1960; VIMONT et al., 1966; DUSAN-PEYRETHON et al., 1967; see DEMENT et al., 1970a, b). In experiments performed in men isolated in caves "beyond time", spontaneous or induced bicircadian rhythms of sleep-waking may occur so that 34–36 hours of continuous waking are followed by 12–14 hours of sleep (SIFFRE and JOUVET, 1971). In these conditions there is a very striking relationship between the duration of the preceding waking and the duration of PS during the sleep that follows, PS being equivalent to one-tenth of the waking time. Thus the duration of PS seems to be related to the preceding waking events. This suggests that some *sequential processes* may occur under the influence of both the pontine pacemaker and the cortical cells so that roughly 1 min of PS corresponds to 10 min of waking in adult man. These sequential ascending events seem to occur at a time when other brain systems are turned off [as shown by the decrease or even the total disappearance of unitary activity in the hippocampus (BELUGOU et al., 1968) or the corpus callosum (BERLUCCHI, 1965)]. It is thus possible, although admittedly speculative, that some process may occur during which phenotypic (epigenetic) information from the preceding waking period (stored in some synaptic contact of the type-II cells) and genotypic coding of the pontine pacemaker are combined in the adult during dreaming. Some electrophysiological aspect of the storage of these waking events might be reflected in the so-called eye movement potentials (EMP) (corollary discharges) which accompany any sensory inputs to the central nervous system or any voluntary output (see p. 244). The fact that drugs acting upon MA may increase the EMP is an indirect proof that these EMP may involve CA neurons. Thus, at least theoretically, there are some biochemical mechanisms involving CA mechanisms which may be available for the storage of these waking events (ROBERTS et al., 1970). It is unlikely that the "subjective" aspect of dreaming will lead us to discover how these interactions take place in the cat. However, the observation of the stereotyped behavior of cats after destruction of the caudal part of the locus coeruleus may help us to study these possible interactions objectively. When one observes the behavior of these cats, it is evident that the resulting motor behavior, which is subject to the "coding" of the pontine "pacemaker", is not a *random motor activity*. Most of the time it is either attack or defense behavior (which certainly has some teleological value). The cats usually act as if they are defending themselves against a much larger predator, like a dog. It remains to be shown whether such behavior is genotypic (i.e. if it could be observed in a blind cat raised in an isolated environment without ever having seen a dog) or whether it results from epigenetic modifications of the primitive coding which exists at birth. If this is the case, then there should be some mechanism which could alter the primitive genotypic "coding" to another

one. Speculative as it is, such a mechanism could be the result of interaction between the cortical neurons and the pontine pacemaker. The mechanism of this interaction is still obscure but there are two possibilities: 1. the cortex may act upon the genome which is responsible for the inborn coding; 2. most probably, it is in the complicated network of the pontine interneurons receiving cortical messages that a new code may appear during dreaming. One may speculate that the process taking place during SWS at the cortical level (and at the molecular level) would also be the priming mechanism for cortical cells which would discharge efferent messages to the pons, when solicited at the beginning of PS by the isolated PGO which always herald PS. In this case a new "syntax", which would initiate a bombardment of some cortical synapses, would serve some teleological function. In the newborn, this genotypic coding may serve to prepare and mature vital instinctive behavior; in the adult, this coding under certain phenotypic corticofugal influences would serve to organize acquired phenotypic schemes of activity.

Thus, dreaming in the adult could be the continuation of the organization of schemes of activity. These schemes would be organized by the pontine pacemaker whose genotypic pattern of discharge would be transformed into a complex "syntax" by corticofugal influences. It is possible, but not yet proved, that such a mechanism may contribute to certain memory processes. PS deprivation in man does not seem to interfere significantly with long-term memory (DEMENT, 1960; WYATT et al., 1971) and studies of the relationship between PS deprivation and learning have given equivocal results (ALBERT et al., 1970; LUCERO, 1970; FISHBEIN et al., 1971). If the hypothesis that PS first plays a role in genotypic coding and secondarily elaborates new schemes of activity is correct, then "nurture would prevail over nature" and the following experiments should give the following results:

1. Homozygote twins should have the same "syntactic" organization of PGO (thus of rapid eye movements) at least immediately after birth.

2. Animals totally isolated from their environment, with the caudal part of the locus coeruleus destroyed, should display, during PS, a repertoire of motor-stereotyped behavior differing from that of control animals exposed to aggression.

3. Long-term deprivation of PS should alter instinctive behavior by suppressing the possibility of phenotypic control of genetic coding: this would explain the "release" of aggressive behavior or hypersexuality (MORDEN et al., 1968; see DEMENT et al., 1970) which is observed after PS deprivation.

But another hypothesis is possible: The genotypic pattern of PGO activity during PS may not be influenced by phenotypic events occurring during waking, If this is the case, the repertoire of the behavior occurring in cats whose caudal locus coeruleus has been destroyed should be similar whatever might be the history of the animal. "Nature would prevail over nurture" and the mammalian brain would be subjected during sleep to a programmed genotypic reorganization. This would explain the striking differences of behavior which may sometimes be observed between individuals raised in the same environment. Dreaming would thus be responsible for the inner core of the "personality" which cannot be altered by environmental factors.

C. Summary

In this speculative essay I consider that waking is the mere cause of sleep. The duration of recuperative SWS is proportional to the log of the duration of waking (at least in adult humans). 5-HT mechanisms trigger "second messenger effects" upon cortical cells to restore the effect of waking at the molecular level. SWS is also responsible for the priming of PS. PS is considered first as a genotypically coded arousal which serves to organize some special plastic type of neurons during ontogeny. In the adult (human) the duration of PS is directly related to the duration of the preceding waking, one minute of PS corresponding to 10 minutes of waking. PS appears to be composed of sequential events during which epigenetic cortical mechanisms might interact or not with the genotypic coding from the pontine pacemaker in order to organize teleologically acquired schemes of activity. A possible experimental approach to this hypothesis is proposed.

VII. Summary and Conclusions

1. Recent histochemical techniques have confirmed the presence of ascending monoaminergic (5-HT, NA and DA) and probably cholinergic systems in the reticular core of the brain stem. Their topography is outlined.

2. The experimental evidence indicating a central neurotransmitter role of 5-HT, CA and ACH is reviewed and the possible strategies and tactics concerning the study of their functions in the regulation of the sleep-waking cycle are discussed.

3. The determinant role of 5-HT in the induction of slow-wave sleep is suggested by converging experimental evidence. The following results are among the most convincing: the inhibition by p-chlorophenylalanine of the first step of 5-HT synthesis leads to total insomnia which may be immediately reversed to sleep by subsequent injection of low doses of 5-HTP which bypass the inhibition of synthesis. Lesion of the 5-HT-containing perikarya of the raphe system induces insomnia which is proportional to the extent of the destruction of the raphe system and to the decrease in 5-HT at the level of the terminals.

4. Catecholaminergic mechanisms appear to be involved, at least in the cat, in the maintenance of tonic cortical arousal. The inhibition of CA synthesis by α-methyl-p-tyrosine can reverse into a state of sedation with cortical

synchronization the excitation and insomnia following destruction of the raphe system or injection of amphetamine. The NA-containing neurons of the anterior part of the locus coeruleus and the group of CA-containing neurons of the mesencephalic tegmentum appear to play a role in the maintenance of cortical arousal. The ventral NA pathway, originating from the medulla, does not seem to be important in waking but is concerned in the rat with self-stimulation or reward. The substantia nigra and the DA nigro-striatal pathway are involved in the maintenance of waking behavior but do not significantly affect the electrical activity of the cortex in the cat.

5. It is possible that cholinergic ascending mechanisms also participate in cortical arousal, whereas local cholinergic cortical mechanisms appear to be responsible for the activation of the cortex during waking. There is little evidence that cholinergic mechanisms have a significant role in slow-wave sleep.

6. Paradoxical sleep seems to involve some "5-HT priming mechanisms" located in the raphe system, whereas its "executive mechanisms", depending upon CA and probably ACH neurons, are located in the region of N. locus coeruleus. The organization of descending and ascending mechanisms of PS is outlined. The PGO activity which occurs during PS depends upon a pontine "pacemaker". The relationship of this activity to MA metabolism is discussed and a hypothetical model of the triggering of PGO activity is presented.

7. A monoamine theory of the sleep-waking cycle is proposed: according to this theory, the brain stem systems of 5-HT neurons are responsible for SWS and for the priming mechanism of PS, whereas CA and ACH systems are involved in the executive mechanisms of PS and in the maintenance of tonic behavioral and EEG arousal. The theory emphasizes the paramount importance of the brain stem, whereas each level of the organization of the nervous system subjected to the interaction of both 5-HT and CA systems is responsible for the behavioral and EEG aspects of the sleep-waking cycle. This theory postulates the existence of two types of insomnia and two types of reduced waking whose characteristics are enumerated and discussed. The monoamine theory is tested against the results of brain lesions upon the sleep-waking cycle.

8. Finally, some speculation was made concerning the possible function of SWS and PS. It is hypothesized that SWS acts mainly upon cortical cells at the molecular level to restore the effects of waking. SWS is also necessary to prepare some interneurons for the action of PS. PS is considered first as a genotypic arousal during ontogeny. In the adult, the duration of PS is related to the duration of waking and the sequential patterns of ponto-geniculo-occipital activity may be considered either genotypically programmed or as resulting from the interaction of epigenetic corticofugal mechanisms stored during waking with the genotypic coding depending upon the pontine pacemaker.

References *

ACHESON, G. H. (ed.): Proceed. of second Symposium on catecholamines. Pharmacol. Rev. **18**, 803 (1966). Baltimore: The Williams & Wilkins.

AGHAJANIAN, G. K., BLOOM, F. E.: Electron-microscopic localization of tritiated norepinephrine in rat brain: effect of drugs. J. Pharmacol. exp. Ther. **156**, 407–417 (1967a).

— — Localization of tritiated serotonin in rat brain by electron microscopic autoradiography. J. Pharmacol. exp. Ther. **156**, 23–31 (1967b).

— — LOVELL, R. A., SHEARD, M. H., FREEDMAN, D. X.: The uptake of 5-hydroxytryptamine-3H from the cerebral ventricles: autoradiographic localization. Biochem. Pharmacol. **15**, 401–408 (1966).

— FOOTE, W. E., SHEARD, M. H.: Action of psychotogenic drugs on single midbrain raphe neurons. J. Pharmacol. exp. Ther. **171**, 178–187 (1970a).

— GRAHAM, A. W., SHEARD, M. H.: Serotonin containing neurons in brain: depression of firing by monoamine oxidase inhibitors. Science **169**, 1100–1102 (1970b).

— ROSECRANS, J. A., SHEARD, M. H.: Serotonin: release in the forebrain by stimulation of midbrain raphe. Science **156**, 402–403 (1967c).

— ROTH, R. H.: Gamma-hydroxybutyrate induced increase in brain dopamine: localization by fluorescence microscopy. J. Pharmacol. exp. Ther. **175**, 131–138 (1970).

— WEISS, T.: LSD blocks the increase in rat brain serotonin turnover induced by an elevated ambient temperature. Nature (Lond.) **220**, 795–796 (1968).

AJURIAGUERRA, J. DE (ed.): Monoamines et système nerveux central. 293 pp. Genève: Georg 1962.

AKERT, K., KOELLA, W. P., HESS, R.: Sleep produced by electrical stimulation of the thalamus. Amer. J. Physiol. **168**, 260–267 (1952).

— MOOR, H., PFENNINGER, K., SANDRI, C.: Contributions of new impregnation methods and freeze etching to the problems of synaptic fine structure. Progr. Brain Res. **31**, 223–240 (1969).

— WASER, P. G. (eds.): Mechanisms of synaptic transmission. Progr. Brain Res. **31** (1969).

AKINDELE, M. O., EVANS, J. I., OSWALD, I.: Monoamine oxidase inhibitors, sleep and mood. Electroenceph. clin. Neurophysiol. **29**, 47–56 (1970).

ALBERICI, M., RODRIGUEZ DE LORES ARNAIZ, G., ROBERTIS, E. DE: Catechol-o-methyltransferase in nerve endings of rat brain. Life Sci. **4**, 1951–1961 (1965).

ALBERT, I., CICALA, G., SIEGEL, J.: The behavioral effects of REM sleep deprivation in rats. Psychophysiology **6**, 550–560 (1970).

ALEKSEYEVA, T. T.: Correlation of nervous and humoral factors in the development of sleep in non-dijointed twins. [Russian.] Zh. vyssh. nerv. Deyat. Pavlova **8**, 835–844 (1958).

ALIVISATOS, S. G., PAPAPHILIS, A. D., UNGAR, F., SETH, P. K.: Chemical nature of binding on serotonin in the central nervous system. Nature (Lond.) **226**, 455–456 (1970).

— UNGAR, F., PARMAR, S. S., SETH, P. K.: Monoamine oxidase dependent upon labeling in vitro of mouse brain by 14 C-Serotonin. Biochem. Pharmacol. **17**, 1993–1997 (1968).

AMIN, A. H., CRAWFORD, I. B., GADDUM, J. H.: The distribution of substance P and serotonin in the central nervous system of the dog. J. Physiol. (Lond.) **126**, 596–618 (1954).

ANDÉN, N. E., BUTCHER, S. G., ENGEL, J.: Central dopamine and noradrenaline receptor activity of the amines formed from m-tyrosine, alpha-methyl-m-tyrosine and alpha-methyldopa. J. Pharm. Pharmacol. **22**, 548–550 (1970).

— CARLSSON, A., DAHLSTRÖM, A., FUXE, K., HILLARP, N. A., LARSSON, K.: Demonstration and mapping out of nigro-neostriatal dopamine neurons. Life Sci. **3**, 523–531 (1964a).

---

* The references have been reviewed up to March 1971.

Andén, N. E., Carlsson, A., Hillarp, N. A., Magnusson, T.: Noradrenaline release by nerve stimulation of the spinal cord. Life Sci. 4, 129–133 (1965a).
— Corrodi, H., Dahlström, A., Fuxe, K., Hökfelt, T.: Effects of tyrosine hydroxylase inhibition on the amine levels of central monoamine neurons. Life Sci. 5, 561–569 (1966a).
— — Fuxe, K., Hökfelt, T.: Evidence for a central 5-hydroxytryptamine receptor stimulation by lysergic acid diethylamide. Brit. J. Pharmacol. 34, 1–8 (1968).
— Dahlström, A., Fuxe, K., Larsson, K.: Functional role of the nigro-neostriatal dopamine neurons. Acta pharmacol. (Kbh.) 24, 255–263 (1966b).
— — — — Mapping out of catecholamine and 5-hydroxytryptamine neurons innervating the telencephalon and diencephalon. Life Sci. 4, 1275–1281 (1965b).
— — — — Olson, L., Ungerstedt, U.: Ascending monoamine neurons to the telencephalon and diencephalon. Acta physiol. scand. 67, 313–327 (1966d).
— — — Olson, L., Ungerstedt, A.: Ascending noradrenalin neurons from the pons and the medulla oblongata. Experientia (Basel) 22, 44–45 (1966c).
— Fuxe, K., Hamberger, G., Hökfelt, T.: A quantitative study on the nigro-neostriatal dopamine neuron system in the rat. Acta physiol. scand. 67, 306–313 (1966e).
— — Henning, M.: Mechanisms of noradrenaline and 5-hydroxytryptamine disappearance induced by alpha-methyl-dopa and alpha-methyl-metatyrosine. Europ. J. Pharmacol. 8, 302–309 (1969).
— — Hökfelt, T.: The importance of the nervous impulse flow for the depletion of the monoamines from central neurones by some drugs. J. Pharm. Pharmacol. 18, 630–632 (1966f).
— — Larsson, K.: Effect of large mesencephalic-diencephalic lesions on the noradrenalin, dopamine and 5-hydroxytryptamine neurons of the central nervous system. Experientia (Basel) 22, 842–844 (1966g).
— Haggendal, T., Magnusson, T., Rosengren, E.: The time course of the disappearance of noradrenaline and 5-hydroxytryptamine in the spinal cord after transection. Acta physiol. scand. 62, 115–119 (1964b).
Andersen, P., Curtis, D. R.: The excitation of thalamic neurones by acetylcholine. Acta physiol. scand. 61, 85–100 (1964).
Angeleri, F., Marchesi, G. F., Quattrini, A.: Effects of chronic thalamic lesions on the electrical activity of the neocortex and on sleep. Arch. ital. Biol. 107, 633–668 (1969).
Antonelli, A. R., Bertaccini, G., Montegazzini, P.: Relationship between mesencephalic-hypothalamic concentration of 5-hydroxytryptamine and cortical activity of cats. J. Neurochem. 8, 157–158 (1961).
Aprison, M. H.: On a proposed theory for the mechanism of action of serotonin in brain. In: Recent adv. biol. psychiat., p. 133–146. New York: Plenum Press 1962.
— Hingtgen, J. H.: Brain acetylcholine and excitation in avoidance behaviour. Biol. Psychiat. 1, 87–89 (1969).
— Kariya, T., Hingtgen, J. N., Toru, M.: Neurochemical correlates of behaviour, changes in acetylcholine, norepinephrine and 5-hydroxytryptamine concentrations in several discrete brain areas of the rat during behavioural excitation. J. Neurochem. 15, 1131–1141 (1968a).
— — — — Neurochemical correlates of behaviour. J. Neurochem. 15, 1131–1139 (1968b).
Arbuthnott, G. W., Crow, T. J., Fuxe, K., Olson, L., Ungerstedt, U.: Depletion of catecholamines in vitro induced by electrical stimulation of central monoamine pathways. Brain Res. 24, 471–483 (1970).
— Fuxe, K., Ungerstedt, U.: Central catecholamine turnover and self stimulation behaviour. Brain Res. 27, 406–413 (1971).
Arduini, A., Pinneo, L. R.: A method for the quantification of tonic activity in the nervous system. Arch. ital. Biol. 100, 415–424 (1962).

ASTIC, L., JOUVET-MOUNIER, D.: Mise en évidence du sommeil paradoxal in utero chez le cobaye. C. R. Acad. Sci. (Paris) 264, 2578–2581 (1969).

AXELROD, J.: The uptake and release of catecholamines and the effect of drugs. In: Biogenic amines. A symposium (H. and W. A. HIMWICH, eds.). Amsterdam: Elsevier 1962.

— Methylation reactions in the formation and metabolism of catecholamines and other biogenic amines. Pharmacol. Rev. 18, 95–113 (1966).

AXELSSON, J.: Catecholamine functions. Ann. Rev. Physiol. 33, 1–31 (1971).

BAK, I. J.: The ultrastructure of the substantia nigra and caudate nucleus of the mouse and the cellular localization of catecholamines. Exp. Brain Res. 3, 40–58 (1967).

BALDISSERA, F., CESA-BIANCHI, M. G., MANCIA, M.: Responses of visual cortex to transcallosal and geniculate stimulations during sleep and wakefulness. Arch. ital. Biol. 104, 247–263 (1966).

BALZANO, E., JEANNEROD, M.: Activité multi-unitaire de structures sous-corticales pendant le cycle veille-sommeil chez le chat. Electroenceph. clin. Neurophysiol. 28, 136–145 (1970).

BANDLER, R.: Agression induced in rats by cholinergic stimulation of hypothalamus. Nature (Lond.) 224, 1035–1036 (1969).

BARCHAS, J.: Relation of melatonin to sleep. Proc. West. Pharmacol. Soc. 11, 22–23 (1968).

BARCHAS, J. D., CIARANELLO, R. D., STEINMAN, A. M.: Epinephrine formation and metabolism in mammalian brain. Biol. Psychiat. 1, 31–48 (1969).

BARD, P., MACHT, M. B.: The behaviour of chronically decerebrate cats. In: Neurological basis of behaviour. A Ciba Foundation Symposium, p. 55–71. London: Churchill 1958.

BARTHOLINI, G., BIRLARD, W. P., PLETSCHER, A., BATES, H. M.: Increase of cerebral catecholamines caused by 3,4 dihydroxyphenylalanine after inhibition of peripheral decarboxylase. Nature 215, 852–853 (1967).

— PRADA, M. DA, PLETSCHER, A.: Decrease of cerebral 5-hydroxytryptamine by 3,4-dihydroxyphenylalanine after inhibition of extra cerebral decarboxylase. J. Pharm. Pharmacol. 20, 228–229 (1968a).

— PLETSCHER, A.: Cerebral accumulation and metabolism of C14 dopa after selective inhibition of peripheral decarboxylase. J. Pharmacol. exp. Ther. 161, 14–20 (1968b).

— — Effect of various decarboxylase inhibitors in the cerebral metabolism of dihydroxyphenylalanine. J. Pharm. Pharmacol. 21, 323–324 (1969).

— — RICHARDS, J.: 6-Hydroxydopamine induced inhibition of brain catecholamine synthesis without ultra structural damage. Experientia (Basel) 26, 598–600 (1970b).

— RICHARDS, J. G., PLETSCHER, A.: Dissociation between biochemical and ultrastructural effects of 6-hydroxydopamine in rat brain. Experientia (Basel) 26, 142–144 (1970a).

BARTLET, A. L.: The influence of chlorpromazine on the metabolism of 5-hydroxytryptamine in the mouse. Brit. J. Pharmacol. Chemother. 24, 497–510 (1965).

BARTOLINI, A., PEPEU, G.: Investigations into the acetylcholine output from the cerebral cortex of the cat in the presence of hyoscine. Brit. J. Pharmacol. Chemother. 31, 66–74 (1967).

BATINI, C., MORUZZI, G., PALESTINI, M., ROSSI, G. F., ZANCHETTI, A.: Effects of complete pontine transections on the sleep-wakefulness rhythm: the midpontine pretrigeminal preparation. Arch. ital. Biol. 97, 1–12 (1959).

BATSEL, H. L.: Electroencephalographic synchronization and desynchronization in the chronic "cerveau isolé" of the dog. Electroenceph. clin. Neurophysiol. 12, 421–430 (1960).

BAUST, W., NIEMCZYK, H.: Studies on the adrenaline-sensitive component the mesencephalic reticular formation. J. Neurophysiol. 26, 692–705 (1963).

— — Further studies on the action of adrenergic drugs on cortical activity. Electroenceph. clin. Neurophysiol. 17, 261–271 (1964).

BAUST, W., NIEMCZYK, H., VIETH, J.: The action of blood pressure on the ascending reticular activating system with special reference to adrenaline induced EEG arousal. Electroenceph. clin. Neurophysiol. **15**, 63–72 (1963).

BAXTER, B. L.: Elicitation of emotional behavior by electrical or chemical stimulation applied at the same loci in cat mesencephalon. Exp. Neurol. **21**, 1–11 (1968).

— Induction of both emotional behavior and a novel form of REM sleep by chemical stimulation applied to cat mesencephalon. Exp. Neurol. **23**, 220–230 (1969).

BEANI, L., BIANCHI, C., SANTINOCETO, L., MARCHETTI, P.: The cerebral acetylcholine release in conscious rabbits with semi-permanently implanted epidural cups. Int. J. Neuropharmacol. **7**, 469–481 (1968).

BEDARD, P., LAROCHELLE, L., PARENT, A., POIRIER, L. J.: The nigrostriatal pathway: a correlative study based on neuroanatomical and neurochemical criteria in the cat and the monkey. Exp. Neurol. **25**, 365–377 (1969a).

— — — — Dopamine and serotonin in the striatum of the cat. Arch. Neurol. (Chic.) **20**, 239–242 (1969b).

BELESLIN, D. B., MYERS, R. D.: The release of acetylcholine and 5-hydroxytryptamine from the mesencephalon of the unanesthetized rhesus monkey. Brain Res. **23**, 437–442 (1970).

BELL, L., IVERSEN, L. L., URETSKY, N. J.: Time course of the effects of 6-hydroxydopamine on catecholamine containing neurones in rat hypothalamus and striatum. Brit. J. Pharmacol. **4**, 790–799 (1970).

BELUGOU, J. L., BENOIT, O., LEYGONIE, F.: Décharges neuronales de l'hippocampe au cours de la veille et du sommeil. J. Physiol. (Paris) **60** (Suppl. 2), 399 (1968).

BENDA, P., PERLES, R.: Etude expérimentale de l'abaissement de la vigilance par le gamma butyrolactone. C. R. Acad. Sci. (Paris) **251**, 1312–1313 (1960).

BENOIT, O.: Activité unitaire du nerf optique, du corps genouillé latéral et de la formation réticulaire durant les différents stades du sommeil. J. Physiol. (Paris) **56**, 259–262 (1964).

— Influences toniques et phasiques exercées par le sommeil sur l'activité de la voie visuelle. J. Physiol. (Paris) **59**, 295–317 (1967).

— Influence des états de sommeil sur l'activité spontanée et la transmission dans les voies et les relais sensitifs et visuels. Thèse de Sciences, Paris, 1971. 235 pp.

BERGER, B. D., STEIN, L.: An analysis of the learning deficits produced by scopolamine. Psychopharmacologia (Berl.) **14**, 271–284 (1969).

BERLUCCHI, G.: Callosal activity in unrestrained, unanesthetized cats. Arch. ital. Biol. **103**, 623–635 (1965).

— MAFFEI, L., MORUZZI, G., STRATA, P.: Mécanismes hypnogènes du tronc de l'encéphale antagonistes du système réticulaire activateur. In: Aspects anatomo-fonctionnels de la physiologie du sommeil (JOUVET, M., ed.), p. 89–105. Paris: Centre National de la Recherche Scientifique 1965.

BESSMAN, S. P., FISHBEIN, W. N.: Gamma hydroxybutyrate, a normal brain metabolite. Nature (Lond.) **200**, 1207 (1963).

BIZZI, E.: Changes in the orthodromic and antidromic response of optic tract during the eye movements of sleep. J. Neurophysiol. **29**, 861–871 (1966a).

— Discharge patterns of single geniculate neurons during the rapid eye movements of sleep. J. Neurophysiol. **29**, 1087–1096 (1966b).

— BROOKS, D. C.: Functional connections between pontine reticular formation and lateral geniculate nucleus during deep sleep. Arch. ital. Biol. **101**, 666–680 (1963).

BLASCHKO, H., BURN, J. H., LANGEMANN, H.: The formation of noradrenaline from dihydroxyphenylserine. Brit. J. Pharmacol. Chemother. **5**, 431–437 (1950).

BLISS, E. L., AILION, J., ZWANZIGER J.: Metabolism of norepinephrine, serotonin and dopamine in rat brain with stress. J. Pharmacol. exp. Ther. **164**, 122–134 (1968).

— ZWANZIGER, J.: Brain amines and emotional stress. J. Psychiat. Res. **4**, 189–199 (1966).

BLOOM, F. E.: Lesions of central norepinephrine terminals with 6-hydroxy-dopamine: chemistry and fine structure. Science 166, 1284–1286 (1969).
— AGHAJANIAN, G. K.: An electron microscopic analysis of large granular synaptic vesicles of the brain in relation to monoamine content. J. Pharmacol. exp. Ther. 159, 261–273 (1968).
— COSTA, E., SALMOIRAGHI, G. C.: Analysis of individual rabbit olfactory bulb neuron responses to the microelectrophoresis of acetylcholine, norepinephrine and serotonin synergists and antagonists. J. Pharmacol. exp. Ther. 146, 16–24 (1964).
— GIARMAN, N. J.: Physiologic and pharmacologic considerations of biogenic amines in the nervous system. Ann. Rev. Pharmacol. 8, 229–259 (1968).
BOAKES, R. J., BRADLEY, P. B., BRIGGS, I., DRAY, A.: Effects of lysergic acid derivatives on 5-hydroxytryptamine excitations of brain stem neurons. Brit. J. Pharmacol. 38, 453–454 (1970a).
— — — — Antagonism of 5-hydroxytryptamine by LSD 25 in the central nervous system: a possible neuronal basis for the actions of LSD 25. Brit. J. Pharmacol. 40, 202–218 (1970b).
— — BROOKES, N., CANDY, J. M., STENCROFT, J. H.: Effects of noradrenaline and its analogues on brain stem neurons. J. Physiol. (Lond.) 201, 20 (1969).
BOBILLIER, P., MOURET, J. R.: Sleep and brain biogenic amines diurnal rhythms in the rat. II. Alterations of brain biogenic amines diurnal rhythms independent of light. Int. J. Neurosciences 2, 271–281 (1971).
BOKUMS, J. A., ELLIOTT, H. W.: Effects of physostigmine in electrical activity of the cat brain. Pharmacology 1, 98–111 (1968).
BONVALLET, M., ALLEN, B. J.: Prolonged spontaneous and evoked reticular activation following discrete bulbar lesions. Electroenceph. clin. Neurophysiol. 15, 969–988 (1963).
— BLOCH, V.: Bulbar control of cortical arousal. Science 133, 1133–1134 (1961).
— DELL, P.: Contrôle bulbaire du système activateur. In: Aspects anatomo-fonctionnels de la physiologie du sommeil (JOUVET, M., ed.). Paris: Centre National de la Recherche Scientifique 1965.
— — HIEBEL, G.: Sinus carotidien et activité électrique cérébrale. C. R. Soc. Biol. (Paris) 147, 1166–1170 (1953).
— — — Tonus sympathique et activité électrique corticale. Electroenceph. clin. Neurophysiol. 6, 119–144 (1954).
BOURNE, G. H. (ed.): The structure and function of nervous tissue. Vol. III Biochemistry and disease. New York: Academic Press 1969.
BOWERS, M. B., HARTMANN, E. L., FREEDMAN, D. X.: Sleep deprivation and brain acetylcholine. Science 153, 1416–1417 (1966).
BRADLEY, P. B., CANDY, J. M.: Iontophoretic release of acetylcholine, noradrenaline, 5-hydroxytryptamine and D-lysergic acid diethylamide from micropipettes. Brit. J. Pharmacol. 40, 194–202 (1970).
— DHAWAN, B. N., WOLSTENCROFT, J. H.: Pharmacological properties of cholinoceptive neurons in the medulla and pons of the cat. J. Physiol. (Lond.) 183, 658–675 (1966).
— ELKES, J.: The effects of some drugs on the electrical activity of the brain. Brain 80, 77–117 (1957).
— KEY, B. J.: The effect of drugs on arousal responses produced by electrical stimulation of the reticular formation of the brain. Electroenceph. clin. Neurophysiol. 10, 97–110 (1958).
— WOLSTENCROFT, J. H.: Actions of drugs on single neurons in the brain stem. Brit. med. Bull. 21, 15–18 (1965).
— — HÖSLI, L., AVANZINO, G. L.: Neuronal basis for the central action of chlorpromazine. Nature (Lond.) 212, 1425–1428 (1966).
BRANCHEY, M., KISSIN, B.: The effect of alpha methylparatyrosine on sleep and arousal in the rat. Psychosom. Sci. 19, 281–282 (1970).
BRAUCHI, J. T., WEST, L. J.: Sleep deprivation. J. Amer. med. Ass. 171, 11–14 (1959).

BREESE, G. R., DENNIS TRAYLOR, T.: Effect of 6-hydroxydopamine on brain norepinephrine and dopamine: evidence for selective degeneration of catecholamine neurons. J. Pharmacol. exp. Ther. 174, 413–420 (1970).

BREMER, F.: The neurophysiological problem of sleep. In: Brain mechanisms and consciousness (ADRIAN, E. D., F. BREMER, and H. H. JASPER, eds.), p. 137–162. Oxford: Blackwell 1954.

— CHATONNET, J.: Acétylcholine et cortex cérébral. Arch. intern. Physiol. 57, 106–109 (1949).

BRODAL, A., TABER, E., WALBERG, F.: The raphe nuclei of the brain stem in the cat. II. Efferent connections. J. comp. Neurol. 114, 239–259 (1960a).

— WALBERG, F., TABER, E.: The raphe nuclei of the brain stem in the cat. III. Afferent connections. J. comp. Neurol. 114, 261–281 (1960b).

BRODIE, B. B., COMER, M. S., COSTA, E., DLABAC, A.: The role of brain serotonin in the mechanism of the central action of reserpine. J. Pharmacol. exp. Ther. 152, 340–350 (1966).

— FINGER, K. F., ORLANS, F. B., QUINN, G. P., SULSER, F.: Evidence that tranquillizing drug action of reserpine is associated with change in brain serotonin and not in brain norepinephrine. J. Pharmacol. exp. Ther. 129, 250–256 (1960).

— OLIN, J. S., KUNTZMAN, R. G., SHORE, P. A.: Possible interrelationship between release of brain norepinephrine and serotonin by reserpine. Science 125, 1293–1294 (1957).

— SHORE, P. A.: On a role for serotonin and norepinephrine as chemical mediators in the central autonomic nervous system. In: Hormones, brain function and behaviour (HOAGLAND, H., ed.), p. 161–176. New York: Academic Press 1957.

BRODY, J. F.: Behavioral effects of serotonin depletion and p-chlorophenylalanine (a serotonin depletor in rats). Psychopharmacologia (Berl.) 17, 14–34 (1970).

BROOKS, D. C.: Localization and characteristics of the cortical waves associated with eye movement in the cat. Exp. Neurol. 22, 603–613 (1969).

— Localization of the lateral geniculate nucleus monophasic waves associated with paradoxical sleep in the cat. Electroenceph. clin. Neurophysiol. 23, 123–133 (1967a).

— Effect of bilateral optic nerve section on visual system monophasic wave activity in the cat. Electroenceph. clin. Neurophysiol. 23, 134–141 (1967).

— BIZZI, E.: Brain stem electrical activity during deep sleep. Arch. ital. Biol. 101, 648–665 (1963).

— GERSHON, M. D.: Eye movement potentials in the oculomotor and visual system: A comparison in reserpine induced waves with those present during wakefulness and rapid eye movement sleep. Brain Res. 27, 223–239 (1971).

BUGUET, A.: Monoamines et sommeils. V. Etude des relations entre les structures monoaminergiques du pont et les pointes ponto-geniculo occipitales du sommeil. Thèse Médecine, Lyon, 1969. Tixier ed. 214 pp.

— PETITJEAN, F., JOUVET, M.: Suppression des pointes PGO du sommeil par lésion ou injection in situ de 6-hydroxydopamine au niveau du tegmentum pontique. C. R. Soc. Biol. (Paris) 164, 2293–2298 (1970).

BURKARD, W. P., JALFRE, M., BLUM, J.: Effect of 6-hydroxydopamine on behaviour and cerebral amine content in rats. Experientia (Basel) 25, 1295–1296 (1969).

BURN, J. H., RAND, M. J.: Acetylcholine in adrenergic transmission. Ann. Rev. Pharmacol. 5, 163–183 (1965).

CALVET, J., CALVET, M. C., SCHERRER, J.: Etude stratigraphique corticale de l'activité E.E.G. spontanée. Eletroenceph. clin. Neurophysiol. 17, 109–125 (1964).

CAPON, A.: Nouvelles recherches sur l'effet d'éveil de l'adrénaline. J. Physiol. (Paris) 51, 424–425 (1959).

CARLI, G., DIETE-SPIFF, K., POMPEIANO, O.: Presynaptic and postsynaptic inhibition on transmission of cutaneous afferent volleys through the cuneate nucleus during sleep. Experientia (Basel) 22, 239–241 (1966).

CARLI, G., ZANCHETTI, A.: A study of pontine lesions suppressing deep sleep in the cat. Arch. ital. Biol. **103**, 751–789 (1965).

CARLSSON, A.: Functional significance of drug induced changes in brain monoamine levels. In: Biogenic amines. A symposium (H. and W. A. HIMWICH, eds.). Amsterdam: Elsevier 1964.

— LINDQVIST, M., FUXE, K., HÖKFELT, T.: Histochemical and biochemical effects of diethyldithiocarbamate on tissue catecholamines. J. Pharm. Pharmacol. **18**, 60–62 (1966).

— — MAGNUSSON, T., WALDECK, B.: On the presence of 3-hydroxytyramine in brain. Science **127**, 471 (1958).

— MEISCH, J. T., WALDECK, B.: On the beta-hydroxylation of (±)-α-methyldopamine *in vivo*. Europ. J. Pharmacol. **5**, 85–92 (1968).

CARR, L. A., MOORE, K. E.: Release of norepinephrine and normetanephrine from cat brain by central nervous system stimulants. Biochem. Pharmacol. **19**, 2671–2675 (1970).

CELESIA, G. G., JASPER, H. H.: Acetylcholine released from cerebral cortex in relation to state of activation. Neurology (Minneap.) **16**, 1053–1064 (1966).

CHASE, T. N., KOPIN, J.: Stimulus-induced release of substances from olfactory bulb using the push-pull cannula. Nature (Lond.) **217**, 484–485 (1968).

CLEMENTE, C. D., STERMAN, M. B., WYRWICKA, W.: Forebrain inhibitory mechanisms: conditioning of basal forebrain induced EEG synchronization and sleep. Exp. Neurol. **7**, 404–417 (1963).

COHEN, B., FELDMAN, M.: Relationship of electrical activity in pontine reticular formation and lateral geniculate body to rapid eye movements. J. Neurophysiol. **31**, 806–818 (1968).

— — DIAMOND, S. P.: Effects of eye movement, brain-stem stimulation, and alertness on transmission through lateral geniculate body of monkey. J. Neurophysiol. **32**, 583–595 (1969).

COHEN, H., FERGUSON, J., HENRIKSEN, S., BARCHAS, J., DEMENT, W.: Reversal of p-chorophenylalanine effects with chlorpromazine. Psychophysiology **6**, 221–222 (1969).

— — — STOLK, J. M., ZARCONE, V. J., BARCHAS, J., DEMENT, W.: Effects of chronic depletion of brain serotonin on sleep and behavior. Proc. Amer. Psychol. Ass. **78**, 831–832 (1970).

COLBURN, N. G., KOPIN, I. J.: Effect of L-DOPA on efflux of cerebral monoamines from synaptosomes. Nature (Lond.) **230**, 331–332 (1971).

COLLIER, B., MITCHELL, J. F.: The central release of acetylcholine during consciousness and after brain lesions. J. Physiol. (Lond.) **188**, 83–99 (1967).

— — Release of acetylcholine from the cerebral cortex during stimulation of the optic pathway. Nature (Lond.) **210**, 424 (1966).

CONNER, R. L., STOLK, J. M., BARCHAS, J. P., LEVINE, S.: Parachlorophenylalanine and habituation to repetitive auditory startle stimuli in rats. Physiol. Behav. **5**, 758–762 (1970).

CONSOLO, S., GARATTINI, S., GHIELMETTI, R., MORSELLI, P., VALZELLI, L.: The hydroxylation of tryptophan *in vivo* by brain. Life Sci. **4**, 625–630 (1965).

CONSTANTINIDIS, J., BARTHOLINI, G., GEISSBÜHLER, F., TISSOT, R.: La barrière capillaire enzymatique pour la DOPA au niveau de quelques noyaux du tronc cérébral du rat. Experientia (Basel) **26**, 381–383 (1970).

— — TISSOT, R., PLETSCHER, A.: Accumulation of dopamine in the parenchyma after decarboxylase inhibition in the capillaries of brain. Experientia (Basel) **24**, 130–132 (1968).

— TORRE, J. C. DE LA, TISSOT, R., GEISSBÜHLER, F.: La barrière capillaire pour la DOPA dans le cerveau et les différents organes. Psychopharmacologia (Berl.) **15**, 75–87 (1969).

— — — HUGGEL, H.: Les monoamines cérébrales lors de l'hibernation chez la chauvesouris. Rev. suisse Zool. **77**, 345–352 (1970).

Cordeau, J. P.: Functional organization of the brain stem reticular formation in relation to sleep and wakefulness. Rev. canad. Biol. 21, 113–125 (1962).
— Monoamines and the physiology of sleep and waking. In: L DOPA and parkinsonism (A. Barbeau and F. H. McCowell, eds.), p. 369–383. Philadelphia: Davis 1970.
— Mancia, M.: Effect of unilateral chronic lesions of the midbrain on the electrocortical activity of the cat. Arch. ital. Biol. 96, 374–379 (1958).
— — Evidence for the existence of an electroencephalographic synchronization mechanism originating in the lower brain stem. Electroenceph. clin. Neurophysiol. 11, 551–564 (1959).
— Moreau, A., Beaulnes, A., Lawrin, C.: EEG and behavioural changes following microinjections of acetylcholine and adrenaline in the brain stem of cats. Arch. ital. Biol. 101, 30–47 (1963).
Corrodi, H., Fuxe, K.: Decrease turnover in central 5-HT nerve terminals induced by antidepressant drugs of the imipramine type. Canad. J. Physiol. Pharmacol. 47, 56–59 (1969).
— — Hökfelt, T.: Replenishment by 5-hydroxytryptophan of the amine stores in the central 5-hydroxytryptamine neurons after depletion induced by reserpine or by an inhibitor of monoamine synthesis. J. Pharm. Pharmacol. 19, 433–438 (1967).
Costa, E.: The role of serotonin in neurobiology. Neurobiology 2, 175–227 (1960).
— Brodie, B. B.: Concept of the neurochemical transducer as an organized molecular unit at sympathetic nerve endings. In: Biogenic amines. A symposium (H. and W. A. Himwich, eds.). Amsterdam: Elsevier 1964.
— Garattini, S. (eds.): Amphetamines and related compounds. New York: Raven Press 1970.
— Giacobini, E. (eds.): Biochemistry of simple neuronal models. New York: Raven Press 1970.
— Pscheidt, G. R., Meter, W. G. van, Himwich, H. E.: Brain concentrations of biogenic amines and EEG patterns of rabbits. J. Pharmacol. exp. Ther. 130, 81–88 (1960).
Costin, A., Hafemann, D. R.: Relationship between oculomotor nucleus and lateral geniculate body monophasic waves. Experientia (Basel) 26, 972–973 (1970).
Couch, J. R.: Responses of neurons in the raphe nuclei to serotonin, norepinephrine and acetylcholine and their correlation with an excitatory synaptic output. Brain Res. 19, 137–150 (1970).
Coulter, J. D., Lester, B. K., Williams, H. L.: Reserpine and sleep. Psychopharmacologia (Berl.) 19, 134 (1971).
Cramer, H., Kuhlo, W.: Effets des inhibiteurs de la monoamine oxydase sur le sommeil et l'électroencéphalogramme chez l'homme. Acta neurol. belg. 67, 658–669 (1967).
Crawford, J. M.: The sensitivity of cortical neurons to acidic aminoacids and acetylcholine. Brain Res. 17, 287–297 (1970).
Cuculic, Z., Bost, K., Himwich, H. E.: An examination of a possible cortical cholinergic link in the EEG arousal reaction. Progr. Brain Res. 28, 27–39 (1968).
Curtis, D. R.: Acetylcholine as a central transmitter. Symposium on function of acetylcholine as a synaptic transmitter. Canad. J. Biochem. Physiol. 41, 2611–2619 (1963).
— Crawford, J. M.: Central synaptic transmission—microelectrophoretic studies. Ann. Rev. Pharmacol. 9, 209–241 (1969).
— Davis, R.: Pharmacological studies upon neurons of the lateral geniculate nucleus of the cat. Brit. J. Pharmacol. 18, 217–246 (1962).
Dagnino, N., Favale, E., Loeb, C., Manfredi, M.: Sensory transmission in the geniculostriate system of the cat during natural sleep and arousal. J. Neurophysiol. 28, 443–457 (1965).
— — — — Seitun, A.: Accelerated synaptic transmission in nucleus ventralis-posterolateralis during deep sleep. Experientia (Basel) 22, 329–330 (1966).
Dahl, D. R.: Short chain fatty acid inhibition of rat brain NA-K adenosine triphosphatase. J. Neurochem. 15, 815–820 (1968).

DAHLSTRÖM, A., FUXE, K.: Evidence for the existence of monoamine neurons in the central nervous system. I. Demonstration of monoamines in the cell bodies of brain stem neurons. Acta physiol. scand. **62**, supp. 232 (1964).

— — Evidence for the existence of monoamine neurons in the central nervous system. II. Experimentally induced changes in the intraneuronal amine levels of bulbospinal neuron systems. Acta physiol. scand. **64**, Suppl. 247 (1965a).

— — Evidence for the existence of an outflow of noradrenaline nerve fibers in the ventral roots of the rat spinal cord. Experientia (Basel) **21**, 409–410 (1965b).

— — KERNELL, D., SEDVALL, G.: Reduction of the monoamine stores in the terminals of bulbospinal neurons following stimulation in the medulla oblongata. Life Sci. **4**, 1207–1213 (1965).

— — OLSON, L., UNGERSTEDT, U.: Ascending systems of catecholamine neurons from the lower brain stem. Acta physiol. scand. **62**, 485–487 (1964).

DALE, H. H.: Pharmacology and nerve endings. Proc. roy. Soc. B **28**, 319–332 (1935).

DAVIS, J. M.: Theories of biological etiology of affective disorders. Int. Rev. Neurobiol. **12**, 145–175 (1970).

DELL, P.: Intervention of an adrenergic mechanism during brain stem reticular activation. In: Adrenergic mechanisms. Ciba foundation symposium (VARRE, J. R., G. E. W. WOLSTENHOLME and C. M. O'CONNOR, eds.), p. 393–409. London: Churchill 1960.

— Reticular homeostasis and cortical reactivity. In: Brain mechanisms (MORUZZI, G., A. FESSARD and H. H. JASPER, eds.), p. 82–103. Amsterdam: Elsevier 1963.

DELORME, F.: Monoamines et sommeils. Etude polygraphique neuropharmacologique et histochimique des états de sommeil chez le chat. Thèse Université de Lyon, Imprimerie LMD 1966. 168 pp.

— FROMENT, J. L., JOUVET, M.: Suppression du sommeil par la p-chloromethamphetamine et p-chlorophenylalanine. C. R. Soc. Biol. (Paris) **160**, 2347–2351 (1966a).

— JEANNEROD, M., JOUVET, M.: Effets remarquables de la réserpine sur l'activité EEG phasique pontogeniculo occipitale. C. R. Soc. Biol. (Paris) **159**, 900–903 (1965).

— RIOTTE, M., JOUVET, M.: Conditions de déclenchement du sommeil paradoxal par les acides gras à chaîne courte chez le chat pontique chronique. C. R. Soc. Biol. (Paris) **160**, 1457–1460 (1966b).

— VIMONT, P., JOUVET, D.: Etude statistique du cycle veille-sommeil chez le chat. C. R. Soc. Biol. (Paris) **158**, 2128–2130 (1964).

DEMENT, W.: The occurrence of low voltage, fast, electroencephalogram patterns during behavioral sleep in the cat. Electroenceph. clin. Neurophysiol. **10**, 291–296 (1958).

— The effect of dream deprivation. Science **131**, 1705–1707 (1960).

— FERGUSON, J., COHEN, H., BARCHAS, J.: Non chemical methods and data using a biochemical model: The REM quanta, p. 275–325. In: Some current issues in psychochemical research strategies in man (MANDELL, A., ed.). New York: Academic Press 1970a.

— RECHTSCHAFFEN, A., GULEVICH, G.: The nature of the narcoleptic sleep attack. Neurology (Minneap.) **16**, 18–34 (1966).

— ZARCONE, V., FERGUSON, J., COHEN, H., PIVIK, T., BARCHAS, J.: Some parallel findings in schizophrenic patients and serotonin depleted cats. In: Schizophrenia. Current concepts and resaerch (SIVA-SANKAR, P. J. D., ed.), p. 775–811. New York: Hicksville 1970b.

DESMEDT, J. E., LA GRUTTA, G.: The effect of selective inhibition of pseudo cholinesterase on the spontaneous and evoked activity of the cat's cerebral cortex. J. Physiol. (Lond.) **136**, 20–40 (1957).

DEUTSCH, J. A., ROCKLIN, K. W.: Amnesia induced by scopolamine and its temporal variations. Nature (Lond.) **216**, 89–90 (1967).

DEWHURST, W. G.: New theory of cerebral amine function and its clinical application. Nature (Lond.) **218**, 1130–1133 (1968).

Dewhurst W. G., Marley, E.: Action of sympathomimetic and allied amines on the central nervous system of the chicken. Brit. J. Pharmacol. Chemother. 25, 705–728 (1965).

Dixit, B. N., Buckley, J. P.: Circadian changes in brain 5-hydroxytryptamine and plasma. Life Sci. 6, 755–759 (1967).

Dominic, J. A., Moore, K. E.: Acute effects of alpha-methyltyrosine on brain catecholamine levels and on spontaneous and amphetamine-stimulated motor activity in mice. Arch. int. Pharmacol. Ther. 178, 166–176 (1969).

Domino, E. F., Stawiski, M.: Effects of the cholinergic antisynthesis agent HC-3 on the awake-sleep cycle of cat. Psychophysiology 7, 315–316 (1971).

— Yamamoto, K.: Nicotine: Effect on the sleep cycle of the cat. Science 150, 637–638 (1965).

— — Dren, A. T.: Role of cholinergic mechanisms in states of wakefulness and sleep. Progr. Brain Res. 28, 113–133 (1968).

Dresse, A.: Importance du système mésencéphalo-télencéphalique noradrénergique comme substratum anatomique du comportement d'autostimulation. Life Sci. 5, 1003–1014 (1966).

Drucker-Colin, R. R., Rojas-Ramirez, J. A., Vera-Trueba, J., Monroy-Ayala, G., Hernandez-Péon, R.: Effect of crossed perfusion of the midbrain reticular formation upon sleep. Brain Res. 23, 269–273 (1970).

Dusan-Peyrethon, D.: Etude dynamique et neuropharmacologique des phénomènes phasiques du sommeil paradoxal. Thèse de Médecine, Lyon: J. Tixier & Fils (ed.) 1968. 111 pp.

— Froment, J. L.: Effets du disulfiram sur les états de sommeil chez le chat. C. R. Soc. Biol. (Paris) 162, 2141–2145 (1968).

— Peyrethon, J., Jouvet, M.: Suppression élective du sommeil paradoxal chez le chat par alpha méthyl-dopa. C. R. Soc. Biol. (Paris) 162, 116–118 (1968).

— — — Etude quantitative des phénomènes phasiques du sommeil paradoxal pendant et après sa déprivation instrumentale. C. R. Soc. Biol. (Paris) 161, 2530–2533 (1967).

Eccleston, D., Padjen, A., Randic, M.: Release of 5-hydroxytryptamine 5-hydroxyindol-3-ylacetic acid in the forebrain by stimulation of midbrain raphe. J. Physiol. (Lond.) 201, 22 pp. (1969).

— Ritchie, I. M., Roberts, M. H. T.: Long term effects of midbrain stimulation on 5-hydroxyindole synthesis in rat brain. Nature (Lond.) 226, 84–85 (1970).

Eränkö, O.: Histochemistry of nervous tissues: catecholamines and cholinesterases. Ann. Rev. Pharmacol. 7, 203–223 (1967).

Erspamer, V. (ed.): 5-hydroxytryptamine and related indole alkylamines. In: Handbook of experimental pharmacology, 928 pp. Berlin-Heidelberg-New York: Springer 1966.

Evans, J. I.: Neurochemical basis for sleep. In: The abnormalities of sleep in man (Gastaut, H., E. Lugaresi, G. Berti Coroni and G. Coccagna ed.). 326 pp. Bologna: Aulogaggi 1968.

— Oswald, I.: Some experiments in the chemistry of narcoleptic sleep. Brit. J. Psychiat. 112, 401–404 (1966).

Evetts, K. D., Uretsky, N. J., Iversen, L. L., Iversen, S. D.: Effects of 6-hydroxydopamine on CNS catecholamines, spontaneous motor activity and amphetamine induced hyperactivity in rats. Nature (Lond.) 225, 961–962 (1970).

Falck, B., Hillarp, N. A., Thieme, G., Thorp, A.: Fluorescence of catecholamines and related compounds condensed with formaldehyde. J. Histochem. Cytochem. 10, 348–354 (1962).

Feldberg, W., Hellon, R. F., Myers, R. D.: Effects on temperature of monoamines injected into the cerebral ventricles of anaesthetized dogs. J. Physiol. (Lond.) 186, 416–424 (1966).

FELDMAN, M., COHEN, B.: Electrical activity in the lateral geniculate body of the alert monkey associated with eye movements. J. Neurophysiol. **31**, 455–467 (1968).

FELDMAN, S. M., WALLER, H. J.: Dissociation of electrocortical activation and behavioural arousal. Nature (Lond.) **196**, 1320–1322 (1962).

FELDSTEIN, A., CHANG, F. H., KUCHARSKI, J. M.: Tryptophol, 5-hydroxytryptophol and 5-methoxytryptophol induced sleep in mice. Life Sci. **9**, 323–329 (1970).

FERGUSON, J., HENRIKSEN, S., COHEN, H., MITCHELL, G., BARCHAS, J., DEMENT, W. C.: Hypersexuality and behavioral changes in cats caused by administration of p-chloro-phenylalanine. Science **168**, 499–501 (1970).

FERRY, C. B.: Cholinergic link hypothesis in adrenergic neuroeffector transmission. Physiol. Rev. **46**, 420–457 (1966).

FIBIGER, H. C., CAMPBELL, B. A.: The effect of para-chlorophenylalanine on spontaneous locomotor activity in the rat. Neuropharmacology **10**, 25–32 (1971).

— LYTLE, L. D., CAMPBELL, B. A.: Cholinergic modulation of adrenergic arousal in the developing rat. J. comp. physiol. Psychol. **72**, 384–390 (1970).

FINLEY, K. H., COBB, S.: The capillary bed of the locus coeruleus. J. comp. Neurol. **73**, 49–58 (1940).

FISHBEIN, W., McGAUGH, J. L., SWARZ, J. R.: Retrograde amnesia: Electroconvulsive shock effects after termination of rapid eye movement sleep deprivation. Science **172**, 80–82 (1971).

FLORIO, V., SCOTTI DE CAROLIS, A., LONGO, V. G.: Observations on the effect of DL-Parachlorophenylalanine on the electroencephalogram. Physiol. Behav. **3**, 861–865 (1968).

FREDERICKSON, C. J., HOBSON, J. A.: Electrical stimulation of the brain stem and subsequent sleep. Arch. ital. Biol. **108**, 564–577 (1970).

FREEDMAN, D. X., GIARMAN, N. J.: Brain amines, electrical activity and behavior. In: EEG and behavior (GLASER, G. H., ed.), p. 198–243. New York: Basic Books 1963.

FRIEDMAN, A. H., WALKER, C. A.: Circadian rhythms in rat mid-brain and caudate nucleus biogenic amine levels. J. Physiol. (Lond.) **197**, 77–85 (1968).

FROMENT, J. L., ESKAZAN, E., JOUVET, M.: Effects du LSD et du methysergide sur les états de sommeil du chat. C. R. Soc. Biol. (Paris) (in press).

FUJIMORI, M., HIMWICH, H. E.: Electroencephalographic analyses of amphetamine and its methoxy derivatives with reference to their sites of EEG alerting in the rabbit brain. Int. J. Neuropharmacol. **8**, 601–615 (1969).

— — The effect of alpha methyl p-tyrosine on sleep in rabbits. Fed. Proc., Abs. 541 (1971).

FULLER, R. W., HINES, C. W.: Inhibition by p-chloramphetamine of the conversion of 5-hydroxytryptamine to 5-hydroxyindole acetic acid in rat brain. J. Pharm. Pharmacol. **22**, 634–635 (1970).

— — MILLS, J.: Lowering of brain serotonin level by chloramphetamines. Biochem. Pharmacol. **14**, 483–489 (1965).

FUXE, K.: Evidence for the existence of monoamine neurons in the central nervous system. IV. Distribution of monoamine nerve terminals in the central nervous system. Acta physiol. scand. **64** (Suppl. 247), 39–85 (1965).

— GUNNE, L. M.: Depletion of the amine stores in brain catecholamine terminals on amygdaloid stimulation. Acta physiol. scand. **62**, 493 (1964).

— HANSON, L. C. F.: Central catecholamine neurons and conditioned avoidance behaviour. Psychopharmacologia (Berl.) **11**, 439 (1967).

— HÖKFELT, T., RITZEW, M., UNGERSTEDT, U.: Studies on uptake of intraventricularly administered tritiated noradrenaline and 5-hydroxytryptamine with combined fluorescence histochemical and autoradiographic techniques. Histochemie **16**, 186–195 (1968).

— OWMAN, C.: Cellular localization of monoamines in the area postrema of certain mammals. J. comp. Neurol. **125**, 337–355 (1965).

Fuxe, K., Ungerstedt, U.: Localization of 5-hydroxytryptamine uptake in rat brain after intraventricular injection. J. Pharm. Pharmacol. 19, 335–337 (1967).
— — Histochemical studies on the distribution of catecholamines and 5-hydroxy-tryptamine after intraventricular injections. Histochemie 13, 16–28 (1968).
Gadea-Ciria, E., Jouvet, M.: Corticofugal control of pontine PGO activity in the cat. Psychophysiology — in press.
Gaillard, J. M., Schaeppi, R., Tissot, R.: Potentialisation des effets centraux de la DOPA après inhibition sélective de la décarboxylase extracérébrale. Arch. int. Pharmacol. Thér. 180, 424–437 (1969).
Gal, E. M., Roggeveen, A. E., Millard, S. A.: Dl (2¹⁴C) p-chlorophenylalanine as an inhibitor of tryptophan 5-hydroxylase. J. Neurochem. 17, 1221–1235 (1970).
Gardiner, J. E.: The inhibition of acetylcholine synthesis in brain by a hemicholinium. Biochem. J. 81, 297–303 (1961).
George, R., Haslett, W. L., Jenden, D. J.: A cholinergic mechanism in the brain-stem reticular formation: induction of paradoxical sleep. Int. J. Pharmacol. 3, 451–552 (1964).
Gerardy, J., Quinaux, N., Maeda, T., Dresse, A.: Analyse des monoamines du locus coeruleus et d'autres structures cérébrales par chromatographie sur couche mince. Arch. int. Pharmacol. Thér. 177, 492–496 (1969).
Gessa, G. L., Crabai, F., Vargiu, L., Spano, P. F.: Selective increase of brain dopamine induced by gamma-hydroxybutyrate: study of the mechanism of action. J. Neurochem. 15, 377–383 (1968).
— Tagliamonte, A., Tagliamonte, P.: Aphrodisiac effect of p-chlorophenylalanine. Science 171, 706 (1971).
Giaquinto, S., Pompeiano, O., Somogyi, I.: Descending inhibitory influences on spinal reflexes during natural sleep. Arch. ital. Biol. 102, 282–308 (1964).
Giarman, N. J.: Neurohumors in the brain. Yale J. Biol. Med. 32, 73–92 (1959).
— Roth, R. H.: Differential estimation of gamma butyrolactone and gamma hydroxy-butyric acid in rat blood and brain. Science 145, 583–584 (1964).
— Schmidt, K. F.: Some neurochemical aspects of the depressant action of 3-butyro-lactone on the central nervous system. Brit. J. Pharmacol. Chemother. 20, 563–568 (1963).
Glowinski, J.: Storage and release of monoamines in the central nervous system. In: Handbook of neurochemistry, vol. 4. New York-London: Plenum Press 1970.
— Axelrod, J., Iversen, L. L.: Regional studies of catecholamines in the rat brain. IV. Effects of drugs on the disposition and metabolism of H³Norepinephrine and H³Dopamine. J. Pharmacol. exp. Ther. 153, 30–42 (1966a).
— — Kopin, I. J., Wurtman, R. J.: Physiological disposition of H³Norepinephrine in the developing rat. J. Pharmacol. exp. Ther. 146, 48–54 (1964).
— Baldessarini, R. J.: Metabolism of norepinephrine in the central nervous system. Pharmacol. Rev. 18, 1201–1238 (1966).
— Iversen, L. L., Axelrod, J.: Storage and synthesis of norepinephrine in the reserpine-treated rat brain. J. Pharmacol. exp. Ther. 151, 385–400 (1966b).
Godin, Y., Mark, J., Mandel, P.: The effects of 4-hydroxybutyric acid on the bio-synthesis of amino acids in the central nervous system. J. Neurochem. 15, 1085–1093 (1968).
Goldstein, M., Lauber, E., McKereghan, M. R.: The inhibition of dopamine-β-hydroxylase by tropolone and other chelating agents. Biochem. Pharmacol. 13, 1103 (1964).
— Nakajima, K.: The effect of disulfiram on catecholamine levels in the brain. J. Pharmacol. exp. Ther. 157, 96–103 (1967).
Gordon, R., Spector, S., Sjoerdsma, A., Udenfriend, S.: Increased synthesis of norepinephrine and epinephrine in the intact rat during exercise and exposure to cold. J. Pharmacol. exp. Ther. 153, 440–448 (1966).

GOTTESMANN, C.: Reserpine et vigilance chez le rat. C. R. Soc. Biol. (Paris) 160, 2056–2061 (1966).

GREEN, H., SAWYER, J. L.: Biochemical pharmacological studies with 5-hydroxy-tryptophan, precursor of serotonin. In: Biogenic amines (HIMWICH ed.). Amsterdam: Elsevier 1964.

GREEN, J. R., HALPERN, M., NIEL, S. VAN: Choline acetylase and acetylcholine esterase changes in chronic isolated cerebral cortex of cat. Life Sci. 9, 481–489 (1970).

GULEVICH, G., DEMENT, W., JOHNSON, L.: Psychiatric and EEG observations on a case of prolonged (264 hours) wakefulness. Arch. gen. Psychiat. 15, 29–35 (1966).

GUMULKA, W., SAMANIN, R., VALZELLI, L., CONSOLO, S.: Behavioural and biochemical effects following the stimulation of the nucleus raphis dorsalis on rats. J. Neurochem. 18, 533–534 (1971).

HAGGENDAL, C. J., DAHLSTRÖM, A. B.: The transport and life-span of amine storage granules in bulbospinal noradrenaline neurons of the rat. J. Pharm. Pharmacol. 21, 55–57 (1969).

HANIG, J. P., AIELLO, E., SEIFTER, J.: Permeability of the blood drain barrier to parenteral 5-hydroxytryptamine in the neonate chick. Europ. J. Pharmacol. 12, 180–182 (1970).

HANIN, I., MASSARELLI, R., COSTA, E.: Acetylcholine concentrations in rat brain: diurnal oscillations. Science 170, 341–342 (1970).

HANSON, L. C. F.: Biochemical and behavioral effects of tyrosine hydroxylase inhibition. Psychopharmacologia (Berl.) 11, 8–18 (1967a).

— Evidence that the central action of (+) amphetamine is mediated via catecholamines. Psychopharmacologia (Berl.) 10, 289–298 (1967b).

— HENNING, M.: Effects of α-methyl-dopa on conditioned behaviour in the cat. Psychopharmacologia (Berl.) 11, 1–8 (1967).

HARTMANN, E.: Reserpine: its effect on the sleep-dream cycle in man. Psychopharmacologia (Berl.) 9, 242–248 (1966).

— On the pharmacology of dreaming sleep (the D state). J. nerv. ment. Dis. 146, 165-173 (1968a).

— The effect of LSD and L tryptophane on the sleep dream cycle in the rat. Psychophysiology 4, 390 (1968b).

— The biochemistry and pharmacology of the D state (dreaming sleep). Exp. Med. Surg. 27, 105-120 (1969a).

— Pharmacological studies of sleep and dreaming chemical and clinical relationships. Biol. Psychiat. 1, 243–258 (1969b).

— The D state and norepinephrine dependant systems. In: Sleep and dreaming, p. 308–328 (E. HARTMANN, ed.). Boston: Little Brown 1970.

— CHUNG, R., CHING-PIAO, C.: L-Tryptophane and sleep. Psychopharmacology 19, 114 (1971).

HARVEY, J. A., HELLER, A., MOORE, R. Y.: The effect of unilateral and bilateral medial forebrain bundle lesions on brain serotonin. J. Pharmacol. exp. Ther. 140, 103–110 (1963).

HASHIMOTO, P. H., MAEDA, T., TORU, K., SHIMIZU, N.: Histochemical demonstration of autonomic regions in the central nervous system of the rabbit by means of a monoamine oxydase staining. Med. J. Osaka Univ. 12, 425–464 (1962).

HAVLIČEK, V.: The effect of dl-3,4-dihydroxyphenylserine (precursor of noradrenaline) on the ECOG of unrestrained rats. Int. J. Neuropharmacol. 6, 83–89 (1967).

— Some aspects of the neurophysiological and neurobiochemical effects of catecholamines on the central nervous system. Verh. dtsch. Ges. exp. Med. 24, 28–34 (1969).

HAZRA, J.: Effect of hemicholinium-3 on slow wave and paradoxical sleep of cat. Europ. J. Pharmacol 11, 395–397 (1970).

HEBB, C.: CNS at the cellular level: identity of transmitter agents. Ann. Rev. Physiol. 32, 165–192 (1970).

HELLER, A., HARVEY, H. A., MOORE, R. Y.: A demonstration of a fall in brain serotonin following central nervous system lesions in the rat. Biochem. Pharmacol. **11**, 859–866 (1962).

HENLEY, K., MORRISON, A.: Release of organized behavior during desynchronized sleep in cats with pontine lesion. Psychophysiology **6**, 245 (1969).

HERMAN, Z. S.: The effects of noradrenaline on rats behaviour. Psychopharmacologia (Berl.) **16**, 369–374 (1970).

HERNANDEZ-PÉON, R.: A neurophysiologic model of dreams and hallucinations. J. nerv. ment. Dis. **141**, 623–651 (1965a).

— A cholinergic hypnogenic limbic forebrain-hindbrain circuit. In: Aspects anatomo-fonctionnels du sommeil (JOUVET, M., ed.), p. 63–88. Paris: Centre National de la Recherche Scientifique 1965b.

— CHAVEZ-IBARRA, G.: Sleep induced by localized electrical or chemical stimulation of the forebrain. Electroenceph. clin. Neurophysiol., Suppl. **24**, 188–198 (1963).

— — MORGANE, P. J., TIMO-IARA, C.: Limbic cholinergic pathways involved in sleep and emotional behavior. Exp. Neurol. **8**, 93–111 (1963).

— O'FLAHERTY, J. J., MAZZUCHELLI-O'FLAHERTY, A. L.: Sleep and other behavioural effects induced by acetylcholinic stimulation of basal temporal cortex and striate structure. Brain Res. **4**, 243–267 (1967).

HERY, F., PUJOL, J. F., LOPEZ, M., MACON, J., GLOWINSKI, J.: Increased synthesis and utilization of serotonin in the central nervous system in the rat during paradoxical sleep deprivation. Brain Res. **21**, 391–403 (1970).

HESS, W. R.: Das Schlafsyndrom als Folge dienzephaler Reizung. Helv. physiol. pharmacol. Acta **2**, 305–344 (1944).

HIEBEL, G., BONVALLET, M., HUBE, P., DELL, P.: Analyse neurophysiologique de l'action centrale de la d-amphetamine (maxiton). Sem. Hôp. Paris **30**, 1880–1885 (1954).

HIMWICH, H. E., ALPERS, H. S.: Psychopharmacologia. Ann. Rev. Pharmacol. **10**, 313–334 (1970).

— HIMWICH, W. A. (eds.): Biogenic amines. Progress in brain research, vol. 8. Amsterdam: Elsevier 1964.

HIMWICH, W. A., SCHADE, J. P. (eds.): Horizons in neuropsychopharmacology. Progress in brain research, vol. 16. Amsterdam: Elsevier 1965.

HISHIKAWA, Y., CRAMER, H., KUHLO, W.: Natural and melatonin induced sleep in young chickens. A behavioral and electrographic study. Exp. Brain Res. **7**, 84–88 (1969).

— NAKAI, K., HIDENOBU, I., KANEKO, Z.: The effect of imipramine, desmethylimipramine and chlorpromazine on the sleep wakefulness cycle of the cat. Electroenceph. clin. Neurophysiol. **19**, 518–521 (1965).

HOBSON, J. A.: The effects of chronic brain stem lesions on cortical and muscular activity during sleep and waking in the cat. Electroenceph. clin. Neurophysiol. **19**, 41–62 (1965).

— ALEXANDER, J., FREDERICKSON, J.: The effect of lateral geniculate lesions on phasic electrical activity of the cortex during desynchronized sleep in the rat. Brain Res. **14**, 607–621 (1969).

HÖKFELT, T.: On the ultrastructural localization of noradrenalin in the central nervous system of the rat. Z. Zellforsch. **79**, 110–118 (1967).

HÖSLI, L., MONNIER, M.: Humoral transmission of sleep and wakefulness. III. Hemodialysis of activating humors during stimulation of the midbrain reticular system. Pflügers Arch. ges. Physiol. **283**, 17–25 (1965).

— TEBECIS, A. K., SCHONWELTER, H. P.: A comparison of the effects of monoamines neurons of the bulbar reticular formation. Brain Res. **25**, 357–370 (1971).

HOFFMAN, J. S., DOMINO, E. F.: Comparative effects of reserpine on the sleep cycle of man and cat. J. Pharmacol. exp. Ther. **170**, 190–198 (1969).

HOLMES, R. L., WOLSTENCROFT, J. H.: Cholinesterase in the medulla and pons of the cat. J. Physiol. (Lond.) **175**, 55 (1964).

HOOK, J. B., MOORE, K. E.: The renal handling of alpha methyltyrosine. J. Pharmacol. exp. Ther. **168**, 310–314 (1969).

HOOPER, G. (ed.): Metabolism of amines in the brain, p. 1–74. London: MacMillan 1969.

HORNYKIEWICZ, O.: Dopamine (3-hydroxytyramine) and brain function. Pharmacol. Rev. **18**, 925–965 (1966).

HOYLAND, J., SHILLITO, E., VOGT, M.: The effect of parachlorophenylalanine on the behaviour of cats. Brit. J. Pharmacol. **40**, 659–667 (1970).

IRIE, M.: Metabolic regulations of human growth hormone secretion. Folia endocr. Jap. **44**, 977–986 (1968).

ISKANDER, T. N., KAELBLING, R.: Catecholamines, a dream sleep model and depression. Amer. J. Psychiat. **127**, 43–50 (1970).

IVERSEN, L. L.: Inhibition of catecholamine uptake by 6-hydroxydopamine in rat brain. Europ. J. Pharmacol. **10**, 408–410 (1970).

— URETSKY, N. J.: Regional effects of 6-hydroxydopamine on catecholamine containing neurons in rat brain and spinal cord. Brain Res. **24**, 364–367 (1970).

IWAMA, I., KAWAMOTO, T., SAKAKURA, H., KASAMATSU, T.: Responsiveness of cat lateral geniculate at pre- and postsynaptic levels during natural sleep. Physiol. Behav. **1**, 45–53 (1966).

JACOBSON, M.: Development, specification and diversification of neuronal connections. In: The neurosciences, 2nd study program (F. O. SCHMITT, ed.), p. 116–129. New York: The Rockefeller University Press 1970.

JASPER, H. H.: Neurochemical mediators of specific and non-specific cortical activation. In: Attention in neurophysiology (EVANS, C. R., and T. B. MULHOLLAND, eds.), p. 317–325. London: Butterworths 1969.

JEANNEROD, M.: Organisation de l'activité électrique phasique du sommeil paradoxal. Etude électrophysiologique et neuropharmacologique. Thèse de Médecine, Lyon: LMD ed. 1965. 90 pp.

— Les phénomènes phasiques du sommeil paradoxical. Rev. lyon. Méd. **18**, 171–180 (1966).

— Relationships between reserpine induced phasic EEG activity and single unit discharge in the pontine reticular formation and lateral geniculate body neurons. Experientia (Basel) **23**, 389–393 (1967).

— DELORME, F., JOUVET, M.: Modulation des réponses évoquées visuelles au cours de l'activité phasique provoquée par la réserpine. C. R. Soc. Biol. (Paris) **159**, 1718–1721 (1965).

— KIYONO, S.: Effets de la réserpine sur la réponse réticulaire aux stimulations sensorielles. Brain Res. **12**, 129–137 (1969a).

— — Décharge unitaire de la formation réticulée pontique et activité phasique ponto-géniculo-occipitale chez le chat sous réserpine. Brain Res. **12**, 112–128 (1969b).

— MOURET, J., JOUVET, M.: Effets secondaires de la déafferentation visuelle sur l'activité électrique phasique ponto-géniculo-occipitale du sommeil paradoxal. J. Physiol. (Paris) **57**, 255–256 (1965b).

— PUTKONEN, P. T. S.: Oculomotor influences on lateral geniculate body neurons. Brain Res. **24**, 125–129 (1970).

— SAKAI, K.: Occipital and geniculate potentials related to eye movements in the unanesthetized cat. Brain Res. **19**, 361–377 (1970).

JEQUIER, E., LOVENBERG, W., SJOERDSMA, A.: Tryptophan hydroxylase inhibition: the mechanism by which p-chlorphenylalanine depletes rat brain serotonin. Molec. Pharmacol. **3**, 274–279 (1967).

JONES, B.: The double role of catecholamines in waking and paradoxical sleep: A neuropharmacological model. PHD thesis, University of Delaware, 1970.

— Hallucinating like behavior elicited in cats by elevation of catecholamines in the central nervous system. Psychophysiology 7, 314 (1971).

— Catecholamine containing neurons in the brain stem of the cat and their role in waking. MA thesis, Lyon: Tixier 1969. 87 pp.

Jones, B., Bobillier, P., Jouvet, M.: Effets de la destruction des neurones conte-
nant des catecholamines du mesencephale sur le cycle veille-sommeil du chat. C. R.
Soc. Biol. (Paris) 163, 176–180 (1969).

Jouvet, D., Delorme, F.: Evolution des signes électriques du sommeil paradoxal au
cours de la narcose au pentobarbital chez le chat. C. R. Soc. Biol. (Paris) 159, 387–390
(1965).

Jouvet, M.: Recherches sur les structures nerveuses et les mécanismes responsables des
différentes phases du sommeil physiologique. Arch. ital. Biol. 100, 125–206 (1962).

— Etude de la dualité des états de sommeil et des mécanismes de la phase paradoxale.
In: Aspects anatomo-fonctionnels de la physiologie du sommeil (M. Jouvet, ed.),
p. 397–449. Paris: Centre National de la Recherche Scientifique 1965a.

— Etude électrophysiologique et neuropharmacologique des états de sommeil. Actualités
pharmacol. 18, 109–173 (1965b).

— Behavioural and EEG effects of paradoxical sleep deprivation in the cat. In: Proceedings of the 23rd international congress of physiological sciences, Tokyo 1965,
vol. 4. Excerpta medica Internat. Congress series, No 87, p. 344–355 (1965c).

— Paradoxical sleep—a study of its nature and mechanisms. Progr. Brain Res. 18,
20–57 (1965d).

— Sommeil et monoamines. Bull. schweiz. Akad. med. Wiss. 22, 287–305 (1966).

— Neurophysiology of the states of sleep. Physiol. Rev. 47, 117–177 (1967a).

— Mechanisms of the states of sleep: a neuropharmacological approach. Res. Publ. Ass.
nerv. ment. Dis. 45, 86–126 (1967b).

— The states of sleep. Sci. Amer. 216, 62–72 (1967c).

— Neuropharmacology of sleep. In: Psychopharmacology. A review of progress (Efron,
D. H., ed.), p. 523–540. Public Health Service Publication No 1836, 1968.

— Neurophysiologische Mechanismen im Schlaf. In: Der Schlaf. Neurophysiologische
Aspekte, S. 103–135. München: J. Ambrosius Barth 1969a.

— Biogenic amines and the states of sleep. Science 163, 32–41 (1969b).

— Some monoaminergic mechanisms controlling sleep and waking. In: Brain and human
behavior (Karczmar, A., and J. Eccles, eds.). Berlin-Heidelberg-New York: Springer
1971 (in press).

— Cier, A., Mounier, D., Valatx, J. L.: Effets du 4 butyrolactone et du 4 hydroxybutyrate de sodium sur l'EEG et le comportement du chat. C. R. Soc. Biol. (Paris)
155, 1313–1316 (1961).

— Delorme, J. F.: Locus coeruleus et sommeil paradoxal. C. R. Soc. Biol. (Paris) 159,
895–899 (1965).

— Michel, F.: Corrélations électromyographiques du sommeil chez le chat décortiqué
et mésencéphalique chronique. C. R. Soc. Biol. (Paris) 153, 422–425 (1959).

— — Courjon, J.: Sur un stade d'activité électrique cérébrale rapide au cours du
sommeil physiologique. C. R. Soc. Biol. (Paris) 153, 1024–1028 (1959).

— Mounier, D.: Effets des lésions de la formation réticulée pontique sur le sommeil
du chat. C. R. Soc. Biol. (Paris) 154, 2301–2305 (1960).

— Vimont, P., Delorme, F.: Suppression élective du sommeil paradoxal chez le chat
par les inhibiteurs de la monoamine oxydase. C. R. Soc. Biol. (Paris) 159, 1595–1599
(1965).

Jouvet-Mounier, D.: Ontogenèse des états de vigilance chez quelques mamifères. Thèse
de Sciences, Lyon: J. Tixier 1968. 231 pp.

— Astic, L., Lacote, D.: Ontogenesis of the states of sleep in rat, cat and guinea pig
during the first postnatal month. Develop. Psychobiol. 2, 216–239 (1970).

Jung, R.: Der Schlaf. In: Physiologie und Pathophysiologie des vegetativen Nervensystems. Hippokrates: Stuttgart 1963.

Kadzielawa, K., Widy-Tyszkiewicz, E.: Electroencephalographic analysis of the central
action of dihydroxyphenylalanine. Electroenceph. clin. Neurophysiol. 28, 259–266
(1970a).

KADZIELAWA, K., WIDY-TYSZKIEWICZ, E.: The influence of various pharmacological agents on the desynchronization produced by dopa in the *cerveau isolé* preparation. Electroenceph. clin. Neurophysiol. **28**, 266–272 (1970b).

KALES, A. (ed.): Sleep. Physiology and pathology. A symposium. Philadelphia: J. B. Lippincott 1969.

KARCZMAR, A. G.: Is the central cholinergic nervous system over exploited? Fed. Proc. **28**, 147–158 (1969).

— LONGO, V. G., SCOTTI DE CAROLIS, A.: A pharmacological model of paradoxical sleep: the role of cholinergic and monoamine systems. Physiol. Behav. **5**, 175–182 (1970).

KATZMAN, R., BJORKLUND, A., OWMAN, CH., STENEVI, U., WEST, K. A.: Evidence for regenerative axon sprouting of central catecholamine neurons in the rat mesencephalon following electrolytic lesion. Brain Res. **25**, 579–596 (1971).

KAWAI, N.: Release of 5-hydroxytryptamine from slices of superior colliculus by optic tract stimulation. Neuropharmacology **9**, 395–397 (1970).

KAWAMURA, H., DOMINO, E. F.: Differential actions on m and n cholinergic agonists on the brainstem activating system. Int. J. Neuropharmacol. **8**, 105–117 (1969).

— MARCHIAFAVA, P. L.: Excitability changes along visual pathways during eye tracking movements. Arch. ital. Biol. **106**, 141–157 (1968).

KETY, S. S.: Neurochemical aspects of emotional behavior. In: Physiological correlates of emotion (BLOCK, P., ed.), p. 61–71. New York: Academic Press 1970.

— EVARTS, E. V., WILLIAMS, H. L. (eds.): Sleep and altered states of consciousness. Res. Publ. Ass. nerv. ment. Dis. **45**, (1967). 591 pp.

— JAVOY, F., THIERRY, A. M., JULOU, L., GLOWINSKI, J.: A sustained effect of electro-convulsive shock on the turnover of norepinephrine in the central nervous system of the rat. Proc. nat. Acad. Sci. (Wash.) **58**, 1249 (1967).

— SAMSON, F. E. (eds.): Neural properties of the biogenic amines. Neurosci. Res. Prog. Bull. **5** (1967).

KEY, G. J., MARLEY, E.: The effect of the sympathomimetic amines on behaviour and electrocortical activity of the chicken. Electroenceph. clin. Neurophysiol. **14**, 90–105 (1962).

KHAZAN, N., BAR, R., SULMAN, F. G.: The effect of cholinergic drugs on paradoxical sleep in the rat. Int. J. Neuropharmacol. **6**, 279–282 (1967).

— BROWN, P.: Differential effects of three tricyclic antidepressants on sleep and REM sleep in the rat. Life Sci. **9**, 279–285 (1970).

— SAWYER, C. H.: Mechanisms of paradoxical sleep as revealed by neurophysiologic and pharmacologic approaches in the rabbit. Psychopharmacologia (Berl.) **5**, 457–462 (1964).

— SULMAN, F. G.: Effect of imipramine on paradoxical sleep in animals with reference to dreaming and enuresis. Psychopharmacologia (Berl.) **10**, 89–96 (1966).

KING, D., JEWETT, R. E.: The effects of alpha methyltyrosine on sleep and brain nor-epinephrine in cat. J. Pharmacol. exp. Ther. **177**, 188–194 (1971).

KIYONO, S., JEANNEROD, M.: Relations entre l'activité géniculée phasique et les mouve-ments oculaires chez le chat normal et sous réserpine. C. R. Soc. Biol. (Paris) **161**, 1607–1611 (1967).

KLAUE, R.: Die bioelektrische Tätigkeit der Großhirnrinde im normalen Schlaf und in der Narkose durch Schlafmittel. J. Psychol. Neurol. (Lpz.) **47**, 510–531 (1937).

KLEIN, M.: Etude polygraphique et phylogénique des états de sommeil. Thèse Université de Lyon, Bosc ed. (1963). 101 pp.

KOE, B. K., WEISSMAN, A.: P-chlorophenylalanine, a specific depletor of brain serotonin. J. Pharmacol. exp. Ther. **154**, 499–516 (1966).

KOELLA, W. P.: Sleep. Its nature and physiological organization. 199 pp. Springfield: Ch. C. Thomas 1967.

— Neurohumoral aspects of sleep control. Biol. Psychiat. **1**, 161–177 (1969a).

— Serotonin and sleep. Exp. Med. Surg. **27**, 157–169 (1969b).

Koella, W. P., Czicman, J.: Mechanism of the EEG synchronizing action of serotonin. Amer. J. Physiol. 211, 926–935 (1966).
— Feldstein, A., Czicman, J. S.: The effect of parachlorophenylalanine on the sleep of cats. Electroenceph. clin. Neurophysiol. 25, 481–490 (1968).
— Smythies, J. R., Bull, D. M., Levy, C. K.: Physiological fractionation of the effect of serotonin on evoked potentials. Amer. J. Physiol. 198, 205–212 (1960).
— Trunca, C. M., Czicman, J. S.: Serotonin: Effect on recruiting responses of the cat. Life Sci. 4, 173–183 (1965).
Koelle, G. B.: The histochemical localizations of cholinesterases in the central nervous system of the rat. J. comp. Neurol. 100, 211–235 (1954).
Kopin, I. J.: Storage and metabolism of catecholamines in the role of monoamine oxidase. Pharmacol. Rev. 16, 179–193 (1964).
— Biochemical aspect of release of norepinephrine and other amines from sympathetic nerve endings. Pharmacol. Rev. 18, 513–523 (1966).
— Fischer, J. E., Musacchio, J. M., Horst, W. D., Weise, V. K.: "False neurochemical transmitters" and the mechanism of sympathetic blockade by monoamine oxidase inhibitors. J. Pharmacol. exp. Ther. 147, 186–194 (1965).
— Weise, V. K., Sedvall, G. C.: Effect of false transmitters in norepinephrine synthesis. J. Pharmacol. exp. Ther. 170, 246–253 (1969).
Kornmüller, A. E., Lux, H. B., Winkel, K., Klee, M.: Neurohumoral ausgelöste Schlafzustände an Tieren mit gekreuztem Kreislauf unter der Kontrolle von EEG-Ableitungen. Naturwissenschaften 48, 503–505 (1961).
Kostowski, W., Giacalone, E.: Stimulation of various forebrain structures and brain 5HT, 5HIAA, and behaviour in rats. Europ. J. Pharmacol. 7, 176–180 (1969).
— — Garattini, S., Valzelli, L.: Studies on behavioral and biochemical changes in rats after lesion of midbrain raphe. Europ. J. Pharmacol. 4, 371–377 (1968).
— — — — Electrical stimulation of midbrain raphe biochemical, behavioral and bioelectrical effects. Europ. J. Pharmacol. 7, 170–176 (1969).
Krnjevic, K.: Central cholinergic pathways. Fed. Proc. 28, 113–121 (1969).
— Randic, M., Straughan, D. W.: Cortical inhibition. Nature (Lond.) 201, 1294–1296 (1964).
— — — Pharmacology of cortical inhibition. J. Physiol. (Lond.) 184, 78–106 (1966a).
— — — An inhibitory process in the cerebral cortex. J. Physiol. (Lond.) 184, 16–49 (1966b).
— Silver, A.: The distribution of "cholinergic" fibres in the cerebral cortex. J. Physiol. (Lond.) 168, 39 (1963).
Kurokawa, M., Machiyama, Y., Kato, M.: Distribution of acetylcholine in the brain during various states of activity. J. Neurochem. 10, 341–348 (1963).
Laborit, H.: Sodium 4-hydroxybutyrate. Int. J. Neuropharmacol. 3, 433–452 (1964).
Laguzzi, R., Petitjean, F., Pujol, J. F., Jouvet, M.: Effets de la 6-hydroxydopamine intraventriculaire sur les états de sommeil et les monoamines cérébrales du chat. C. R. Soc. Biol. (Paris) (in press).
Lajtha, A. (ed.): Handbook of neurochemistry. New York: Plenum Press 1969.
Laverty, R., Sharman, D. F., Vogt, M.: Action of 2,4,5-trihydroxyphenylethylamine on the storage and release of noradrenaline. Brit. J. Pharmacol. Chemother. 24, 549–561 (1965).
— Taylor, K. M.: Effects of intraventricular 6-hydroxydopamine on rat behaviour and brain catecholamine metabolism. Brit. J. Pharmacol. 40, 836–846 (1970).
Ledebur, I. X., Tissot, R.: Modification de l'activité électrique centrale du lapin sous l'effet de micro-injections de précurseurs des monoamines dans les structures somnogènes bulbaires et pontiques. Electroenceph. clin. Neurophysiol. 20, 370–381 (1966).
Le Gassicke, J., Ashcroft, J. W., Eccleston, D., Evans, J. I., Oswald, I., Ritson, E. B.: The clinical state, sleep and amine metabolism of a tranylcypromine ("Parnate") addict. Brit. J. Psychiat. 111, 357–365 (1965).

LEGENDRE, R., PIÉRON, H.: Recherches sur le besoin de sommeil consécutif à une veille prolongée. Z. allg. Physiol. **14**, 235–262 (1913).

LEWIS, P. R., SHUTE, C. C. D.: The cholinergic limbic system: projections to hippocampal formation, medial cortex, nuclei of the ascending cholinergic reticular system, and the subfornical organ and supra-optic crest. Brain **90**, 521–541 (1967).

— — SILVER, A.: Confirmation from choline acetylase analyses of a massive cholinergic innervation to the rat hippocampus. J. Physiol. (Lond.) **191**, 215 (1967).

LICHTENSTEIGER, W., LANGEMANN, H.: Uptake of exogenous catecholamines by monoamine-containing neurons of the central nervous system: uptake of catecholamines by Arcuato-infundibular neurons. J. Pharmacol. exp. Ther. **151**, 400–409 (1966).

LIN, R. C., COSTA, E., NEFF, N. H., WANG, C. T., NAGAI, S. H.: In vivo measurement of 5-hydroxytryptamine turnover rate in the rat brain from the conversion of C^{14} tryptophan to $C^{14}5$ hydroxytryptamine. J. Pharmacol. exp. Ther. **170**, 232–239 (1969).

LINDSLEY, D. B., SCHREINER, L. H., KNOWLES, W. B., MAGOUN, H. V.: Behavioral and EEG changes following chronic brain stem lesions in the cat. Electroenceph. clin. Neurophysiol. **2**, 483–498 (1950).

LIPPMANN, W., LLOYD, K.: Dopamine-beta-hydroxylase inhibition by dimethyldithiocarbamate and related compounds. Biochem. Pharmacol. **18**, 2501–2517 (1969).

LOIZOU, L. A.: Projections of the nucleus locus coeruleus in the albino rat. Brain Res. **15**, 563–566 (1969).

— Uptake of monoamines into central neurons and the blood brain barrier in the infant rat. Brit. J. Pharmacol. **40**, 800–813 (1970).

— SALT, P.: Regional changes in monoamines of the rat brain during postnatal development. Brain Res. **20**, 467–470 (1970).

LOIZZO, A., LONGO, V. G.: A pharmacological approach to paradoxical sleep. Physiol. Behav. **3**, 91–99 (1968).

LONGO, V. G.: Mechanism of the behavioral and electroencephalographic effects of atropine and related compounds. Pharmacol. Rev. **18**, 965–996 (1966).

LOUP, M., CADILHAC, J.: Le développement des neurones à monoamines du cerveau chez le chaton. C. R. Soc. Biol. (Paris) **164**, 1582–1587 (1970).

LUBY, E. D., FROHMAN, C. E., GRISELL, J. L., LENZO, J. E., GOTTLIEB, J. S.: Sleep deprivation: Effects on behaviour thinking, motor performance and biological energy transfer system. Psychosom. Med. **22**, 182–192 (1960).

LUCE, G. G.: Current research on sleep and dreams. Publ. Hlth Serv. Publ. No 1389 (1966).

LUCERO, M. A.: Lengthening on REM sleep duration consecutive to learning in the rat. Brain Res. **20**, 319–322 (1970).

MACCHITELLI, F. J., FISCHETTI, D., MONTARARELLI, N.: Changes in behavior and electrocortical activity in the monkey following administration of 5-hydroxytryptophan. Psychopharmacologia (Berl.) **9**, 447–456 (1966).

MACON, J. B., SOKOLOFF, L., GLOWINSKI, J.: Feedback control of rat brain 5-hydroxytryptamine synthesis. J. Neurochem. **18**, 323–332 (1971).

MACPHAIL, E. M., MILLER, N. E.: Cholinergic brain stimulation in cats: failure to obtain sleep. J. comp. Physiol. Psychol. **65**, 499–503 (1968).

MAEDA, T.: Histochemical consideration on the relationship between monoamine oxidase and biogenic amines in the brain. Advanc. Neurol. **13**, 812–820 (1970) [in Japanese].

— ABE, T., SHIMIZU, N.: Histochemical demonstration of aromatic monoamine in the locus coeruleus of the mammalian brain .Nature (Lond.) **188**, 326–327 (1960).

— DRESSE, A.: Recherches sur le développement du locus coeruleus. Etude des catecholamines au microscope de fluorescence. Acta neurol. belg. **69**, 5–10 (1969).

— GEREBTZOFF, M. A.: Recherches sur le développement du locus coeruleus. 2. Etude histoenzymologique. Acta neurol. belg. **69**, 11–19 (1969).

— SHIMIZU, N.: Ascending amine fibers from locus coeruleus and other pontine amine neurons to the forebrain of the rat. Brain Res. **36**, 19–35 (1972).

Malcolm, L. J., Bruce, I. S. C., Burke, W.: Excitability of the lateral geniculate nucleus in the alert, non-alert and sleeping cat. Exp. Brain Res. **10**, 283–297 (1970a).
— Watson, J. A., Burke, W.: PGO waves as unitary events. Brain Res. **24**, 130–133 (1970b).
Mancia, M.: EEG and behavioural changes owing to splitting of the brain stem in cats. Electroenceph. clin. Neurophysiol. **27**, 487–503 (1969).
— Desiraju, T., Chhina, G. S.: The monkey split brain stem. Effects on the sleep wakefulness cycle. Electroenceph. clin. Neurophysiol. **24**, 409–416 (1968).
Mandel, P., Godin, U.: Approches biochimiques au problème du sommeil. In: Aspects anatomo-fonctionnels de la physiologie du sommeil (Jouvet, M., ed.), p. 13–36. Paris: Centre National de la Scientifique Recherche 1965a.
— — Sur l'intervention possible de l'acide gamma aminobutyrique dans le phénomène du sommeil. C. R. Soc. Biol. (Paris) **158**, 2475–2477 (1965b).
Mandell, A. J., Mandell, M. P.: Biochemical aspects of rapid eye movement sleep. Amer. J. Psychiat. **122**, 391–402 (1965).
— Spooner, C. E.: Psychochemical research studies in man. Science **162**, 1442–1453 (1968).
— — An N.N. indole transmethylation theory of the mechanism of MAOI-indole amino-acid load behavioral activation. In: Schizophrenia, current concepts and research (Siva-Sankar, ed.), p. 496–505. New York 1969.
— — Winters, W. D., Cruiksmank, M., Sabbott, I. M.: Imipramine antagonism of the CNS effects of norepinephrine, behavioral and biochemical correlates. Int. J. Neuropharmacol. **8**, 235–245 (1969).
Manshardt, J., Wurtman, R. J.: Daily rhythm in the noradrenalin content of rat hypothalamus. Nature (Lond.) **217**, 574–575 (1968).
Mantegazzini, P., Glasser, A.: Action de la DL-3-4-dioxyphenylalanine (dopa) et de la dopamine sur l'activité électrique du chat „cerveau isolé". Arch. ital. Biol. **98**, 367–374 (1960).
— Pepej, G.: Increase of cortical acetylcholine induced by midbrain hemisection in the cat. J. Physiol. (Lond.) **173**, 20 (1964).
Marantz, R., Rechtschaffen, A.: Effect of alpha-methyltyrosine on sleep in the rat. Percept. motor skills **25**, 805–808 (1967).
— — Lovell, R. A., Whitehead, P. K.: Effect of alpha-methyltyrosine on the recovery from paradoxical sleep deprivation in the rat. Commun. Behav. Biol. **2**, 161–164 (1968).
Marcus, R. J., Winters, W. D., Mori, K., Spooner, C. E.: EEG and behavioral comparison of the effects of gamma-hydroxybutyrate, gamma-butyrolactone and short chain fatty acids in the rat. Int. J. Neuropharmacol. **6**, 175–187 (1967).
Marczynski, T. J., Yamaguchi, N., Ling, G. M., Grodzinska, L.: Sleep induced by the administration of melatonin to the hypothalamus in unrestrained cats. Experientia (Basel) **20**, 435–437 (1964).
Mark, J., Heiner, L., Mandel, P., Godin, U.: Norepinephrine turnover in brain and stress reactions in rats during paradoxical sleep deprivation. Life Sci. **8**, 1085–1093 (1969).
Marley, E.: Behavioural and electrophysiological effects of catecholamines. Pharmacol. Rev. **18**, 753–768 (1966).
— Key, B. J.: Maturation of the electrocorticogram and behaviour in the kitten and guinea pig and the effect of some sympathomimetic amines. Electroenceph. clin. Neurophysiol. **15**, 620–636 (1963).
— Stephenson, J. D.: Effects of catecholamines infused into the brain of young chickens. Brit. J. Pharmacol. **40**, 639–658 (1970).
Masai, H., Kusunoki, T., Ishibashi, H.: A histochemical study on the fundamental plan of the central nervous system. Experientia (Basel) **21**, 572 (1965).
Matsumoto, J., Jouvet, M.: Effets de réserpine, DOPA et 5HTP sur les 2 états de sommeil. C. R. Soc. Biol. (Paris) **158**, 2137–2140 (1964).

MATSUMOTO, J., WATANABE, S.: Paradoxical sleep: Effects of adrenergic blocking agents. Proc. Jap. Acad. **43**, 680–683 (1967).

MATSUZAKI, M.: Differential effects of sodium butyrate and physostigmine upon the activities of para-sleep in acute brain stem preparations. Brain Res. **13**, 247–265 (1969).

— KASAHARA, M.: Induction of para-sleep by cholinesterase inhibitors in the mesencephalic cat. Proc. Jap. Acad. **42**, 989 (1966).

— OKADA, Y., SHUTO, S.: Cholinergic actions related to paradoxical sleep induction in the mesencephalic cat. Experientia (Basel) **23**, 1029–1031 (1967).

— TAKAGI, H.: Sleep induced by sodium butyrate in the cat. Brain Res. **4**, 206–222 (1967a).

— — Para-sleep induction by sodium butyrate in acute brain stem preparations. Brain Res. **4**, 223–242 (1967b).

— — TOKIZANE, T.: Paradoxical phase of sleep. Its artificial induction in the cat by sodium butyrate. Science **146**, 1328–1329 (1964).

McCARLEY, R. W., HOBSON, J. A.: Cortical unit activity in desynchronized sleep. Science **167**, 901–904 (1970).

McGINTY, D. J.: Somnolence, recovery and hyposomnia following ventromedial diencephalic lesions in the rat. Electroenceph. clin. Neurophysiol. **26**, 70–80 (1969).

— STERMAN, M. B.: Sleep suppression after basal forebrain lesions in the cat. Science **160**, 1253–1255 (1968).

McINTOSH, F. C.: The spontaneous and evoked release of acetylcholine from the cerebral cortex. Canad. J. Biochem. **41**, 2555–2571 (1963).

— OBORIN, P. E.: Release of acetylcholine from intact cerebral cortex. XIX. Int. Physiol. Congr. Montreal 1953. Abstr., p. 580–581.

McLENNAN, H.: Synaptic transmission. Philadelphia: Saunders 1970. p. 1–173.

— The release of acetylcholine and of 3-hydroxytyramine from the caudate nucleus. J. Physiol. (Lond.) **174**, 152–161 (1964).

METER, W. G. VAN, AYALA, G. F.: EEG effect of intracarotid or intravertebral arterial administration of d-amphetamine in rabbits with basilar artery ligation. Electroenceph. clin. Neurophysiol. **13**, 382–384 (1961).

MICHEL, F.: L'encéphale dédoublé chez le chat. J. Physiol. (Paris) **59**, 266 (1967).

— JEANNEROD, M., MOURET, J., RECHTSHAFFEN, A., JOUVET, M.: Sur les mécanismes de l'activité de pointes au niveau du système visuel au cours de la phase paradoxale du sommeil. C. R. Soc. Biol. (Paris) **158**, 103–106 (1964).

— ROFFWARG, H. P.: Chronic split brain stem preparation: effect on the sleep-waking cycle. Experientia (Basel) **23**, 126–128 (1967).

MIKITEN, T., NIEBYL, P., HENDLEY, C.: EEG desynchronization during behavioural sleep associated with spike discharges from the thalamus of the cat. Fed. Proc. **20**, 327 (1961).

MIZUNO, N., NAKAMURA, Y.: Direct hypothalamic projections to the locus coeruleus. Brain Res. **19**, 160–162 (1970).

MOIR, A. T. B., ECCLESTON, D.: The effects of precursor loading in the cerebral metabolism of 5-hydroxyindoles. J. Neurochem. **15**, 1093–1109 (1968).

MONNIER, M.: Le sommeil expérimental et sa transmission humorale. Actualités neurophysiol. **5**, 203–237 (1964).

— FALLERT, M.: Neurophysiologische und biochemische Mechanismen der Schlafsteuerung. Schweiz. med. Wschr. **97**, 866–875 (1967a).

— — Neurophysiologische und biochemische Regulationsmechanismen der Wachfunktion. Wien. klin. Wschr. **79**, 509–515 (1967b).

— GRABER, S.: Action de DOPA et du blocage de la monoaminoxydase par l'iproniazid sur le cerveau. Sonderdruck Schweiz. Arch. Neurol. Neurochir. Psychiat. **92**, 410–414 (1963).

— HÖSLI, L.: Dyalisis of sleep and waking factors in blood of rabbit. Science **146**, 796–798 (1964).

Monnier, M., Hösli, L.: Humoral regulation of sleep and wakefulness by hypnogenic and activating dialysable factors. Progr. Brain Res. **18**, 118–123 (1965).
— Koller, T., Graber, S.: Humoral influences of induced sleep and arousal upon electrical brain activity of animals with crossed circulation. Exp. Neurol. **8**, 264–277 (1963).
— — Hösli, L.: Humoral mechanism in experimental sleep. In: Aspects anatomo-fonctionnels de la physiologie du sommeil (Jouvet, M., ed.), p. 37–50. Paris: Centre National de la Recherche Scientifique 1965.
— Romanowski, W.: Les systèmes cholinoceptifs cérébraux—actions de l'acétylcholine, de la physostigmine, pilocarpine et de GABA. Electroenceph. clin. Neurophysiol. **14**, 486–500 (1962).
— Tissot, R.: Action de la réserpine et de ses médiateurs (5-HTP-serotonine et DOPA-noradrenaline) sur le comportement et le cerveau du lapin. Helv. physiol. pharmacol. Acta **16**, 255–267 (1958).
Monti, J. M.: Effect of recurrent stimulation of the brain stem reticular formation on REM sleep in cats. Exp. Neurol. **28**, 484–493 (1970).
Moore, K. E.: Effects of disulfiram and diethyldithiocarbamate on spontaneous locomotor activity and brain catecholamine levels in mice. Biochem. Pharmacol. **18**, 1627—1634 (1969).
— Wright, P. F., Bert, J. K.: Toxicologic studies with α-methyltyrosine, an inhibitor of tyrosine hydroxylase. J. Pharmacol. exp. Ther. **155**, 506–516 (1967).
Moore, R. Y., Heller, R. A.: Monoamine levels and neuronal degeneration in rat brain following lateral hypothalamic lesions. J. Pharmacol. exp. Ther. **156**, 12–23 (1967).
— Wong, S. L. R., Heller, A.: Regional effects of hypothalamic lesions on brain serotonin. Arch. Neurol. (Chic.) **13**, 346–354 (1965).
Morden, B., Conner, R., Mitchell, G., Dement, W., Levine, S.: Effect of rapid eye movement sleep deprivation on shock-induced fighting. Physiol. Behav. **3**, 425–432 (1968).
Morgane, P. J.: Chemical mapping of hypnogenic and arousal systems in the brain. Psychophysiology **6**, 219 (1969).
— Degeneration studies of the rostral projections of the raphe nuclei in the cat. Psychophysiology **7**, 317 (1971).
Morrison, A. R., Pompeiano, O.: Vestibular influences during sleep. IV. Functional relations between vestibular nuclei and lateral geniculate nucleus during desynchronized sleep. Arch. ital. Biol. **104**, 425–459 (1966).
Moruzzi, G.: Active processes in the brain stem during sleep. In: The Harvey lectures. Series 58, p. 233–297. New York: Academic Press 1963.
— The historical development of the deafferentation hypothesis of sleep. Proc. Amər. phil. Soc. **108**, 19–28 (1964).
— Magoun, H. W.: Brain stem reticular formation and activation of the E.E.G. Electroenceph. clin. Neurophysiol. **1**, 455–473 (1949).
Mouret, J. R., Bobillier, P.: Sleep and brain biogenic amine diurnal rhythms in the rat. I. Alterations of sleep diurnal rhythms independent of light. Int. J. Neurosciences **2**, 265–269 (1971).
— — Jouvet, M.: Insomnia following parachlorophenylalanine in the rat. Europ. J. Pharmacol. **5**, 17–22 (1968a).
— Froment, J. L., Bobillier, P., Jouvet, M.: Etude neuropharmacologique et biochimique des insomnies provoquées par la p-chlorophenylalanine. J. Physiol. (Paris) **59**, 463–464 (1967).
— Jeannerod, M., Jouvet, M.: L'activité électrique du système visuel au cours de la phase paradoxale du sommeil chez le chat. J. Physiol. (Paris) **55**, 305–306 (1963).
— Vilppula, A., Frachon, N., Jouvet, M.: Effets d'un inhibiteur de la monoamineoxydase sur le sommeil du rat. C. R. Soc. Biol. (Paris) **162**, 914–917 (1968b).

MYERS, R. D.: Emotional and autonomic responses following hypothalamic chemical stimulation. Canad. J. Psychol. **18**, 6–14 (1964).

— Transfusion of cerebrospinal fluid and tissue bound chemical factors between the brains of conscious monkeys: a new neurobiological essay. Physiol. Behav. **2**, 373–379 (1967).

— The role of hypothalamic transmitter factors in the control of body temperature. In: Physiological and behavioral temperature regulation (J. D. HARDY et al., eds.), p. 648–666. Springfield: Ch. Thomas 1970.

— BELESLIN, D. B.: The spontaneous release of 5-hydroxytryptamine and acetylcholine within the diencephalon of the unanaesthetized rhesus monkey. Exp. Brain Res. **11**, 539–552 (1970).

— YAKSH, T. L.: Feeding and temperature responses in the unrestrained rat after injections of cholinergic and aminergic substances into the cerebral ventricles. Physiol. Behav. **3**, 917–929 (1968).

— — Control of body temperature in the unanaesthetized monkey by cholinergic and aminergic systems in the hypothalamus. J. Physiol. (Lond.) **202**, 483–501 (1969).

NAQUET, R., DENAVIT, M., LANOIR, J., ALBE-FESSARD, D.: Altérations transitoires ou définitives de zones diencéphaliques chez le chat. Leurs effets sur l'activité électrique corticale et le sommeil. In: Aspects anatomo-fonctionnels de la physiologie du sommeil (JOUVET, M., ed.). Paris: Centre National de la Recherche Scientifique 1965.

NAUTA, W. J. H.: Hypothalamic regulation of sleep in rats. An experimental study. J. Neurophysiol. **9**, 285–316 (1946).

— Hippocampal projections and related neural pathways to the midbrain of the cat. Brain **81**, 319–340 (1958).

— KOELLA, W. P., QUARTON, G. C.: Sleep, wakefulness, dreams and memory. Neurosci. Res. Program Bull. **4**, 5–103 (1966).

NEAL, M. J., HEMSWORTH, B. A., MITCHELL, J. F.: The excitation of central cholinergic mechanisms by stimulation of the auditory pathway. Life Sci. **7**, 757–765 (1968).

NEMOZ, C., HENRY, P., JOUVET, M., SITE, J.: Neuroanatomie biochimique. Utilisation d'un ordinateur pour le traitement des données recueillies après lésion du tronc cérébral chez le chat. Bull. Centre Calcul Université Lyon (SIMEP, ed.) 1970. 95 pp.

NG, K. Y., CHASE, T. N., COLBURN, R. W., KOPIN, I. J.: L-Dopa-induced release of cerebral monoamines. Science **170**, 76–77 (1970).

OKADA, F.: The maturation of the circadian rhythm of brain serotonin in the rat. Life Sci. **10**, 77–87 (1971).

OLDS, J.: Hypothalamic substrate of reward. Physiol. Rev. **42**, 554–604 (1962).

— YUWILER, A., OLDS, M. E., YUN, C.: Neurohumors in hypothalamic substrates of reward. Amer. J. Physiol. **207**, 242–255 (1964).

OLIVIER, A., PARENT, A., SIMARD, H., POIRIER, L. J.: Cholinesterase striatopallidal and striatonigral efferents in the cat and the monkey. Brain Res. **18**, 273–282 (1970).

OLSON, L., FUXE, K.: On the projections from the locus coeruleus noradrenaline neurons: the cerebellar innervation. Brain Res. **28**, 165–171 (1971).

OSWALD, I.: Drugs and sleep. Pharmacol. Rev. **20**, 273–303 (1968).

— — — — THACORE, V. R.: Some experiments in the chemistry of normal sleep. Brit. J. Psychiat. **112**, 391–399 (1966).

PAPPENHEIMER, J. R., MILLER, T. B., GOODRICH, C. A.: Sleep-promoting effects of cerebrospinal fluid from sleep-deprived goats. Proc. nat. Acad. Sci. (Wash.) **58**, 513–518 (1967).

PARENT, A., POIRIER, L. J.: The medial forebrain bundle (MFB) and ascending monoaminergic pathways in the cat. Canad. J. Physiol. Pharmacol. **47**, 781–787 (1969).

PAVLOV, I. P.: Conditioned reflexes. Oxford: Oxford University Press 1927. XV+ 430 pp.

PEPEU, G., BARTOLINI, A.: Effect of some psychopharmacological agents on acetyl-choline release from the cerebral cortex of the cat. Boll. Soc. ital. Biol. sper. **43**, 1409–1411 (1967).

— MANTEGAZZINI, P.: Midbrain hemisection: Effect on cortical acetylcholine in the cat. Science **145**, 1069–1070 (1964).

PERENIN, M. T., JEANNEROD, M.: Lésions internucléaires: effets sur la motricité oculaire pendant l'éveil et le sommeil paradoxal chez le chat. Brain Res. **32**, 299–310 (1971).

PETITJEAN, F.: Etude de l'hypersomnie expérimentale consécutive à une lésion ponto-mésencéphalique. Thèse 3e cycle Université de Lyon, 1970. 133 pp.

— JOUVET, M.: Hypersomnie et augmentation de l'acide 5-hydroxy-indolacétique céré-bral par lésion isthmique chez le chat. C. R. Soc. Biol. (Paris) **164**, 2288–2293 (1970).

PEYRETHON, J., DUSAN-PEYRETHON, D.: Etude polygraphique du cycle veille-sommeil d'un Téléostéen (Tinca tinca). C. R. Soc. Biol. (Paris) **161**, 2533–2537 (1967).

— — Etude polygraphique du cycle veille-sommeil chez trois genres de reptiles. C. R. Soc. Biol. (Paris) **163**, 181–186 (1969).

PHILLIS, J. W., CHONG, G. C.: Acetylcholine release from the cerebral and cerebellar cortices: its role in cortical arousal. Nature (Lond.) **207**, 1253–1255 (1965).

— TEBECIS, A. K., YORK, D. H.: The inhibitory action of monoamines on lateral geni-culate neurones. J. Physiol. (Lond.) **190**, 563–583 (1967a).

— — — A study of cholinoreceptive cells in the lateral geniculate nucleus. J. Physiol. (Lond.) **192**, 695–713 (1967b).

— — — The responses of thalamic neurons to iontophoretically applied monoamines. J. Physiol. (Lond.) **192**, 715–740 (1967c).

PIERON, H.: Le problème physiologique du sommeil. 520 pp. Paris: Masson 1913.

PIN, C., JONES, B., JOUVET, M.: Topographie des neurones monoaminergiques du tronc cérébral du chat: étude par histofluorescence. C. R. Soc. Biol. (Paris) **162**, 2136–2141 (1968).

PLETSCHER, A., DAPRADA, M., BARTHOLINI, G., BURKARD, W. P., BRUDERER, H.: Two types of monoamine liberation by chlorinated aralkylamines. Life Sci. **4**, 2301–2309 (1965).

— — BURKARD, W. P., BARTHOLINI, G., STEINER, F. A., BRUDERER, H., BIGLER, F.: Aralkylamines with different effetcs on the metabolism of aromatic monoamines. J. Pharmacol. exp. Ther. **154**, 64–73 (1966).

POLC, P., MONNIER, M.: An activating mechanism in the ponto-bulbar raphe system of the rabbit. Brain Res. **22**, 47–63 (1970).

POMPEIANO, O.: Mechanism of sensorimotor integration during sleep. Progr. Physiol. Psychol. **3**, 1–179 (1970).

PORTIG, P. J., SHARMAN, D. F., VOGT, M.: Release by tubocurarine of dopamine and homovanillic acid from the superfused caudate nucleus. J. Physiol. (Lond.) **194**, 565–572 (1968).

PRAAG, H. M. VAN: Indoleamines and the central nervous system. Psychiat. Neurol. Neurochir. (Amst.) **73**, 9–36 (1970).

PUIZILLOUT, J., TERNAUX, J. P.: Persistance d'un endormement vago-aortique après destruction chirurgicale et pharmacologique des noyaux du raphé. J. Physiol. (Paris) **63**, 272' (1971).

PUJOL, J. F.: Contribution à l'étude des modifications de la régulation du métabolisme des monoamines centrales pendant le sommeil et la veille. Thèse de Doctorat ès Sciences, Paris, 1970. 192 pp.

— BOBILLIER, P., BUGUET, A., JONES, B., JOUVET, M.: Biosynthèse de la sérotonine cérébrale: étude neurophysiologique et biochimique après p-chlorophénylalanine et destruction du système du raphé. C. R. Acad. Sci. (Paris) **268**, 100–102 (1969).

— BUGUET, A., FROMENT, J. L., JONES, B., JOUVET, M.: The central metabolism of serotonin in the cat during insomnia: A neurophysiological and biochemical study after p-chlorophenylalanine or destruction of the raphe system. Brain Res. **29**, 195–212 (1971).

PUJOL, J. F., MOURET, J., JOUVET, M., GLOWINSKI, J.: Increased turnover of cerebral norepinephrine during rebound of paradoxical sleep in the rat. Science **159**, 112–114 (1968).

QUAY, W. B.: Differences in circadian rhythms in 5-hydroxytryptamine according to brain region. Amer. J. Physiol. **215**, 1448–1452 (1968).

RALL, T. W., GILMAN, A. G.: The role of cyclic AMP in the nervons system. Neurosci. Res. Program Bull. **8**, 221–323 (1970).

RAMANAMURTHY, P. S. U., SRIKANTIA, S. G.: Effects of leucine on brain serotonin. J. Neurochem. **17**, 27–32 (1970).

RANDIC, M., SIMINOFF, R., STRAUGHAN, D.: Acetylcholine depression of cortical neurons. Exp. Neurol. **9**, 236–243 (1964).

RANSON, S. W.: Somnolence caused by hypothalamic lesion in the monkey. Arch. Neurol. Psychiat. (Chic.) **41**, 1–23 (1939).

RECH, R. M., BORYS, H. K., MOORE, K. E.: Alterations in behavior and brain catecholamine levels in rats treated with alpha-methyltyrosine. J. Pharmacol. exp. Ther. **153**, 412–420 (1966).

RECHTSCHAFFEN, A., LOVELL, R. A., FREEDMAN, D. W., WHITEHEAD, P. K., ALDRICH, M.: Effect of p-chlorophenylalanine on sleep in rats. Psychophysiology **6**, 223 (1969).

REIS, D. J., CORVELLI, A., CONNERS, J.: Circadian and ultradian rhythms of serotonin regionally in cat brain. J. Pharmacol. exp. Ther. **167**, 328–334 (1969).

— FUXE, K.: Brain norepinephrine: evidence that neuronal release is essential for sham rage behavior following brain stem transection in cat. Proc. nat. Acad. Sci. (Wash.) **64**, 108–112 (1969).

— GUTNICK, E.: Daily segmental rhythms of norepinephrine in spinal cord of the cat. Amer. J. Physiol. **218**, 1707–1710 (1970).

— MIURA, M., WEINBREN, M., GUNNE, L. M.: Brain catecholamines relation to defense reaction evoked by acute brainstem transection in cat. Science **156**, 1768–1770 (1967).

— MOORHEAD, D. T., MESLINO, N.: Dopa induced excitement in the cat. Arch. Neurol. (Chic.) **22**, 31–39 (1970).

— WEINBREN, M., CORVELLI, A.: A circadian rhythm of norepinephrine regionally in cat brain: its relationship to environmental lighting and to regional diurnal variations in brain serotonin. J. Pharmacol. exp. Ther. **164**, 135–146 (1968).

— WURTMAN, R. J.: Diurnal changes in brain noradrenaline. Life Sci. **7**, 91–98 (1968).

REITE, M., PEGRAM, G. V., STEPHENS, L. M., BIXLER, E. C., LEWIS, O. L.: The effect of reserpine and monoamine oxidase inhibitors on paradoxical sleep in the monkey. Psychopharmacologia (Berl.) **14**, 12–17 (1969).

RENAULT, J.: Monoamines et sommeils. Role du système du raphé et de la sérotonine cérébrale dans l'endormissement. Thèse de Médecine, Université Lyon, 1967 (TIXIER ed.). 140 pp.

RIBEREAU-GAYON, P., SAPIS, J. C.: Sur la présence dans le vin de tyrosol, de tryptophol d'alcool phenylethylique et de gamma butyrolactone, produits secondaires de la fermentation alcoolique. C. R. Acad. Sci. (Paris) **261**, 1915–1916 (1965).

RICCI, G. F., ZAMPARO, L.: Electrocortical correlates of avoidance conditioning in the monkey. Their modifications with atropine and amphetamine. In: Pharmacology of conditioning, learning and retention (MICHELSON, M. Y., and V. G. LONGO, eds.), p. 269–283. Praha: Czechoslovak medical press 1965.

RINALDI, F., HIMWICH, H. E.: Cholinergic mechanism involved in function of mesodiencephalic activating system. Arch. Neurol. Psychiat. (Chic.) **73**, 396–402 (1955).

RINGLE, D. A., HERNDON, B. L.: Plasma dialysates from sleep deprived rabbits and their effect on the electrocorticogram of rats. Pflügers Arch. **303**, 344–349 (1968).

— — Effects on rats of CSF from sleep-deprived rabbits. Pflügers Arch. **306**, 320–328 (1969).

RIZZOLI, A. A., AGOSTI, S., GALZIGNA, L.: Interaction between cerebral amines and 4-hydroxybutyrate in the induction of sleep. J. Pharm. Pharmacol. **21**, 465–466 (1969).

Rizzoli, A. A., Galzigna, L.: Molecular mechanism of unconscious state induced by butyrate Biochem. Pharmacol. **19**, 2727–2736 (1970).

Robertis, E. de, Pellegrino de Iraldi, A., Rodriguez de Lores Arnaiz, G., Zieher, L. M.: Synaptic vesicles from the rat hypothalamus. Isolation and norepinephrine content. Life Sci. **4**, 193–201 (1965).

— Rodriguez de Lores Arnaiz, G., Salganicoff, L., Pellegrino de Iraldi, A., Zieher, L. M.: Isolation of synaptic vesicles and structural organization of the acetylcholine system within brain nerve endings. J. Neurochem. **10**, 225–235 (1963).

Roberts, M. H. T., Straughan, D. W.: Excitation and depression of cortical neurons by 5-hydroxytryptamine. J. Physiol. (Lond.) **193**, 269–295 (1967).

Roberts, R. B., Flexner, J. B., Flexner, L. B.: Some evidence for the involvement of adrenergic sites in the neurons trace. Proc. nat. Acad. Sci. (Wash.) **66**, 310–313 (1970).

Robinson, N.: Histochemistry of rat brain stem monoamine oxidase during maturation. J. Neurochem. **15**, 1151–1158 (1968).

Rocaboy, J. C., Le Cam, A., Samperez, S., Jouan, P.: Sur la mise en évidence d'une association macromoléculaire de la 5-hydroxytryptamine-^{14}C dans les centres nerveux supérieurs du rat. C. R. Acad. Sci. (Paris) **269**, 247–250 (1969).

Roffwarg, H. P., Muzio, J. N., Dement, W. C.: Ontogenetic development of the human sleep-dream cycle. Science **152**, 604–619 (1966).

Roll, S. K.: Intracranial self-stimulation and wakefulness. Effect of manipulating ambient brain catecholamines. Science **168**, 1370–1372 (1970).

Rossi, G. F.: Sleep-inducing mechanisms in the brain stem. Electroenceph. clin. Neurophysiol., Suppl. **24**, 113–132 (1963).

— Favale, E., Hara, T., Giussani, A., Sacco, G.: Researches on the nervous mechanisms underlying deep sleep in the cat. Arch. ital. Biol. **99**, 270–292 (1961).

Roth, G. I., Walton, P. L., Yamamoto, W. S.: Area postrema-abrupt EEG synchronization following close intra-arterial perfusion with serotonin. Brain Res. **23**, 223–233 (1970).

— Yamamoto, W. S.: The microcirculation of the area postrema in the rat. J. comp. Neurol. **133**, 329–341 (1968).

Roth, R. H., Giarman, N. J.: Conversion in vivo of gamma-aminobutyric to gamma-hydroxybutyric acid in the rat. Biochem. Pharmacol. **18**, 247–250 (1969).

— — Natural occurrence of gamma-hydroxybutyrate in mammalian brain. Biochem. Pharmacol. **19**, 1087–1095 (1970).

— Suhr, Y.: Mechanism of the gamma-hydroxybutyrate induced increase in brain dopamine and its relationship to "sleep". Biochem. Pharmacol. **19**, 3001–3012 (1970).

Rothballer, A. B.: Studies on the adrenaline-sensitive component of the reticular activating system. Electroenceph. clin. Neurophysiol. **8**, 603–621 (1956).

— The effect of phenylephrine, methamphetamine, cocaine, and serotonin upon the adrenaline sensitive component of the reticular activating system. Electroenceph. clin. Neurophysiol. **9**, 409–417 (1957).

— The effects of catecholamines on the central nervous system. Pharmacol. Rev. **11**, 494–547 (1959).

Rougeul, A., Verdeaux, J., Letalle, A.: Effets électrographiques et comportementaux de divers hallucinogènes chez le chat libre. Rev. neurol. **120**, 391–394 (1969).

Roussel, B.: Monoamines et sommeils: Suppression du sommeil paradoxal et diminution de la noradrénaline cérébrale par les lésions des noyaux locus coeruleus. Thèse Médecine, Lyon: Tixier ed. 1967. 141 pp.

— Buguet, A., Bobillier, P., Jouvet, M.: Locus coeruleus—sommeil paradoxal et noradrénaline cérébrale. C. R. Soc. Biol. (Paris) **161**, 2537–2541 (1967).

Routtenberg, A.: The two arousal hypothesis: reticular formation and limbic system. Psychol. Rev. **75**, 51–80 (1968).

— Malsbury, C.: Brainstem pathways of reward. J. comp. physiol. Psychol. **68**, 22–31 (1969).

ROUTTENBERG, A., SLADEK, J., BONDAREFF, W.: Histochemical fluorescence after application of neurochemicals to caudate nucleus and septal area in vivo. Science **161**, 272–274 (1968).

RUSSEL, R. W.: Behavioral aspects of cholinergic transmission. Fed. Proc. **28**, 121–131 (1969).

RUSSELL, G. V.: The nucleus locus coeruleus (dorso lateralis tegmenti). Tex. Rep. Biol. Med. **13**, 939–988 (1955).

SABELLI, H. C., GIARDINA, W. J.: Tryptaldehydes (indoleacetaldehydes) in serotoninergic sleep in newly hatched chicks. Arzneimittel-Forsch. **20**, 74–80 (1970).

— — ALIVISATOS, S. G., SETH, P. K., UNGAR, F.: Aldehydes of brain amines affect central nervous system. Nature (Lond.) **223**, 73–74 (1969).

SAKAKURA, H.: Spontaneous and evoked unitary activities of cat lateral geniculate neurons in sleep and wakefulness. Jap. J. Physiol. **18**, 23–43 (1968).

— IWAMA, K.: Unitary recording from cat lateral geniculate during natural sleep. Proc. Jap. Acad. **42**, 418–423 (1966).

SALMOIRAGHI, G. C., BLOOM, F. E., COSTA, E.: Adrenergic mechanisms in rabbit olfactory bulb. Amer. J. Physiol. **207**, 1417–1425 (1964).

— COSTA, E., BLOOM, F. E.: Pharmacology of central synapses. Ann. Rev. Pharmacol. **5**, 213–235 (1965).

— STEFANIS, C. N.: Patterns of central neurons responses to suspected transmitters. Arch. ital. Biol. **103**, 705–725 (1965).

SAMSON, F. E., DAHL, N., DAHL, D. N.: A study on the narcotic action of the short chain fatty acids. J. clin. Invest. **35**, 1291–1298 (1956).

SANDERS-BUSH, E., SULSER, F.: P-chloroamphetamine: in vivo investigations on the mechanism of action of the selective depletion of cerebral serotonin. J. Pharmacol. exp. Ther. **175**, 419–426 (1970).

SANER, A., THOENEN, H.: Model experiments on the molecular mechanism of action of 6-hydroxydopamine. Molec. Pharmacol. **7**, 147–155 (1970).

SASSIN, J. F., PARKER, D. C., JOHNSON, L. C., ROSSMAN, L. G., MACE, J. W., GOTLIN, R. W.: Effects of slow wave sleep deprivation on human growth hormone release in sleep: preliminary study. Life Sci. **8**, 1291–1299 (1969a).

— — MACE, J. W., GOTLIN, R. W., JOHNSON, L. C., ROSSMAN, L. G.: Human growth hormone release. Relation to slow wave sleep and sleep-waking cycle. Science **165**, 513–515 (1969b).

SATINSKY, D.: Pharmacological responsiveness of lateral geniculate nucleus neurons. Int. J. Neuropharmacol. **6**, 387–399 (1967).

SATTERLEE, W. G., SERPICK, A., BIANCHINE, J. R.: The carcinoid syndrome: chronic treatment with para-chlorophenylalanine. Ann. intern. Med. **72**, 919–923 (1970).

SCHAIN, R. J., COPENHAVER, J. H., CARVER, M. J.: Inhibition by phenylalanine of the entry of 5-hydroxy-tryptophan-1-C[14] into cerebrospinal fluid. Proc. Soc. exp. Biol. (N.Y.) **118**, 184–186 (1965).

SCHECKEL, C. L., BOFF, E.: Behavioral stimulation in rats associated with a selective release of drain norepinephrine. Arch. int. Pharmacodyn. **152**, 479–490 (1964).

— PAZERY, L. M.: Hyperactive states related to the metabolism of norepinephrine and similar biochemicals. Ann. N.Y. Acad. Sci. **159**, 939–958 (1969).

SCHEVING, L. E., HARRISON, W. H., GORDON, P., PAULY, J. E.: Daily fluctuation (circadian and ultradian) in biogenic amines of the rat brain. Amer. J. Physiol. **124**, 166–173 (1968).

SCHILDKRAUT, J. J.: The catecholamine hypothesis of affective disorders: a review of supporting evidence. Amer. J. Psychiat. **122**, 509–523 (1965).

— Neuropsychopharmacology and the affective disorders. 111 pp. Boston: Little Brown & Co. 1970.

— KETY, S.: Biogenic amines and emotion. Science **156**, 21–30 (1967).

SCHMIDT, J., KRUG, M., MAIER, R., POHLE, W., MATTHIES, H.: Action of noradrenaline, serotonin and acetylcholine on the impulse activity of neurons of the pontine formatio reticularis of the rat. Acta biol. med. germ. 18, 703–713 (1967).

SCHMITT, F. O.: Macromolecular specificity and biological memory (SCHMITT, F. O., ed.) Cambridge: MIT Press 1962.

SCHNEDORF, J. F., IVY, A. C.: An examination of the hypnotoxin theory of sleep. Amer. J. Physiol. 125, 491–505 (1939).

SCHNEIDERMAN, N., MONNIER, M., HÖSLI, L.: Humoral transmission of sleep. IV. Cerebral and visceral effects of sleep dialysate. Pflügers Arch. ges. Physiol. 288, 65–81 (1966)i

SCHOENFELD, R. I., SEIDEN, L. S.: Effect of alpha-methyltyrosine on operant behavior and brain catecholamine levels. J. Pharmacol. exp. Ther. 167, 319–327 (1969).

SEGAL, D. S., MANDELL, A. J.: Behavioral activation of rats during intraventricular infusion of norepinephrine. Proc. nat. Acad. Sci. (Wash.) 66, 289–293 (1970).

— WHALEN, R. E.: Effect of chronic administration of p-chlorophenylalanine on sexual receptivity of the female rat. Psychopharmacologia (Berl.) 16, 434–438 (1970).

SEIDEN, L. S., HANSON, L. C. F.: Reversal of the reserpine-induced suppression of the conditioned avoidance response in the cat by L-DOPA. Psychopharmacologia (Berl.) 6, 239–245 (1964).

— PETERSON, D. D.: Blockade of L-Dopa reversal of reserpine-induced conditioned avoidance response suppression by disulfiram. J. Pharmacol. exp. Ther. 163, 84–90 (1968).

SHARPLESS, S. K., ROTHBALLER, A. B.: Humoral factors released from intracranial sources during stimulation of reticular formation. Amer. J. Physiol. 200, 909–915 (1961).

SHASKAN, E. G., SNYDER, S. H.: Kinetics of serotonin accumulation into slices from rat brain: relationship to catecholamine uptake. J. Pharmacol. exp. Ther. 175, 404–418 (1970).

SHEARD, M. H.: The effect of p-chlorophenylalanine on behaviour in rats. Relation to brain serotonin and 5-hydroxyindolacetic acid. Brain Res. 15, 524–528 (1969).

— AGHAJANIAN, G. K.: Neural release of brain serotonin and body temperature. Nature (Lond.) 216, 495–496 (1967).

— — Stimulation of the midbrain raphe: Effect on serotonin metabolism. J. Pharmacol. exp. Ther. 163, 425–431 (1968).

SHIMIZU, A., HIMWICH, H. E.: The effects of amphetamine on the sleep-wakefulness cycle of developing kittens. Psychopharmacologia (Berl.) 13, 161–169 (1968).

SHIMIZU, N., IMAMOTO, K.: Fine structure of the locus coeruleus in the rat. Arch. Histol. jap. 31, 229–246 (1970).

— MORIKAWA, N.: Histochemical study of monoamine oxidase in the developing rat brain. Nature (Lond.) 184, 650–651 (1959).

— — OKADA, M.: Histochemical studies of monoamine oxidase of the brain of rodents. Z. Zellforsch. 49, 389–400 (1959).

SHUTE, C. C. D.: The distribution of cholinergic and monoaminergic neurones in the brain. J. Physiol. (Lond.) 201, 2 (1969).

— LEWIS, P. R.: Cholinesterase-containing systems of the brain of the rat. Nature (Lond.) 199, 1160–1164 (1963).

— — Cholinergic and monoaminergic systems of the brain. Nature (Lond.) 212, 710–711 (1966).

— — The ascending cholinergic reticular system: neocortical, olfactory and subcortical projections. Brain 90, 497–520 (1967).

SIFFRE, M., JOUVET, M.: Induction of bicircadian periods by light in isolated man. Psychophysiology — in press.

SIGGINS, G. R., HOFFER, B. J., BLOOM, F. E.: Studies on norepinephrine containing afferents to Purkinje cells of rat cerebellum. III. Evidence for mediation of norepinephrine effects by cyclic 3'5' adenosine monophosphate. Brain Res. 25, 535–553 (1971).

SIMPSON, B. A., IVERSEN, S. D.: Effects of substantia nigra lesions on the locomotor and stereotype responses to amphetamine. Nature (Lond.) **230**, 30–32 (1971).

SINGER, G., HO, A., GERSHON, S.: Changes in activity of choline acetylase in central nervous system of rat after intraventricular administration of noradrenaline. Nature (Lond.) **230**, 152–153 (1971).

SINHA, A. K., HENRIKSON, S., DEMENT, W., BARCHAS, J. D.: Changes in brain amine content in sleep and wakefulness. Amer. J. Physiol. (in press).

SJOERDSMA, A., ENGELMAN, K., SPECTOR, S., UDENFRIEND, S.: Inhibition of catecholamine synthesis in man with alpha-methyl-tyrosine, an inhibitor of tyrosine hydroxylase. Lancet **II**, 1092–1094, **1965**.

— LOVENBERG, W., ENGELMAN, K., CARPENTER, W. T., WYATT, R. J., GESSA, G. L.: Serotonin now: clinical implications of inhibiting its synthesis with parachlorophenylalanine. Ann. intern. Med. **73**, 607–629 (1970).

SLATER, P.: The effects of triethylcholine and hemicholinium-3 in the acetylcholine content of rat brain. Int. J. Neuropharmacol. **7**, 421–429 (1968).

SMYTHIES, J. R. (ed.): The mode of action of psychotomimetic drugs. Neurosci. Res. Progr. Bull. **8** (1970).

SNYDER, S. H., GLOWINSKI, J., AXELROD, J.: The storage of norepinephrine and some of its derivatives in brain synaptosomes. Life Sci. **4**, 797–809 (1965).

SODERBERG, U.: The effect of blood-borne influences of the cerebral cortex in wakefulness and sleep. In: Proceedings of the 22nd international Congress of physiological Sciences. Leyden 1962. Excerpta Medica internat. congress series, p. 457–462 (1962).

SOURKES, T. L.: Action of alpha methyldopa in the brain. Brit. med. Bull. **21**, 66–69 (1965).

— POIRIER, L.: Influence of the substantia nigra on the concentration of 5-hydroxytryptamine and dopamine of the striatum. Nature (Lond.) **207**, 202–203 (1965).

SPAFFORD, D. C.: A study of the influence of the neurohumor serotonin on hibernation in the golden mantled-ground squirrel *Citellus lateralis*. MA thesis. University of California, Riverside 1970.

SPECTOR, S., SJOERDSMA, A., UDENFRIEND, S.: Blockade of endogenous norepinephrine synthesis by alpha-methyl-tyrosine, an inhibitor of tyrosine hydroxylase. J. Pharmacol. exp. Ther. **147**, 86–96 (1965).

SPOONER, C. E., MANDELL, A. J., WINTERS, W. D., SABBOT, I. M., CRUIKSHANK, M. K.: Pharmacological and biochemical correlates of 5-hydroxytryptamine entry into the central nervous system during maturation. Proc. West. Pharmacol. Soc. **11**, 98–105 (1968).

— WINTERS, W. D.: Evidence for a direct action of monoamines on the chick central nervous system. Experientia (Basel) **21**, 256–258 (1965).

— — Neuropharmacological profile of the young chick. Int. J. Neuropharmacol. **5**, 217–236 (1966).

— — Intra-arterial blood pressure recording in the unrestrained chick during wakefulness and sleep. Arch. int. Pharmacodyn. **161**, 1–6 (1966).

— — The influence of centrally active amine induced blood pressure changes on the electroencephalogram and behavior. Int. J. Neuropharmacol. **6**, 109–119 (1967).

SQUIRES, R. D., JACOBSON, F. H.: Chronic deficits of temperature regulation produced in cat by preoptic lesions. Amer. J. Physiol. **214**, 549–553 (1968).

STEIN, L., WISE, C. D.: Release of norepinephrine from hypothalamus and amygdala by rewarding medial forebrain bundle stimulation and amphetamine. J. comp. physiol. Psychol. **67**, 189–198 (1969).

— — Mechanism of the facilitating effects of amphetamines on behaviour. In: Psychotomimetic drugs (EFRON, D. H., ed.), p. 123–145. New York: Raven Press 1970.

— — Possible etiology of schizophrenia: Progressive damage to the noradrenergic reward system by 6-hydroxydopamine. Science **171**, 1032–1037 (1971).

Steiner, F. A.: Influence of microelectrophoretically applied acetylcholine on the responsiveness of hippocampal and lateral geniculate neurones. Pflügers Arch. Europ. J. Physiol. **303**, 173–181 (1968).

Sterman, M. B., Clemente, C. D.: Forebrain inhibitory mechanisms: cortical synchronization induced by basal forebrain stimulation. Exp. Neurol. **6**, 91–102 (1962a).

— — Forebrain inhibitory mechanisms: sleep patterns induced by basal forebrain stimulation in the behaving cat. Exp. Neurol. **6**, 103–117 (1962b).

— Knauss, T., Lehmann, D., Clemente, C. D.: Circadian sleep and waking patterns in the laboratory cat. Electroenceph. clin. Neurophysiol. **19**, 509–517 (1965).

Stolk, J., Barchas, J., Dement, W., Schauberg, S.: Brain catecholamine metabolism following P.chlorophenylalanine treatment. Pharmacologist **11**, 258 (1969).

Stolk, J. M., Rech, R. H.: Antagonism of D-amphetamine by alpha-methyl-p-tyrosine: behavioral evidence for the participation of catecholamine stores and synthesis in the amphetamine stimulant response. Neuropharmacology **9**, 249–265 (1970).

Stone, E. A., Dicara, L. V.: Activity level and accumulation of tritiated norepinephrine in rat brain. Life Sci. **8**, 433–441 (1969).

Svensson, T. H.: The effect of inhibition of catecholamine synthesis on desamphetamine induced central stimulation. Europ. J. Pharmacol. **12**, 161–167 (1970).

Sweet, C. P., Hobson, J. A.: The effects of posterior hypothalamic lesions on behavioral and electrographic manifestations of sleep and waking in cats. Arch. ital. Biol. **106**, 283–293 (1968).

Szerb, J. C.: The effect of tertiary and quaternary atropine cortical acetylcholine output and on the electroencephalogram in cats. Canad. J. Physiol. Pharmac. **42**, 303–314 (1964).

— Cortical acetylcholine release and electroencephalographic arousal. J. Physiol. (Lond.) **192**, 329–345 (1967).

— Malik, H., Hunter, E. G.: Relationship between acetylcholine content and release in the cat's cerebral cortex. Canad. J. Physiol. Pharmac. **48**, 780–790 (1970).

Taber, E., Brodal, A., Walberg, F.: The raphe nuclei of the brain stem of the cat. I. Normal topography and cytoarchitecture and general discussion. J. comp. Neurol. **114**, 161–188 (1960).

Tabushi, K., Himwich, H. E.: The acute effects of reserpine on the sleep-wakefulness cycle in rabbits. Psychopharmacologia (Berl.) **16**, 240–252 (1969).

Tagliamonte, A., Tagliamonte, P., Perez-Cruet, J., Gessa, G. L.: Tryptophan increased by treatment with drugs which stimulate synthesis of serotonin. Nature (Lond.) **229**, 125–126 (1971).

Takaki, H., Matsuzaki, M.: Sleep state and its induction by sodium butyrate in acute "encephale isolé" and "isolated midbrain pons-medulla" preparations. Jap. J. Physiol. **18**, 380–390 (1968).

Tebecis, A. K.: Studies on cholinergic transmission in the medial geniculate nucleus. Brit. J. Pharmacol. **38**, 138–147 (1970).

Tenen, S. S.: The effects of p-chlorophenylalanine, a serotonin depletor on avoidance acquisition, pain sensitivity and related behavior in the rat. Psychopharmacologia (Berl.) **10**, 204–220 (1967).

Teuber, H. L., in: La fonction du regard. Colloque de l'Institut National de la Santé et de la Recherche Médicale. Dubois, Poulsen, A., C. G. Lairy and A. Remond, edit., p. 187–200. Paris 1971.

Thierry, A. M., Blanc, G., Glowinski, J.: Preferential utilization of newly synthetized norepinephrine in the brain stem of stressed rats. Europ. J. Pharmacol. **10**, 139–142 (1970).

— — — Effect of stress on the disposition of catecholamines localized in various intraneuronal storage forms in the brain stem of the rat. J. Neurochem. **18**, 449–462 (1971a).

— — — Dopamine-norepinephrine: Another regulatory step of norepinephrine synthesis in central noradrenergic neurons. Europ. J. Pharmacol. **14**, 303–307 (1971b).

Thiérry, A. M., Javoy, F., Glowinski, J., Kety, S.: Effects of stress on the metabolism of norepinephrine, dopamine and serotonin in the central nervous system of the rat. J. Pharmacol. exp. Ther. 163, 163–171 (1968).

Thoenen, H., Mueller, R. A., Axelrod, J.: Phase difference in the induction of tyrosine hydroxylase in cell body and nerve terminals of sympathetic neurones. Proc. nat. Acad. Sci. (Wash.) 65, 58–62 (1970).

Thomas, J., Benoit, O.: Individualisation d'un sommeil à ondes lentes et activité phasique. Brain Res. 5, 221–235 (1967).

Tissot, R.: Monoamines et régulations thymiques, p. 87–153. In: Confrontations psychiatriques, vol. 6. Paris: Specia 1970.

Torda, C.: Effect of brain serotonin depletion on sleep in rats. Brain Res. 6, 375–377 (1967).

— Effects of changes of brain norepinephrine content of sleep cycle in rat. Brain Res. 10, 200–207 (1968).

Toru, M., Hingtgen, J. N., Aprison, M. H.: Acetylcholine concentrations in brain areas of rats during three states of avoidance behavior: normal, depression and excitation. Life Sci. 5, 181–188 (1966).

Trendelenburg, U.: Mechanisms of supersensitivity and subsensitivity to sympathomimetic amines. Pharmacol. Rev. 18, 629–640 (1966).

Tsuchiya, K., Toru, M., Kobayashi, T.: Sleep deprivation-changes of monoamines and acetylcholine. Life Sci. 8, 867–873 (1969).

Udenfriend, S., Weissbach, H., Bogdanski, D. G.: Increase in tissue serotonin following administration of its precursor. 5HTP. J. biol. Chem. 224, 803–810 (1957).

Ungerstedt, U.: 6-Hydroxydopamine induced degeneration of central monoamine neurons. Europ. J. Pharmacol. 5, 107–111 (1969).

— Noradrenaline pathways in the rat brain. Principal structures. Science (in press).

— Arbuthnott, G. W.: Quantitative recording rotational behaviour in rats after 6-hydroxydopamine lesions of the nigrostriatal dopamine system. Brain Res. 21, 485–493 (1970).

Uretsky, N. J., Iversen, L. L.: Effects of 6-hydroxydopamine on catecholamine containing neurones in the rat brain. J. Neurochem. 17, 269–278 (1970).

— Simmonds, M. A., Iversen, L. L.: Changes in the retention and metabolism of H-l-norepinephrine in rat brain in vivo after 6-hydroxydopamine pretreatment. J. Pharmacol. exp. Ther. 176, 489–496 (1971).

Ursin, R.: Sleep stage relations within the sleep cycles of the cat. Brain Res. 19, 91–99 (1970).

Valatx, J. L., Jouvet, D., Jouvet, M.: Evolution électroencéphalographique des différents états de sommeil chez le chaton. Electroenceph. clin. Neurophysiol. 17, 218–233 (1964).

Valleala, P.: The temporal relation of unit discharge in visual cortex and activity of the extraocular muscles during sleep. Arch. ital. Biol. 105, 1–15 (1967).

Velasco, M., Lindsley, D. B.: Role of orbital cortex in regulation of thalamocortical electrical activity. Science 149, 1375–1377 (1965).

Velluti, R., Hernandez-Péon, R.: Atropine blockade within a cholinergic hypnogenic circuit. Exp. Neurol. 8, 20–29 (1963).

Villablanca, J.: Electroencephalogram in the permanently isolated forebrain of the cat. Science 138, 44–45 (1962).

— Ocular behavior in the chronic cerveau isolé cat. Brain Res. 2, 99–102 (1966a).

— Behavioral and polygraphic study of "sleep" and wakefulness in chronic decreebrate cats. Electroenceph. clin. Neurophysiol. 21, 562–577 (1966b).

— Effects of atropine, eserine and adrenaline in cats with mesencephalic transection and in the "isolated hemisphere" cat preparation. Arch. Biol. Med. Exp. 3, 118–129 (1966c).

Vimont-Vicary, P.: La suppression des différents états de sommeil. Etude comporte-mentale, EEG, et Neuropharmacologique chez le chat. Thèse Université de Lyon, L.M.D. ed. (1966). 95 pp.

— Jouvet-Mounier, D., Delorme, F.: Effects EEG et comportementaux des privations de sommeil paradoxal chez le chat. Electroenceph. clin. Neurophysiol. **20**, 439–449 (1966).

Vogt, M.: The concentration of sympathin in different parts of the cerebral nervous system under normal conditions and after the administration of drugs. J. Physiol. (Lond.) **123**, 451–481 (1954).

Votava, Z.: Pharmacology of the central cholinergic synapses. Ann. Rev. Pharmacol. **7**, 223–241 (1967).

Wada, J. A., McGeer, E. G.: Central aromatic amines and behavior. Arch. Neurol. (Chic.) **14**, 129–143 (1966).

— Terao, A.: Effect of parachlorophenylalanine on basal forebrain stimulation. Exp. Neurol. **28**, 501–506 (1970).

Wallach, M. B., Winters, W. D., Mandell, A. J., Spooner, C. E.: A correlation of EEG, reticular multiple unit activity and gross behavior following various anti-depressant agents in the cat. Electroenceph. clin. Neurophysiol. **27**, 563–573 (1969a).

— — — — Effects of antidepressant drugs on wakefulness and sleep in the cat. Electro-enceph. clin. Neurophysiol. **27**, 574–581 (1969b).

Walsh, J. T., Cordeau, J. P.: Responsiveness in the visual system during various phases of sleep and waking. Exp. Neurol. **11**, 80–104 (1965).

Weill-Malherbe, H., Axelrod, J., Tomchick, R.: Blood brain barier for adrenaline. Science **129**, 1226–1227 (1959).

Weiner, N.: Regulation of norepinephrine biosynthesis. Ann. Rev. Pharmacol. **10**, 273–291 (1970).

Weiss, B. L., Aghajanian, G. K.: Activation of brain serotonin metabolism by heat: role of midbrain raphe neurons. Brain Res. **26**, 37–48 (1971).

Weiss, B., Heller, A.: Methodological problems in evaluating the role of cholinergic mechanisms in behavior. Fed. Proc. **28**, 135–147 (1969).

Weiss, E., Bordwell, B., Seeger, M., Lee, J., Dement, W., Barchas, J.: Changes in brain serotonin and 5 hydroxy-indole-3 acetic acid in REM sleep deprived rats. Psychophysiology **5**, 209 (1968).

Weissman, A., Koe, B. K.: Behavioral effects of L-alpha-methyl-tyrosine hydroxylase inhibitor s. Life Sci. **4**, 1037–1049 (1965).

Weitzman, E. D., McGregor, P., Moore, C., Jacoby, J.: The effect of alpha-methyl-paratyrosine on sleep patterns of the monkey. Life Sci. **8**, 751–758 (1969).

— Rapport, M. M., McGregor, P., Jacoby, J.: Sleep patterns of the monkey and brain se-rotonin concentration: Effect of p-chlorophenylalanine. Science **160**, 1361–1363 (1968).

Welch, A. S., Welch, B. L.: Reduction of norepinephrine in the lower brainstem by psychological stimulus. Proc. nat. Acad. Sci. (Wash.) **60**, 478–482 (1968).

Wende, Ch. van der, Johnson, J. C.: Interaction of serotonin with the catecholamine and norepinephrine oxidation. Biochem. Pharmacol. **19**, 1991–2000 (1970a).

— — Interaction of serotonin with the catecholamins. II. Activation and inhibition of adrenochrome formation. Biochem. Pharmacol. **19**, 2001–2007 (1970b).

Werman, R.: Criteria for identification of a central nervous system transmitter. Comp. Biochem. Physiol. **18**, 745–767 (1966).

West, L. J., Janszen, H. H., Lester, B. K., Cornelisson, F. S.: The psychosis of sleep deprivation. Ann. N.Y. Acad. Sci. **96**, 66–70 (1962).

White, R. P., Samson, F. E.: Effects of fatty acids anions on the electroencephalogram of unanesthetized rabbits. Amer. J. Physiol. **186**, 271–274 (1956).

Wikler, A.: Pharmacologic dissociation of behaviour and EEG "sleep patterns" in dogs Morphine, N. allylmorphine and atropine. Proc. Soc. exp. Biol. (N.Y.) **79**, 261–265 (1952).

WILLIAMS, H. L., LUBIN, A., GOODNOW, J. J.: Impaired performance with acute sleep loss. Psychol. Monogr. 73, 1–26 (1959).

WINTERS, W. D., SPOONER, C. E., A neurophysiological comparison of gamma-hydroxybutyrate with pentobarbital in cats. Electroenceph. clin. Neurophysiol. 18, 287–296 (1965).

WISE, C. D., RUELIUS, H. W.: The binding of serotonin in brain: a study in vitro of the influence of physicochemical factors and drugs. Biochem. Pharmacol. 17, 617–633 (1968).

— STEIN, L.: Facilitation of brain self-stimulation by central administration of norepinephrine. Science 163, 299–301 (1969).

WOOD, J. G.: Electron microscopic localization of amines in central nervous tissue. Nature (Lond.) 209, 1131–1133 (1966).

WOODS, J. W.: Behavior of chronic decerebrate rats. J. Neurophysiol. 27, 635–645 (1964).

WURTMAN, R. J. (ed.): Brain monoamines and endocrine function. Neurosci. Res. Prog. Bull. 9, No 2, 297 (1971).

WYATT, R. J., ENGELMAN, K., KUPFER, D. J., FRAM, D. H., SJOERDSMA, A., SNYDER, F.: Effects of L-tryptophan (a natural sedative) on human sleep. Lancet 1970 II, 842–846.

— — — SJOERDSMA, A., SNYDER, F.: Effects of parachlorophenylalanine on sleep in man. Electroenceph. clin. Neurophysiol. 27, 529–532 (1969).

— FRAM, D. H., KUPFER, D. J., SNYDER, F.: Total prolonged drug-induced REM sleep suppression im anxious-depressed patients. Arch. gen. Psychiat. 24, 145–156 (1971).

— KUPFER, D. J., SCOTT, J., ROBINSON, D. S., SNYDER, F.: Longitudinal studies of the effect of monoamine oxidase inhibitors on sleep in man. Psychopharmacologia (Berl.) 15, 236–244 (1969).

WYRWICKA, W., STERMAN, M. B., CLEMENTE, C. D.: Conditioning of induced electroencephalographic sleep patterns in the cat. Science 137, 616–618 (1962).

YAMAGUCHI, N., MARCZINSKI, T. J., LINGI, M.: The effects of electrical and chemical stimulation of the preoptic region and some nonspecific thalamic nuclei in unrestrained, waking animals. Electroenceph. clin. Neurophysiol. 15, 145–166 (1963).

YAMAMOTO, K. I., DOMINO, E. F.: Nicotine induced EEG and behavioral arousal. Int. J. Neuropharmacol. 4, 359–373 (1965).

— — Cholinergic agonist-antagonist interactions on neocortical and limbic EEG activation. Int. J. Neuropharmacol. 6, 357–375 (1967).

YORK, D. H.: Possible dopaminergic pathway from substantia nigra to putamen Brain. Res. 20, 233–251 (1970).

ZANCHETTI, A.: Brain stem mechanisms of sleep. Anesthesiology 28, 81–99 (1967).

ZIGMOND, M. J., WURTMAN, R. J.: Daily rhythm in the accumulation of brain catecholamines synthesized from incubating H^3 tyrosine. J. Pharmacol. exp. Ther. 172, 416–423 (1970).

Author Index

Page numbers in *italics* refer to bibliography

Subject Index

Ergebnisse der Physiologie

Biologischen Chemie und experimentellen Pharmakologie

Reviews of Physiology

Biochemistry and Experimental Pharmacology

Herausgeber / Editors

R. H. Adrian, Cambridge · E. Helmreich, Würzburg
H. Holzer, Freiburg · R. Jung, Freiburg · K. Kramer, München
O. Krayer, Boston · F. Lynen, München · P. A. Miescher, Genf
H. Rasmussen, Philadelphia · A. E. Renold, Genf
U. Trendelenburg, Würzburg · W. Vogt, Göttingen
H. H. Weber, Heidelberg

Sonderdruck aus Band 64

G. Moruzzi
The Sleep-Waking Cycle
With 39 Figures

Nicht im Handel

Springer-Verlag Berlin Heidelberg GmbH 1972

Contents

Ergebnisse der Physiologie

Biologischen Chemie und experimentellen Pharmakologie

Reviews of Physiology

Biochemistry and Experimental Pharmacology

Herausgeber / Editors

R. H. Adrian, Cambridge · E. Helmreich, Würzburg
H. Holzer, Freiburg · R. Jung, Freiburg · K. Kramer, München
O. Krayer, Boston · F. Lynen, München · P. A. Miescher, Genf
H. Rasmussen, Philadelphia · A. E. Renold, Genf
U. Trendelenburg, Würzburg · W. Vogt, Göttingen
H. H. Weber, Heidelberg

Sonderdruck aus Band 64

M. Jouvet
The Role of Monoamines and Acetylcholine-Containing
Neurons in the Regulation of the Sleep-Waking Cycle
With 24 Figures

Nicht im Handel

Springer-Verlag Berlin Heidelberg GmbH 1972

Contents

Brain and Human Behavior

Edited by
Alexander G. Karczmar
and Sir John C. Eccles

With 162 figures
X, 475 pages. 1972
Cloth DM 98,—
US $ 30.50

Twenty experts in the interdisciplinary field of neurosciences contributed to this book. The contents were structured beforehand to fully illustrate the status and the problems of neurosciences and the contributors, who include three Nobel Prize winners as well as British, American, French, German, Argentinian and Russian scientists, were selected for their ability to cover the wide range of topics. The topics range from molecular and subcellular organization of the neurons to their function in field systems and as scanners of the environment, to their involvement in processes of learning and memory, and extending into the behavior of the organism with a view to its genetic and sociological interactions. This then led on to the perennial problems of the relationship of mind to its substrate and of behavior either as a mechanism or as an expression of free will. This text presents examples of advanced techniques in these various areas; it should help to develop overall understanding of the current state of behavioral theory and practice; and it emphasizes the philosophical implications of this subject.

Published earlier:

John C. Eccles: The Physiology of Synapses
The subject of this book is the research work for which Professor Eccles was awarded the Nobel prize
With 101 figures. XII, 316 pages. 1964. Cloth DM 40,—; US $ 12.50

Studies in Physiology
Presented to John C. Eccles. Edited by D. R. Curtis and A. K. McIntyre with the collaboration of numerous experts
With 80 figures. VIII, 276 pages. 1965. Cloth DM 28,—; US $ 8.80

Brain and Conscious Experience
Study Week September 28 to October 4, 1964, of the Pontificia Academia Scientiarum. **Edited by Sir John C. Eccles**
With 147 figures. XXII, 591 pages. 1966. Cloth DM 74,—; US $ 23.10

John C. Eccles: Facing Reality
Philosophical Adventures by a Brain Scientist
With 36 figures. XI, 210 pages. 1970 (Heidelberg Science Library, Vol. 13)
DM 22,60; US $ 7.10
Distribution rights for U.K., Commonwealth, and the Traditional British Market (excluding Canada): Longman Group Ltd., Harlow/Essex

Springer-Verlag
Berlin
Heidelberg
New York

München · London
Paris · Tokyo · Sydney

Experimental Brain Research
Experimentelle Hirnforschung
Expérimentation Cérébrale
Editorial Board: O. Creutzfeldt, D. R. Curtis, P. Dell, J. C. Eccles, R. Jung, D. M. MacKay, D. Ploog, J. Szentágothai, V. P. Whittaker, V. J. Wilson
Subscription Information. 1972, Volumes 15—16 (4 issues each) —
DM 128,—; US $ 38.40 per volume, plus postage

Universitätsdruckerei H. Stürtz AG, Würzburg